# The Mathematical Experience

# The
# Mathematical
# Experience

Philip J. Davis
Reuben Hersh

With an Introduction by Gian-Carlo Rota

HOUGHTON MIFFLIN COMPANY BOSTON

*Library of Congress Cataloging in Publication Data*

Davis, Philip J., date
  The mathematical experience.

  Reprint. Originally published: Boston: Birkhäuser, 1981.
  Bibliography: p.
  Includes index.
  1. Mathematics—Philosophy. 2. Mathematics—History. 3. Mathematics—Study and teaching. I. Hersh, Reuben, date. II. Title.
  QA8.4.D37  1982      510       81-20304
  ISBN 0-395-32131-X (pbk.)        AACR2

Printed in the United States of America

AL10 9 8 7 6 5 4 3 2 1

Reprinted by arrangement with Birkhäuser Boston

Houghton Mifflin Company paperback 1982

*For my parents,*
*Mildred and Philip Hersh*

* * * *

*For my brother,*
*Hyman R. Davis*

# Contents

*Contents*

# Preface

THE OLDEST MATHEMATICAL tablets we have date from 2400 B.C., but there is no reason to suppose that the urge to create and use mathematics is not coextensive with the whole of civilization. In four or five millennia a vast body of practices and concepts known as mathematics has emerged and has been linked in a variety of ways with our day-to-day life. What is the nature of mathematics? What is its meaning? What are its concerns? What is its methodology? How is it created? How is it used? How does it fit in with the varieties of human experience? What benefits flow from it? What harm? What importance can be ascribed to it?

These difficult questions are not made easier by the fact that the amount of material is so large and the amount of interlinking is so extensive that it is simply not possible for any one person to comprehend it all, let alone sum it up and compress the summary between the covers of an average-sized book. Lest we be cowed by this vast amount of material, let us think of mathematics in another way. Mathematics has been a human activity for thousands of years. To some small extent, everybody is a mathematician and does mathematics consciously. To buy at the market, to measure a strip of wall paper or to decorate a ceramic pot with a regular pattern is doing mathematics. Further, everybody is to some small extent a philosopher of mathematics. Let him only exclaim on occasion: "But figures can't lie!" and he joins the ranks of Plato and of Lakatos.

In addition to the vast population that uses mathematics on a modest scale, there are a small number of people who are professional mathematicians. They practice mathematics, foster it, teach it, create it, and use it in a wide variety of situations. It should be possible to explain to nonprofessionals just what these people are doing, what they say they are doing, and why the rest of the world should support them at it. This, in brief, is the task we have set for ourselves. The book is not intended to present a systematic, self-contained discussion of a specific corpus of mathematical material, either recent or classical. It is intended rather to capture the inexhaustible variety presented by the mathematical experience. The major strands of our exposition will be the substance of mathematics, its history, its philosophy, and how mathematical knowledge is elicited. The book should be regarded not as a compression but rather as an impression. It is not a mathematics book; it is a book about mathematics. Inevitably it must contain some mathematics. Similarly, it is not a history or a philosophy book, but it will discuss mathematical history and philosophy. It follows that the reader must bring to it some slight prior knowledge of these things and a seed of interest to plant and water. The general reader with this background should have no difficulty in getting through the major portion of the book. But there are a number of places where we have brought in specialized material and directed our exposition to the professional who uses or produces mathematics. Here the reader may feel like a guest who has been invited to a family dinner. After polite general conversation, the family turns to narrow family concerns, its delights and its worries, and the guest is left up in the air, but fascinated. At such places the reader should judiciously and lightheartedly push on.

For the most part, the essays in this book can be read independently of each other.

Some comment is necessary about the use of the word "I" in a book written by two people. In some instances it will be obvious which of the authors wrote the "I." In any case, mistaken identity can lead to no great damage, for each author agrees, in a general way, with the opinions of his colleague.

# Acknowledgements

SOME OF THE MATERIAL of this book was excerpted from published articles. Several of these have joint authorship: "Non-Cantorian Set Theory" by Paul Cohen and Reuben Hersh and "Non-Standard Analysis" by Martin Davis and Reuben Hersh both appeared in the *Scientific American*. "Nonanalytic Aspects of Mathematics" by Philip J. Davis and James A. Anderson appeared in the *SIAM Review*. To Professors Anderson, Cohen, and M. Davis and to these publishers, we extend our grateful acknowledgement for permission to include their work here.

Individual articles by the authors excerpted here include "Number," "Numerical Analysis," and "Mathematics by Fiat?" by Philip J. Davis which appeared in the *Scientific American*, "The Mathematical Sciences," M.I.T. Press, and the *Two Year College Mathematical Journal* respectively; "Some Proposals for Reviving the Philosophy of Mathematics" and "Introducing Imre Lakatos" by Reuben Hersh, which appeared in *Advances in Mathematics* and the *Mathematical Intelligencer*, respectively.

We appreciate the courtesy of the following organizations and individuals who allowed us to reproduce material in this book: The Academy of Sciences at Göttingen, Ambix, Dover Publishers, *Mathematics of Computation*, M.I.T. Press, *New Yorker Magazine*, Professor A. H. Schoenfeld, and John Wiley and Sons.

The section on Fourier analysis was written by Reuben Hersh and Phyllis Hersh. In critical discussions of philosophical questions, in patient and careful editing of rough drafts, and in her unfailing moral support of this

Acknowledgements

project, Phyllis Hersh made essential contributions which it is a pleasure to acknowledge.

The following individuals and institutions generously allowed us to reproduce graphic and artistic material: Professors Thomas Banchoff and Charles Strauss, the Brown University Library, the Museum of Modern Art, The Lummus Company, Professor Ron Resch, Routledge and Kegan Paul, Professor A. J. Sachs, the University of Chicago Press, the Whitworth Art Gallery, the University of Manchester, the University of Utah, Department of Computer Science, the Yale University Press.

We wish to thank Professors Peter Lax and Gian-Carlo Rota for encouragement and suggestions. Professor Gabriel Stoltzenberg engaged us in a lively and productive correspondence on some of the issues discussed here. Professor Lawrence D. Kugler read the manuscript and made many valuable criticisms. Professor José Luis Abreu's participation in a Seminar on the Philosophy of Mathematics at the University of New Mexico is greatly appreciated.

The participants in the Seminar on Philosophical Issues in Mathematics, held at Brown University, as well as the students in courses given at the University of New Mexico and at Brown, helped us crystallize our views and this help is gratefully acknowledged. The assistance of Professor Igor Najfeld was particularly welcome.

We should like to express our appreciation to our colleagues in the History of Mathematics Department at Brown University. In the course of many years of shared lunches, Professors David Pingree, Otto Neugebauer, A. J. Sachs, and Gerald Toomer supplied us with the "three I's": information, insight, and inspiration. Thanks go to Professor Din-Yu Hsieh for information about the history of Chinese mathematics.

Special thanks to Eleanor Addison for many line drawings. We are grateful to Edith Lazear for her careful and critical reading of Chapters 7 and 8 and her editorial comments.

We wish to thank Katrina Avery, Frances Beagan, Joseph M. Davis, Ezoura Fonseca, and Frances Gajdowski for their efficient help in the preparation and handling of the manuscript. Ms. Avery also helped us with a number of classical references.

P. J. Davis
R. Hersh

# Introduction

DEDICATED TO MARK KAC
*"oh philosophie alimentaire!"*
–Sartre

AT THE TURN OF THE CENTURY, the Swiss historian Jakob Burckhardt, who, unlike most historians, was fond of guessing the future, once confided to his friend Friedrich Nietzsche the prediction that the Twentieth Century would be "the age of oversimplification".

Burckhardt's prediction has proved frighteningly accurate. Dictators and demagogues of all colors have captured the trust of the masses by promising a life of bread and bliss, to come right after the war to end all wars. Philosophers have proposed daring reductions of the complexity of existence to the mechanics of elastic billiard balls; others, more sophisticated, have held that life is language, and that language is in turn nothing but strings of marble-like units held together by the catchy connectives of Fregean logic. Artists who dished out in all seriousness checkerboard patterns in red, white, and blue are now fetching the highest bids at Sotheby's. The use of such words as "mechanically" "automatically" and "immediately" is now accepted by the wizards of Madison Avenue as the first law of advertising.

Not even the best minds of Science have been immune to the lure of oversimplification. Physics has been driven by the search for one, only one law which one day, just around the corner, will unify all forces: gravitation and electricity and strong and weak interactions and what not. Biologists are now mesmerized by the prospect that the secret of life may be gleaned from a double helix dotted with large molecules. Psychologists have prescribed in turn sexual release, wonder drugs and primal screams as the cure for common depression, while preachers would counter with the less expensive offer to join the hosannahing chorus of the born-again.

It goes to the credit of mathematicians to have been the slowest to join this movement. Mathematics, like theology and all free creations of the Mind, obeys the inexorable laws of the imaginary, and the Pollyannas of the day are of little help in establishing the truth of a conjecture. One may pay lip service to Descartes and Grothendieck when they wish that geometry be reduced to algebra, or to Russell and Gentzen when they command that mathematics become logic, but we know that some mathematicians are more endowed with the talent of drawing pictures, others with that of juggling symbols and yet others with the ability of picking the flaw in an argument.

Nonetheless, some mathematicians have given in to the simplistics of our day when it comes to the understanding of the nature of their activity and of the standing of mathematics in the world at large. With good reason, nobody likes to be told what he is really doing or to have his intimate working habits analyzed and written up. What might Senator Proxmire say if he were to set his eyes upon such an account? It might be more rewarding to slip into the Senator's hands the textbook for Philosophy of Science 301, where the author, an ambitious young member of the Philosophy Department, depicts with impeccable clarity the ideal mathematician ideally working in an ideal world.

We often hear that mathematics consists mainly in "proving theorems". Is a writer's job mainly that of "writing sentences"? A mathematician's work is mostly a tangle of guesswork, analogy, wishful thinking and frustration, and proof, far from being the core of discovery, is more often than not a way of making sure that our minds are not playing tricks. Few people, if any, had dared write this out loud before Davis and Hersh. Theorems are not to mathematics what successful courses are to a meal. The nutritional analogy is misleading. To master mathematics is to master an intangible view, it is to acquire the skill of the virtuoso who cannot pin his performance on criteria. The theorems of geometry are not related to the field of Geometry as elements are to a set. The relationship is more subtle, and Davis and Hersh give a rare honest description of this relationship.

After Davis and Hersh, it will be hard to uphold the *Glasperlenspiel* view of mathematics. The mystery of mathematics, in the authors' amply documented account, is that conclusions originating in the play of the mind do find striking practical applications. Davis and Hersh have chosen to describe the mystery rather than explain it away.

Making mathematics accessible to the educated layman, while keeping high scientific standards, has always been considered a treacherous navigation between the Scylla of professional contempt and the Charybdis of public misunderstanding. Davis and Hersh have sailed across the Strait under full sail. They have opened a discussion of the mathematical experience that is inevitable for survival. Watching from the stern of their ship, we breathe a sigh of relief as the vortex of oversimplification recedes into the distance.

GIAN-CARLO ROTA
*August 9, 1980*

*"The knowledge at which geometry aims is the knowledge of the eternal."*

PLATO, REPUBLIC, VII, 527

*"That sometimes clear . . . and sometimes vague stuff . . . which is . . . mathematics."*

IMRE LAKATOS, 1922–1974

*"What is laid down, ordered, factual, is never enough to embrace the whole truth: life always spills over the rim of every cup."*

BORIS PASTERNAK, 1890–1960

# Overture

UP TILL ABOUT five years ago, I was a normal mathematician. I didn't do risky and unorthodox things, like writing a book such as this. I had my "field"—partial differential equations—and I stayed in it, or at most wandered across its borders into an adjacent field. My serious thinking, my real intellectual life, used categories and evaluative modes that I had absorbed years before, in my training as a graduate student. Because I did not stray far from these modes and categories, I was only dimly conscious of them. They were part of the way I saw the world, not part of the world I was looking at.

My advancement was dependent on my research and publication in my field. That is to say, there were important rewards for mastering the outlook and ways of thought shared by those whose training was similar to mine, the other workers in the field. Their judgment would decide the value of what I did. No one else would be qualified to do so; and it is very doubtful that anyone else would have been interested in doing so. To liberate myself from this outlook—that is, to recognize it, to become aware that it was only one of many possible ways of looking at the world, to be able to put it on or off by choice, to compare it and evaluate it with other ways of looking at the world—none of this was required by my job or my career. On the contrary, such unorthodox and dubious adventures would have seemed at best a foolish waste of precious time—at worst, a disreputable dabbling with shady and suspect ventures such as psychology, sociology, or philosophy.

The fact is, though, that I have come to a point where my wonderment and fascination with the meaning and purpose, if any, of this strange activity we call mathematics is equal to, sometimes even stronger than, my fascination with actually *doing* mathematics. I find mathematics an infinitely complex and mysterious world; exploring it is an addiction from which I hope never to be cured. In this, I am a mathematician like all others. But in addition, I have developed a second half, an Other, who watches this mathematician with amazement, and is even more fascinated that such a strange creature and such a strange activity have come into the world, and persisted for thousands of years.

I trace its beginnings to the day when I came at last to teach a course called Foundations of Mathematics. This is a course intended primarily for mathematics majors, at the upper division (junior or senior) level. My purpose in teaching this course, as in the others I had taught over the years, was to learn the material myself. At that time I knew that there was a history of controversy about the foundations. I knew that there had been three major "schools"; the logicists associated with Bertrand Russell, the formalists led by David Hilbert, and the constructivist school of L. E. J. Brouwer. I had a general idea of the teaching of each of these three schools. But I had no idea which one I agreed with, if any, and I had only a vague idea of what had become of the three schools in the half century since their founders were active.

I hoped that by teaching the course I would have the opportunity to read and study about the foundations of mathematics, and ultimately to clarify my own views of those parts which were controversial. I did not expect to become a researcher in the foundations of mathematics, any more than I became a number theorist after teaching number theory.

Since my interest in the foundations was philosophical rather than technical, I tried to plan the course so that it

2

could be attended by interested students with no special requirements or prerequisites; in particular, I hoped to attract philosophy students, and mathematics education students. As it happened there were a few such students; there were also students from electrical engineering, from computer science, and other fields. Still, the mathematics students were the majority. I found a couple of good-looking textbooks, and plunged in.

In standing before a mixed class of mathematics, education, and philosophy students, to lecture on the foundations of mathematics, I found myself in a new and strange situation. I had been teaching mathematics for some 15 years, at all levels and in many different topics, but in all my other courses the job was not to talk about mathematics, it was to *do* it. Here my purpose was not to do it, but to talk about it. It was different and frightening.

As the semester progressed, it became clear to me that this time it was going to be a different story. The course was a success in one sense, for there was a lot of interesting material, lots of chances for stimulating discussions and independent study, lots of things for me to learn that I had never looked at before. But in another sense, I saw that my project was hopeless.

In an ordinary mathematics class, the program is fairly clear cut. We have problems to solve, or a method of calculation to explain, or a theorem to prove. The main work to be done will be in writing, usually on the blackboard. If the problems are solved, the theorems proved, or the calculations completed, then teacher and class know that they have completed the daily task. Of course, even in this ordinary mathematical setting, there is always the possibility or likelihood of something unexpected happening. An unforeseen difficulty, an unexpected question from a student, can cause the progress of the class to deviate from what the instructor had intended. Still, one knew where one was supposed to be going; one also knew that the main thing was what you wrote down. As to spoken words, either from the class or from the teacher, they were important insofar as they helped to communicate the import of what was written.

3

In opening my course on the foundations of mathematics, I formulated the questions which I believed were central, and which I hoped we could answer or at least clarify by the end of the semester.

What is a number? What is a set? What is a proof? What do we know in mathematics, and how do we know it? What is "mathematical rigor"? What is "mathematical intuition"?

As I formulated these questions, I realized that I didn't know the answers. Of course, this was not surprising, for such vague questions, "philosophical" questions, should not be expected to have clearcut answers of the kind we look for in mathematics. There will always be differences of opinion about questions such as these.

But what bothered me was that I didn't know what my own opinion was. What was worse, I didn't have a basis, a criterion on which to evaluate different opinions, to advocate or attack one view point or another.

I started to talk to other mathematicians about proof, knowledge, and reality in mathematics and I found that my situation of confused uncertainty was typical. But I also found a remarkable thirst for conversation and discussion about our private experiences and inner beliefs.

This book is part of the outcome of these years of pondering, listening, and arguing.

# 1
## THE MATHEMATICAL LANDSCAPE

# What is
# Mathematics?

A NAIVE DEFINITION, adequate for the dictionary and for an initial understanding, is that *mathematics is the science of quantity and space.* Expanding this definition a bit, one might add that mathematics also deals with the symbolism relating to quantity and to space.

This definition certainly has a historical basis and will serve us for a start, but it is one of the purposes of this work to modify and amplify it in a way that reflects the growth of the subject over the past several centuries and indicates the visions of various schools of mathematics as to what the subject ought to be.

The sciences of quantity and of space in their simpler forms are known as *arithmetic* and *geometry*. Arithmetic, as taught in grade school, is concerned with numbers of various sorts, and the rules for operations with numbers—addition, subtraction, and so forth. And it deals with situations in daily life where these operations are used.

Geometry is taught in the later grades. It is concerned in part with questions of spatial measurements. If I draw such a line and another such line, how far apart will their end points be? How many square inches are there in a rectangle 4 inches long and 8 inches wide? Geometry is also concerned with aspects of space that have a strong aesthetic appeal or a surprise element. For example, it tells us that in any parallelogram whatsoever, the diagonals bisect one another; in any triangle whatsoever, the three medians intersect in a common point. It teaches us that a floor can be tiled with equilateral triangles or hexagons, but not with regular pentagons.

But geometry, if taught according to the arrangement laid out by Euclid in 300 B.C., has another vitally significant aspect. This is its presentation as a deductive science. Beginning with a number of elementary ideas which are assumed to be self-evident, and on the basis of a few definite rules of mathematical and logical manipulation, Euclidean geometry builds up a fabric of deductions of increasing complexity.

What is stressed in the teaching of elementary geometry is not only the spatial or visual aspect of the subject but the methodology wherein hypothesis leads to conclusion. This deductive process is known as *proof*. Euclidean geometry is the first example of a formalized deductive system and has become the model for all such systems. Geometry has been the great practice field for logical thinking, and the study of geometry has been held (rightly or wrongly) to provide the student with a basic training in such thinking.

Although the deductive aspects of arithmetic were clear to ancient mathematicians, these were not stressed either in teaching or in the creation of new mathematics until the 1800s. Indeed, as late as the 1950s one heard statements from secondary school teachers, reeling under the impact of the "new math," to the effect that they had always thought geometry had "proof" while arithmetic and algebra did not.

With the increased emphasis placed on the deductive aspects of all branches of mathematics, C. S. Peirce in the middle of the nineteenth century, announced that "mathematics is the science of making necessary conclusions." Conclusions about what? About quantity? About space? The content of mathematics is not defined by this definition; mathematics could be "about" anything as long as it is a subject that exhibits the pattern of assumption-deduction-conclusion. Sherlock Holmes remarks to Watson in *The Sign of Four* that "Detection is, or ought to be, an exact science and should be treated in the same cold and unemotional manner. You have attempted to tinge it with romanticism, which produces much the same effect as if you

7

worked a love-story or an elopement into the fifth proposition of Euclid." Here Conan Doyle, with tongue in cheek, is asserting that criminal detection might very well be considered a branch of mathematics. Peirce would agree.

The definition of mathematics changes. Each generation and each thoughtful mathematician within a generation formulates a definition according to his lights. We shall examine a number of alternate formulations before we write Finis to this volume.

### Further Readings. See Bibliography

A. Alexandroff; A. Kolmogoroff and M. Lawrentieff; R. Courant and H. Robbins; T. Danzig [1959]; H. Eves and C. Newsom; M. Gaffney and L. Steen; N. Goodman; E. Kasner and J. Newman; R. Kershner and L. Wilcox; M. Kline [1972]; A. Kolmogoroff; J. Newman [1956]; E. Snapper; E. Stabler; L. Steen [1978]

# Where is Mathematics?

W HERE IS THE PLACE of mathematics? Where does it exist? On the printed page, of course, and prior to printing, on tablets or on papyri. Here is a mathematical book—take it in your hand; you have a palpable record of mathematics as an intellectual endeavor. But first it must exist in people's minds, for a shelf of books doesn't create mathematics. Mathematics exists on taped lectures, in computer memories and printed circuits. Should we say also that it resides in mathematical machines such as slide rules and cash registers and, as some believe, in the arrangement of the stones at Stonehenge? Should we say that it resides in the genes of the sunflower plant if that plant brings forth seeds arranged in Bernoullian spirals and transmits mathematical information from generation to generation? Should we say that mathematics exists on a wall if a lampshade casts a parabolic shadow on that wall? Or do we believe that all these are mere shadow manifestations of the real mathematics which, as some philosophers have asserted, exists eternally and independently of this actualized universe, independently of all possible actualizations of a universe?

What is knowledge, mathematical or otherwise? In a correspondence with the writer, Sir Alfred Ayer suggests that one of the leading dreams of philosophy has been "to agree on a criterion for deciding what there is," to which we might add, "and for deciding where it is to be found."

# The Mathematical Community

**T**HERE IS HARDLY a culture, however primitive, which does not exhibit some rudimentary kind of mathematics. The mainstream of western mathematics as a systematic pursuit has its origin in Egypt and Mesopotamia. It spread to Greece and to the Graeco-Roman world. For some 500 years following the fall of Rome, the fire of mathematical creativeness was all but extinguished in Europe; it is thought to have been preserved in Persia. After some centuries of inactivity, the flame appeared again in the Islamic world and from there mathematical knowledge and enthusiasm spread through Sicily and Italy to the whole of Europe.

A rough timetable would be

| | | |
|---|---|---|
| Egyptian: | 3000 B.C. to | 1600 B.C. |
| Babylonian: | 1700 B.C. to | 300 B.C. |
| Greek: | 600 B.C. to | 200 B.C. |
| Graeco-Roman: | 150 A.D. to | 525 A.D. |

| Islamic: | 750 A.D. to 1450 A.D. |
|---|---|
| Western: | 1100 A.D. to 1600 A.D. |
| Modern: | 1600 A.D. to present. |

Other streams of mathematical activity are the Chinese, the Japanese, the Hindu, and the Inca-Aztec. The interaction between western and eastern mathematics is a subject of scholarly investigation and conjecture.

At the present time, there is hardly a country in the world which is not creating new mathematics. Even the emerging nations, so called, all wish to establish up-to-date university programs in mathematics, and the hallmark of excellence is taken to be the research activity of their staffs.

In contrast to the relative isolation of early oriental and western mathematics from each other, the mathematics of today is unified. It is worked and transmitted in full and open knowledge. Personal secrecy like that practiced by the Renaissance and Baroque mathematicians hardly exists. There is a vast international network of publications; there are national and international open meetings and exchanges of scholars and students.

*François Viète*
*1540–1603*

In all honesty, though, it should be admitted that restriction of information has occurred during wartime. There is also considerable literature on mathematical cryptography, as practiced by the professional cryptographers, which is not, for obvious reasons, generally available.

In the past mathematics has been pursued by people in various walks of life. Thomas Bradwardine (1325) was Archbishop of Canterbury. Ulugh Beg with his trigonometric tables was the grandson of Tamerlane. Luca Pacioli (1470) was a monk. Ferrari (1548) was a tax assessor. Cardano (1550) was professor of medicine. Viète (1580) was a lawyer in the royal privy council. Van Ceulen (1610) was a fencing master. Fermat (1635) was a lawyer. Many mathematicians earned part of their living as protegés of the Crown: John Dee, Kepler, Descartes, Euler; some even had the title of "Mathematicus." Up to about 1600, a mathematician could earn a few pounds by casting horoscopes or writing amulets for the wealthy.

*René Descartes*
*1596–1650*

10

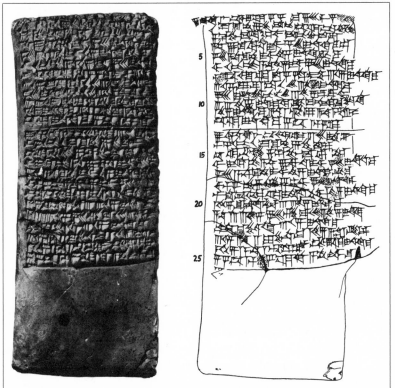

[1] *9 (gín) is the (total expenses in) silver of a kilá; I added the length and the width, and (the result is) 6;30 (GAR); ½ GAR is [its depth],*
[2] *10 gín (volume) the assignment, 6 še (silver) the wages. What are the length (and) its width?*
[3] *When you perform (the operations), take the reciprocal of the wages,*
[4] *multiply by 9 gín, the (total expenses in) silver, (and) you will get 4,30;*
[5] *multiply 4,30 by the assignment, (and) you will get 45;*
[6] *take the reciprocal of its depth, multiply by 45, (and) you will get 7;30;*
[7] *halve the length and the width which I added together, (and) you will get 3;15;*
[8] *square 3;15, (and) you will get 10;33,45;*
[9] *subtract 7;30 from 10;33,45, (and)*
[10] *you will get 3;3,45; take its square root, (and)*
[11] *you will get 1;45; add it to the one, subtract it from the other, (and)*
[12] *you will get the length (and) the width. 5 (GAR) is the length; 1½ GAR is the width.*

*Courtesy: Prof. A. J. Sachs, from Neugebauer and Sachs, "Mathematical Cuneiform Texts"*

Explanation: The drawing is a contemporary version of the symbols on the clay tablet. A line by line translation of the first twelve lines is given. The notation 3;3,45 used in the translation means $3 + \frac{3}{60} + \frac{45}{3600} = 3.0625$. In modern terms, the problem posed by this tablet is: given $x + y$ and $xy$, find $x$ and $y$. Solution:

$$\begin{aligned} x = \\ y = \end{aligned} \frac{x+y}{2} \pm \sqrt{\left(\frac{x+y}{2}\right)^2 - xy}$$

These days there is nothing to prevent a wealthy person from pursuing mathematics full or part time in isolation, as in the era when science was an aristocrat's hobby. But this kind of activity is now not at a sufficiently high voltage to sustain invention of good quality. Nor does the church (or the monarchy) support mathematics as it once did.

For the past century, universities have been our principal sponsors. By releasing part of his time, the university encourages a lecturer to engage in mathematical research. At present, most mathematicians are supported directly or indirectly by the university, by corporations such as IBM, or by the federal government, which in 1977 spent about $130,000,000 on mathematics of all sorts.

To the extent that all children learn some mathematics, and that a certain small fraction of mathematics is in the common language, the mathematics community and the community at large are identical. At the higher levels of practice, at the levels where new mathematics is created and transmitted, we are a fairly small community. The combined membership list of the American Mathematical Society, the Mathematical Association of America, and the Society for Industrial and Applied Mathematics for the year 1978 lists about 30,000 names. It is by no means necessary for one to think of oneself as a mathematician to operate at the highest mathematical levels; one might be a physicist, an engineer, a computer scientist, an economist, a geographer, a statistician or a psychologist. Perhaps the American mathematical community should be reckoned at 60 or 90 thousand with corresponding numbers in all the developed or developing countries.

Numerous regional, national, and international meetings are held periodically. There is lively activity in the writing and publishing of books at all levels, and there are more than 1600 individual technical journals to which it is appropriate to submit mathematical material.

These activities make up an international forum in which mathematics is perpetuated and innovated; in which discrepancies in practice and meaning are thrashed out.

### Further Readings. See Bibliography

R. Archibald; E. Bell; B. Boos and M. Niss; C. Boyer; F. Cajori; J. S. Frame; R. Gillings; E. Husserl; M. Kline [1972]; U. Libbrecht; Y. Mikami; J. Needham; O. Neugebauer; O. Neugebauer and A. Sachs; D. Struik; B. Van der Waerden

# The Tools
# of the Trade

WHAT AUXILIARY TOOLS or equipment are necessary for the pursuit of mathematics? There is a famous picture showing Archimedes poring over a problem drawn in the sand while Roman soldiers lurk menacingly in the background. This picture has penetrated the psyche of the profession and has helped to shape its external image. It tells us that mathematics is done with a minimum of tools—a bit of sand, perhaps, and an awful lot of brains.

Some mathematicians like to think that it could even be done in a dark closet by a solitary man drawing on the resources of a brilliant platonic intellect. It is true that mathematics does not require vast amounts of laboratory equipment, that "Gedankenexperimente" (thought-experiments) are largely what is needed. But it is by no means fair to say that mathematics is done totally in the head.

Perhaps, in very ancient days, primitive mathematics, like the great epics and like ancient religions, was transmitted by oral tradition. But it soon became clear that to do mathematics one must have, at the very least, instruments of writing or recording and of duplication. Before the invention of printing, there were "scribe factories" for the wholesale replication of documents.

The ruler and compass are built into the axioms at the foundation of Euclidean geometry. Euclidean geometry can be defined as the science of ruler-and-compass constructions.

Arithmetic has been aided by many instruments and devices. Three of the most successful have been the abacus, the slide rule, and the modern electronic computer. And, the logical capabilities of the computer have already relegated its arithmetic skills to secondary importance.

In the beginning, we used to count computers. There were four: one in Philadelphia, one in Aberdeen, one in

*Astrolabe, 1568.*

Cambridge, and one in Washington. Then there were ten. Then, suddenly, there were two hundred. The last figure heard was thirty–five thousand. The computers proliferated, and generation followed generation, until now the fifty dollar hand-held job packs more computing power than the hippopotamian hulks rusting in the Smithsonian: the ENIACS, the MARKS, the SEACS, and the GOLEMS. Perhaps tomorrow the $1.98 computer will flood the drugstores and become a throwaway object like a plastic razor or a piece of Kleenex.

Legend has it that in the late 1940s when old Tom Watson of the IBM corporation learned of the potentialities of the computer he estimated that two or three of them would take care of the needs of the nation. Neither he nor anyone else foresaw how the mathematical needs of the nation would rise up miraculously to fill the available computing power.

14

The relationship of computers to mathematics has been far more complex than laymen might suspect. Most people assume that anyone who calls himself a professional mathematician uses computing machines. In truth, compared to engineers, physicists, chemists, and economists, most mathematicians have been indifferent to and ignorant of the use of computers. Indeed, the notion that creative mathematical work could ever be mechanized seems, to many mathematicians, demeaning to their professional self-esteem. Of course, to the applied mathematician, working along with scientists and engineers to get numerical answers to practical questions, the computer has been an indispensable assistant for many years.

When programmed appropriately, the computer also has the ability to perform many symbolic mathematical operations. For example, it can do formal algebra, formal calculus, formal power series expansions and formal work in differential equations. It has been thought that a program like FORMAC or MACSYMA would be an invaluable aid to the applied mathematician. But this has not yet been the case, for reasons which are not clear.

In geometry, the computer is a drawing instrument of much greater power than any of the linkages and templates of the traditional drafting room. Computer graphics show beautifully shaded and colored pictures of "objects" which are only mathematically or programatically defined. The viewer would swear that these images are projected photographs of real objects. But he would be wrong; the "objects" depicted have no "real world" existence. In some cases, they could not possibly have such existence.

On the other hand, it is still sometimes more efficient to use a physical model rather than attempting a computer graphics display. A chemical engineering firm, with whose practice the writer is familiar, designs plants for the petrochemical industry. These plants often have reticulated piping arrangements of a very complicated nature. It is standard company practice to build a scaled, color-coded model from little plastic Tinker Toy parts and to work in a significant way with this physical model.

The computer served to intensify the study of numerical

analysis and to wake matrix theory from a fifty-year slumber. It called attention to the importance of logic and of the theory of discrete abstract structures. It led to the creation of new disciplines such as linear programming and the study of computational complexity.

Occasionally, as with the four-color problem (see Chapter 8.), it lent a substantial assist to a classical unsolved problem, as a helicopter might rescue a Conestoga wagon from sinking in the mud of the Pecos River. But all these effects were marginal. Most mathematical research continued to go on just as it would have if the computing machine did not exist.

Within the last few years, however, computers have had a noticeable impact in the field of pure mathematics. This may be the result of the arrival of a generation of mathematicians who learned computer programming in high school and to whom a computer terminal is as familiar as a telephone or a bicycle. One begins to see a change in mathematical research. There is greater interest in constructive and algorithmic results, and decreasing interest in purely existential or dialectical results that have little or no computational meaning. (See Chapter 4 for further discussion of these issues.) The fact that computers are available affects mathematics by luring mathematicians to move in di-

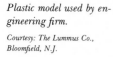

*Plastic model used by engineering firm.*

Courtesy: The Lummus Co., Bloomfield, N.J.

16

rections where the computer can play a part. Nevertheless, it is true, even today, that most mathematical research is carried on without any actual or potential use of computers.

**Further Readings. See Bibliography**

D. Hartree; W. Meyer zur Capellen; F. J. Murray [1961]; G. R. Stibitz; M. L. Dertouzos and J. Moses; H. H. Goldstine [1972] [1977]; I. Taviss; P. Henrici [1974]; J. Traub

# How Much Mathematics Is Now Known?

THE MATHEMATICS BOOKS at Brown University are housed on the fifth floor of the Sciences Library. In the trade, this is commonly regarded as a fine mathematical collection, and a rough calculation shows that this floor contains the equivalent of 60,000 average-sized volumes. Now there is a certain redundancy in the contents of these volumes and a certain deficiency in the Brown holdings, so let us say these balance out. To this figure we should add, perhaps, an equal quantity of mathematical material in adjacent areas such as engineering, physics, astronomy, cartography, or in new applied areas such as economics. In this way, we arrive at a total of, say, 100,000 volumes.

One hundred thousand volumes. This amount of knowledge and information is far beyond the comprehension of any one person. Yet it is small compared to other collections, such as physics, medicine, law, or literature. Within the lifetime of a man living today, the *whole* of mathematics was considered to be essentially within the grasp of a devoted student. The Russian-Swiss mathematician Alex-

17

ander Ostrowski once said that when he came up for his qualifying examination at the University of Marburg (around 1915) it was expected that he would be prepared to deal with any question in any branch of mathematics.

The same assertion would not be made today. In the late 1940s, John von Neumann estimated that a skilled mathematician might know, in essence, ten percent of what was available. There is a popular saying that knowledge always adds, never subtracts. This saying persists despite such shocking assessments as that of A. N. Whitehead who observed that Europe in 1500 knew less than Greece knew at the time of Archimedes. Mathematics builds on itself; it is aggregative. Algebra builds on arithmetic. Geometry builds on arithmetic and on algebra. Calculus builds on all three. Topology is an offshoot of geometry, set theory, and algebra. Differential equations builds on calculus, topology, and algebra. Mathematics is often depicted as a mighty tree with its roots, trunk, branches, and twigs labelled according to certain subdisciplines. It is a tree that grows in time.

*John von Neumann*
*1903–1957*

Constructs are enlarged and filled in. New theories are created. New mathematical objects are delineated and put under the spotlight. New relations and interconnections are found, thereby expressing new unities. New applications are sought and devised.

As this occurs, what is old and true is retained—at least in principle. Everything that once was mathematics remains mathematics—at least in principle. And so it would appear that the subject is a vast, increasing organism, with branch upon branch of theory and practice. The prior branch is prerequisite for the understanding of the subsequent branch. Thus, the student knows that in order to study and understand the theory of differential equations, he should have had courses in elementary calculus and in linear algebra. This serial dependence is in contrast to other disciplines, such as art or music. One can like or "understand" modern art without being familiar with baroque art; one can create jazz without any grounding in seventeenth century madrigals.

18

But while there is much truth in the view of mathematics as a cumulative science, this view as presented is somewhat naive. As mathematical textures are built up, there are concomitantly other processes at work which tend to break them down. Individual facts are found to be erroneous or incomplete. Theories become unpopular and are neglected. Work passes into obscurity and becomes grist for the mill of antiquarians (as with, say, prosthaphaeresic multiplication*). Other theories become saturated and are not pursued further. Older work is seen from modern perspectives and is recast, reformulated, while the older formulations may even become unintelligible (Newton's original writings can now be interpreted only by specialists). Applications become irrelevant and forgotten (the aerodynamics of Zeppelins). Superior methods are discovered and replace inferior ones (vast tables of special functions for computation are replaced by the wired-in approximations of the digital computer). All this contributes to a diminution of the material that must be held in the forefront of the mathematical consciousness.

There is also a loss of knowledge due to destruction or deterioration of the physical record. Libraries have been destroyed in wars and in social upheavals. And what is not accomplished by wars may be done by chemistry. The paper used in the early days of printing was much finer than what is used today. Around 1850 cheap, wood-pulp paper with acid-forming coatings was introduced, and the self-destructive qualities of this combination, together with our polluted atmosphere, can lead to the crumbling of pages as a book is read.

How many mathematics books should the Ph.D. candidate in mathematics know? The average candidate will take about fourteen to eighteen semester courses of undergraduate mathematics and sixteen graduate courses. At one book per course, and then doubling the answer for collateral and exploratory reading, we arrive at a figure of

* I.e., multiplication carried out by the addition of trigonometric functions.

19

about sixty to eighty volumes. In other words, two shelves of books will do the trick. This is a figure well within the range of human comprehension; it has to be.

Thus we can think of our 60,000 books as an ocean of knowledge, with an average depth of sixty or seventy books. At different locations within this ocean—that is, at different subspecialties within mathematics—we can take a depth sample, the two-foot bookshelf that would represent the basic education of a specialist in that area. Dividing 60,000 books by sixty, we find there should be at least 1,000 distinct subspecialties. But this is an underestimate, for many books would appear on more than one subspecialty's basic booklist. The coarse subdivision of mathematics, according to the AMS (MOS) Classification Scheme of 1980, is given in Appendix *B*. The fine structure would show mathematical writing broken down into more than 3,000 categories.

In most of these 3,000 categories, new mathematics is being created at a constantly increasing rate. The ocean is expanding, both in depth and in breadth.

### Further Readings. See Bibliography

J. von Neumann; C. S. Fisher

# Ulam's Dilemma

WE CAN USE THE TERM "Ulam's dilemma" for the situation which Stanislaw Ulam has described vividly in his autobiography, *Adventures of a Mathematician.*

"At a talk which I gave at a celebration of the twenty-fifth anniversary of the construction of von Neumann's computer in Princeton a few years ago, I suddenly started estimating silently in my mind how many theorems are published yearly in mathematical journals. I made a quick mental calculation and came to a number like one hundred thousand theorems per year. I mentioned this and my audience gasped. The next day two of the younger mathematicians in the audience came to tell me that, impressed by this enormous figure, they undertook a more systematic and detailed search in the Institute library. By multiplying the number of journals by the number of yearly issues, by the number of papers per issue and the average number of theorems per paper, their estimate came to nearly two hundred thousand theorems a year. If the number of theorems is larger than one can possibly survey, who can be trusted to judge what is 'important'? One cannot have survival of the fittest if there is no interaction. It is actually impossible to keep abreast of even the more outstanding and exciting results. How can one reconcile this with the view that mathematics will survive as a single science? In mathematics one becomes married to one's own little field. Because of this, the judgment of value in mathematical research is becoming more and more difficult, and most of us are becoming mainly technicians. The variety of objects worked on by young scientists is growing exponentially. Perhaps one should not call it a pollution of thought; it is possibly a mirror of the prodigality of nature which produces a million species of different insects."

All mathematicians recognize the situation that Ulam describes. Only within the narrow perspective of a particular specialty can one see a coherent pattern of development. What are the leading problems? What are the most important recent developments? It is possible to answer such questions within a narrow specialty such as, for example, "nonlinear second-order elliptic partial differential equations."

But to ask the same question in a broader context is almost useless, for two distinct reasons. First of all, there will rarely be any single person who is in command of recent work in more than two or three areas. An overall evaluation demands a synthesis of the judgments of many different people and some will be more critical, some more sympathetic. But even if this difficulty were not present, even if we had judges who knew and understood current research in all of mathematics, we would encounter a second difficulty: we have no stated criteria that would permit us to evaluate work in widely separated fields of mathematics. Consider, say, the two fields of nonlinear wave propagation and category-theoretic logic. From the viewpoint of those working in each of these areas, discoveries of great importance are being made. But it is doubtful if any one person knows what is going on in both of these fields. Certainly ninety-five percent of all professional mathematicians understand neither one nor the other.

Under these conditions, accurate judgment and rational planning are hardly possible. And, in fact, no one attempts to decide (in a global sense, inclusive of all mathematics) what is important, what is ephemeral.

Richard Courant wrote, many years ago, that the river of mathematics, if separated from physics, might break up into many separate little rivulets and finally dry up altogether. What has happened is rather different. It is as if the various streams of mathematics have overflowed their banks, run together, and flooded a vast plain, so that we see countless currents, separating and merging, some of them quite shallow and aimless. Those channels that are still deep and swift-flowing are easy to lose in the general chaos.

Spokesmen for federal funding agencies are very explicit in denying any attempt to evaluate or choose between one area of mathematics and another. If more research proposals are made in area $x$ and are favorably refereed, then more will be funded. In the absence of anyone who feels he has the right or the qualification to make value judgments, decisions are made "by the market" or "by public opinion." But democratic decision-making is supposed to be carried out with controversy and debate to create an informed electorate. In mathematical value judgments, however, we have virtually no debate or discussion, and the vote is more like the economic vote of the consumer who decides to buy or not to buy some commodity. Perhaps classical market economics and modern merchandising theory could shed some light on what will happen. There is no assurance of survival of the fittest, except in the tautological sense that whatever does in fact survive has thereby proved itself fittest—by definition!

Can we try to establish some rational principles by which one could sort through 200,000 theorems a year? Or should we simply accept that there is no more need to choose among theorems than to choose among species of insects? Neither course is entirely satisfactory. Nonetheless decisions are made every day as to what should be published and what should be funded. No one outside the profession is competent to make these decisions; within the profession, almost no one is competent to make them in any context broader than a narrow specialty. There are some exceptional mthematicians whose range of expertise includes several major specialties (for example, probability, combinatorics, and linear operator theory). By forming a committee of such people as these, one can constitute an editorial board for a major journal, or an advisory panel for a federal funding agency. How does such a committee reach its decision? Certainly not by debating and agreeing on fundamental choices of what is most valuable and important in mathematics today.

We find that our judgment of what is valuable in mathematics is based on our notion of the nature and purpose of mathematics itself. What is it to know something in mathematics? What sort of meaning is conveyed by mathematical statements? Thus, unavoidable problems of daily mathematical practice lead to fundamental questions of epistemology and ontology, but most professionals have learned to bypass such questions as irrelevant.

In practice, each member of the panel has a vital commitment to his own area (however skeptical he may be about everybody else) and the committee follows the political principle of nonaggression, or mutual indifference. Each "area" or "field" gets its quota, no one has to justify his own field's existence, and everyone tolerates the continued existence of various other "superfluous" branches of mathematics.

### Further Readings. See Bibliography

B. Boos and M. Niss; S. Ulam; Anon. "Federal Funds . . ."

# How Much
# Mathematics
# Can There Be?

WITH BILLIONS OF BITS of information being processed every second by machine, and with 200,000 mathematical theorems of the traditional, hand-crafted variety produced annually, it is clear that the world is in a Golden Age of mathematical production. Whether it is also a golden age for new mathematical ideas is another question altogether.

It would appear from the record that mankind can go on and on generating mathematics. But this may be a naive assessment based on linear (or exponential) extrapolation, an assessment that fails to take into account diminution due to irrelevance or obsolescence. Nor does it take into account the possibility of internal saturation. And it certainly postulates continuing support from the community at large.

The possibility of internal saturation is intriguing. The argument is that within a fairly limited mode of expression or operation there are only a very limited number of recognizably different forms, and while it would be possible to proliferate these forms indefinitely, a few prototypes adequately express the character of the mode. Thus, although it is said that no two snowflakes are identical, it is generally acknowledged that from the point of view of visual enjoyment, when you have seen a few, you have seen them all.

In mathematics, many areas show signs of internal exhaustion—for example, the elementary geometry of the circle and the triangle, or the classical theory of functions of a complex variable. While one can call on the former to provide five-finger exercises for beginners and the latter for applications to other areas, it seems unlikely that either will ever again produce anything that is both new and startling within its bounded confines.

It seems certain that there is a limit to the amount of living mathematics that humanity can sustain at any time. As new mathematical specialties arise, old ones will have to be neglected.

All experience so far seems to show that there are two inexhaustible sources of new mathematical questions. One source is the development of science and technology, which make ever new demands on mathematics for assistance. The other source is mathematics itself. As it becomes more elaborate and complex, each new, completed result becomes the potential starting point for several new investigations. Each pair of seemingly unrelated mathematical specialties pose an implicit challenge: to find a fruitful connection between them.

Although each special field in mathematics can be expected to become exhausted, and although the exponential growth in mathematical production is bound to level off sooner or later, it is hard to foresee an end to all mathematical production, except as part of an end to mankind's general striving for more knowledge and more power. Such an end to striving may indeed come about some day. Whether this end would be a triumph or a tragedy, it is far beyond any horizon now visible.

### Further Readings. See Bibliography

C. S. Fisher; J. von Neumann

# Appendix A

| | |
|---|---|
| 1637 | **Descartes.** Analytic geometry. Theory of equations. |
| 1650 | **Pascal.** Conics. Probability theory. |
| 1680 | **Newton.** Calculus. Theory of equations. Gravity. Planetary motion. Infinite series. Hydrostatics and dynamics. |
| 1682 | **Leibniz.** Calculus. |
| 1700 | **Bernoulli.** Calculus; probability. |
| 1750 | **Euler.** Calculus. Complex variables. Applied mathematics. |
| 1780 | **Lagrange.** Differential equations. Calculus of variations. |
| 1805 | **Laplace.** Differential equations. Planetary theory. Probability. |
| 1820 | **Gauss.** Number theory. Differential geometry. Algebra. Astronomy. |
| 1825 | **Bolyai, Lobatchevksy.** Non-Euclidean geometry. |
| 1854 | **Riemann.** Integration theory. Complex variables. Geometry. |
| 1880 | **Cantor.** Theory of infinite sets. |
| 1890 | **Weierstrass.** Real and complex analysis. |
| 1895 | **Poincaré.** Topology. Differential equations. |
| 1899 | **Hilbert.** Integral equations. Foundations of mathematics. |
| 1907 | **Brouwer.** Topology. Constructivism. |
| 1910 | **Russell, Whitehead.** Mathematical logic. |

Appendix A, continued.

---

### Brief Chronology of Ancient Chinese Mathematics

**"Chou Pei Suan Ching".** 300 BC (?) (Sacred Book of Arithmetic). Astronomical calculations, right triangles, fractions.

**"Chiu-chang Suan-shu"** (250 BC) (Arithmetic in Nine Sections.)

**Lui Hui** (250) **"Hai-tao Suan-ching"** (Sea Island Arithmetic Classic)

Anonymous, 300. **"Sun-Tsu Suan Ching"** (Arithmetic Classic of Sun-Tsu).

**Tsu Ch'ung-chih** (430–501). **"Chui-Shu".** (Art of Mending) $\pi \approx 355/113$.

**Wang Hs'iao-t'ung** (625) **"Ch'i-ku Suan-ching"** (Continuation of Ancient Mathematics) Cubic equations.

**Ch'in Chiu-shao** (1247). **"Su-shu Chiu-chang"** (Nine Chapters of Mathematics) Equations of higher degree. Horner's method.

**Li Yeh** (1192–1279). **"T'se-yuan Hai Ching"** (The Sea Mirror of the Circle Measurements) Geometric problems leading to equations of higher degree.

**Chu Shih-chieh** (1303). **"Szu-yuen Yu-chien".** (The Precious Mirror of the Four Elements) Pascal's triangle. Summation of series.

**Kuo Shou-ching** (1231–1316). "Shou-shih" calendar. Spherical trigonometry.

**Ch'eng Tai-wei** (1593). **"Suan-fa T'ung-tsung"** (A Systematized Treatise on Arithmetic). Oldest extant work that discusses the abacus.

**Ricci and Hsu** (1607). **"Chi-ho Yuan-pen"** (Elements of Geometry) Translation of Euclid.

---

# Appendix B

## THE CLASSIFICATION OF MATHEMATICS. 1868 AND 1979 COMPARED

**Subdivisions of the *Jahrbuch über die Fortschritte der Mathematik*, 1868.**

History and Philosophy

Algebra

Number Theory

Probability

Series

Differential and Integral Calculus

Theory of Functions

Analytic Geometry

Synthetic Geometry

Mechanics

Mathematical Physics

Geodesy and Astronomy

THIRTY-EIGHT SUBCATEGORIES

### The Classification of Mathematics, 1979 (From the *Mathematical Reviews*)

General
History and biography

Logic and foundations
Set theory
Combinatorics, graph theory
Order, lattices, ordered algebraic structures
General mathematical systems

Number theory
Algebraic number theory, field theory and polynomials
Commutative rings and algebras
Algebraic geometry
Linear and multilinear algebra; matrix theory
Associative rings and algebras

Nonassociative rings and
  algebras
Category theory,
  homological algebra

Group theory and
  generalizations
Topological groups, Lie
  groups

Functions of real variables
Measure and integration
Functions of a complex
  variable
Potential theory
Several complex variables
  and analytic spaces
Special functions
Ordinary differential
  equations
Partial differential
  equations
Finite differences and
  functional equations

Sequences, series,
  summability
Approximations and
  expansions
Fourier analysis
Abstract harmonic analysis
Integral transforms,
  operational calculus
Integral equations
Functional analysis
Operator theory
Calculus of variations and
  optimal control

Geometry
Convex sets and geometric
  inequalities
Differential geometry
General topology

Algebraic topology
Manifolds and cell
  complexes
Global analysis, analysis on
  manifolds

Probability theory and
  stochastic processes
Statistics
Numerical analysis
Computer science
General applied
  mathematics

Mechanics of particles and
  systems
Mechanics of solids
Fluid mechanics, acoustics
Optics, electromagnetic
  theory

Classical thermodynamics,
  heat transfer
Quantum mechanics
Statistical physics, structure
  of matter
Relativity
Astronomy and
  astrophysics
Geophysics

Economics, operations
  research, programming,
  games
Biology and behavioral
  sciences
Systems, control
Information and
  communication, circuits,
  automata

APPROXIMATELY 3400
  SUBCATEGORIES

30

# 2

## VARIETIES OF MATHEMATICAL EXPERIENCE

# The Current
# Individual and
# Collective
# Consciousness

"The whole cultural world, in all its forms, exists through tradition."
"Tradition is the forgetting of the origins."

Edmund Husserl, *"The Origin of Geometry"*

THERE IS A LIMITED amount of knowledge, practice, and aspiration which is currently manifested in the thoughts and activities of contemporary mathematicians. The mathematics that is frequently used or is in the process of emerging is part of the current consciousness. This is the material which—to use a metaphor from computer science—is in the high speed memory or storage cells. What is done, created, practiced, at any given moment of time can be viewed in two distinct ways: as part of the larger cultural and intellectual consciousness and milieu, frozen in time, or as part of a changing flow of consciousness.

What was in Archimedes' head was different from what was in Newton's head and this, in turn, differed from what was in Gauss's head. It is not just a matter of "more," that Gauss knew more mathematics than Newton who, in turn, knew more than Archimedes. It is also a matter of "different." The current state of knowledge is woven into a network of different motivations and aspirations, different interpretations and potentialities.

Archimedes, Newton, and Gauss all knew that in a triangle the sum of the angles adds to 180°. Archimedes knew this as a phenomenon of nature as well as a conclusion deduced on the basis of the axioms of Euclid. Newton knew the statement as a deduction and as application, but he might also have pondered the question of whether the statement is so true, so bound up with what is right in the universe, that God Almighty could not set it aside. Gauss knew that the statement was sometimes valid and sometimes invalid depending on how one started the game of deduction, and he worried about what other strange contradictions to Euclid could be derived on a similar basis.

Take a more elementary example. Counting and arithmetic can be and have been done in a variety of ways: by stones, by abacuses, by counting beads, by finger reckoning, with pencil and paper, with mechanical adding machines, with hand-held digital computers. Each of these modes leads one to a slightly different perception of, and a different relationship to, the integers. If there is an outcry today against children doing their sums by computer, the criers are correct in asserting that things won't be the same as they were when one struggled with pencil and paper arithmetic and all its nasty carryings and borrowings. They are wrong in thinking that pencil and paper arithmetic is ideal, and that what replaces it is not viable.

To understand the mathematics of an earlier period requires that we penetrate the contemporary individual and collective consciousness. This is a particularly difficult task because the formal and informal mathematical writings that come down to us do not describe the network of consciousness in any detail. It is unlikely that the meaning of mathematics could be reconstructed on the basis of the printed record alone. The sketches that follow are intended to give some insight into the inner feelings that can lie behind mathematical engagement.

# The Ideal
# Mathematician

W

E WILL CONSTRUCT a portrait of the "ideal mathematician." By this we do not mean the perfect mathematician, the mathematician without defect or limitation. Rather, we mean to describe the most mathematician-like mathematician, as one might describe the ideal thoroughbred greyhound, or the ideal thirteenth-century monk. We will try to construct an impossibly pure specimen, in order to exhibit the paradoxical and problematical aspects of the mathematician's role. In particular, we want to display clearly the discrepancy between the actual work and activity of the mathematician and his own perception of his work and activity.

The ideal mathematician's work is intelligible only to a small group of specialists, numbering a few dozen or at most a few hundred. This group has existed only for a few decades, and there is every possibility that it may become extinct in another few decades. However, the mathematician regards his work as part of the very structure of the world, containing truths which are valid forever, from the beginning of time, even in the most remote corner of the universe.

He rests his faith on rigorous proof; he believes that the difference between a correct proof and an incorrect one is an unmistakable and decisive difference. He can think of no condemnation more damning than to say of a student, "He doesn't even know what a proof is." Yet he is able to give no coherent explanation of what is meant by rigor, or what is required to make a proof rigorous. In his own work, the line between complete and incomplete proof is always somewhat fuzzy, and often controversial.

To talk about the ideal mathematician at all, we must have a name for his "field," his subject. Let's call it, for instance, "non-Riemannian hypersquares."

34

He is labeled by his field, by how much he publishes, and especially by whose work he uses, and by whose taste he follows in his choice of problems.

He studies objects whose existence is unsuspected by all except a handful of his fellows. Indeed, if one who is not an initiate asks him what he studies, he is incapable of showing or telling what it is. It is necessary to go through an arduous apprenticeship of several years to understand the theory to which he is devoted. Only then would one's mind be prepared to receive his explanation of what he is studying. Short of that, one could be given a "definition," which would be so recondite as to defeat all attempts at comprehension.

The objects which our mathematician studies were unknown before the twentieth century; most likely, they were unknown even thirty years ago. Today they are the chief interest in life for a few dozen (at most, a few hundred) of his comrades. He and his comrades do not doubt, however, that non-Riemannian hypersquares have a real existence as definite and objective as that of the Rock of Gibraltar or Halley's comet. In fact, the proof of the existence of non-Riemannian hypersquares is one of their main achievements, whereas the existence of the Rock of Gibraltar is very probable, but not rigorously proved.

It has never occurred to him to question what the word "exist" means here. One could try to discover its meaning by watching him at work and observing what the word "exist" signifies operationally.

In any case, for him the non-Riemannian hypersquare exists, and he pursues it with passionate devotion. He spends all his days in contemplating it. His life is successful to the extent that he can discover new facts about it.

He finds it difficult to establish meaningful conversation with that large portion of humanity that has never heard of a non-Riemannian hypersquare. This creates grave difficulties for him; there are two colleagues in his department who know something about non-Riemannian hypersquares, but one of them is on sabbatical, and the other is much more interested in non-Eulerian semirings. He goes to conferences, and on summer visits to colleagues, to meet people who talk his language, who can appreciate his work and whose recognition, approval, and admiration are the only meaningful rewards he can ever hope for.

At the conferences, the principal topic is usually "the decision problem" (or perhaps "the construction problem" or "the classification problem") for non-Riemannian hypersquares. This problem was first stated by Professor Nameless, the founder of the theory of non-Riemannian hypersquares. It is important because Professor Nameless stated it and gave a partial solution which, unfortunately, no one but Professor Nameless was ever able to understand. Since Professor Nameless' day, all the best non-Riemannian hypersquarers have worked on the problem, obtaining many partial results. Thus the problem has acquired great prestige.

Our hero often dreams he has solved it. He has twice convinced himself during waking hours that he had solved it but, both times, a gap in his reasoning was discovered by other non-Riemannian devotees, and the problem remains open. In the meantime, he continues to discover new and interesting facts about the non-Riemannian hypersquares. To his fellow experts, he communicates these results in a casual shorthand. "If you apply a tangential mollifier to the left quasi-martingale, you can get an estimate better than quadratic, so the convergence in the Bergstein theorem turns out to be of the same order as the degree of approximation in the Steinberg theorem."

This breezy style is not to be found in his published writings. There he piles up formalism on top of formalism. Three pages of definitions are followed by seven lemmas and, finally, a theorem whose hypotheses take half a page to state, while its proof reduces essentially to "Apply Lemmas 1–7 to definitions A–H."

His writing follows an unbreakable convention: to conceal any sign that the author or the intended reader is a human being. It gives the impression that, from the stated definitions, the desired results follow infallibly by a purely mechanical procedure. In fact, no computing machine has ever been built that could accept his definitions as inputs. To read his proofs, one must be privy to a whole subculture of motivations, standard arguments and examples, habits of thought and agreed-upon modes of reasoning. The intended readers (all twelve of them) can decode the formal presentation, detect the new idea hidden in lemma 4, ignore the routine and uninteresting calculations of lemmas 1, 2, 3, 5, 6, 7, and see what the author is doing and why he does it. But for the noninitiate, this is a cipher that will never yield its secret. If (heaven forbid) the fraternity of non-Riemannian hypersquarers should ever die out, our hero's writings would become less translatable than those of the Maya.

The difficulties of communication emerged vividly when the ideal mathematician received a visit from a public information officer of the University.

*P.I.O.*     I appreciate your taking time to talk to me. Mathematics was always my worst subject.

*I.M.:*      That's O.K. You've got your job to do.

*P.I.O.*     I was given the assignment of writing a press release about the renewal of your grant. The usual thing would be a one-sentence item, "Professor X received a grant of $Y$ dollars to continue his research on the decision problem for non-Riemannian hypersquares." But I thought it would be a good challenge for me to try and give people a better idea about what your work really involves. First of all, what is a hypersquare?

*I.M.*       I hate to say this, but the truth is, if I told you what it is, you would think I was trying to put you down and make you feel stupid. The definition is really somewhat technical, and it just wouldn't mean anything at all to most people.

*P.I.O.*     Would it be something engineers or physicists would know about?

*I.M.*       No. Well, maybe a few theoretical physicists. Very few.

*P.I.O.*     Even if you can't give me the real definition, can't you give me some idea of the general nature and purpose of your work?

*I.M.:*      All right, I'll try. Consider a smooth function $f$ on a measure space $\Omega$ taking its value in a sheaf of germs equipped with a convergence structure of saturated type. In the simplest case . . .

*P.I.O.:*      Perhaps I'm asking the wrong questions. Can you tell me something about the applications of your research?

*I.M.:*      Applications?

*P.I.O.:*      Yes, applications.

*I.M.:*      I've been told that some attempts have been made to use non-Riemannian hypersquares as models for elementary particles in nuclear physics. I don't know if any progress was made.

*P.I.O.:*      Have there been any major breakthroughs recently in your area? Any exciting new results that people are talking about?

*I.M.:*      Sure, there's the Steinberg-Bergstein paper. That's the biggest advance in at least five years.

*P.I.O.:*      What did they do?

*I.M.:*      I can't tell you.

*P.I.O.:*      I see. Do you feel there is adequate support in research in your field?

*I.M.:*      Adequate? It's hardly lip service. Some of the best young people in the field are being denied research support. I have no doubt that with extra support we could be making much more rapid progress on the decision problem.

*P.I.O.:*      Do you see any way that the work in your area could lead to anything that would be understandable to the ordinary citizen of this country?

*I.M.:*      No.

*P.I.O.:*      How about engineers or scientists?

*I.M.:*      I doubt it very much.

*P.I.O.:*      Among pure mathematicians, would the majority be interested in or acquainted with your work?

*I.M.:*      No, it would be a small minority.

*P.I.O.:*      Is there anything at all that you would like to say about your work?

*I.M.:*      Just the usual one sentence will be fine.

*P.I.O.:*   Don't you want the public to sympathize with your work and support it?

*I.M.:*   Sure, but not if it means debasing myself.

*P.I.O.:*   Debasing yourself?

*I.M.:*   Getting involved in public relations gimmicks, that sort of thing.

*P.I.O.:*   I see. Well, thanks again for your time.

*I.M.:*   That's O.K. You've got a job to do.

*Alfred Tarski*
*1902–*

Well, a public relations officer. What can one expect? Let's see how our ideal mathematician made out with a student who came to him with a strange question.

*Student:*   Sir, what is a mathematical proof?

*I.M.:*   You don't know *that*? What year are you in?

*Student:*   Third-year graduate.

*I.M.:*   Incredible! A proof is what you've been watching me do at the board three times a week for three years! That's what a proof is.

*Student:*   Sorry, sir, I should have explained. I'm in philosophy, not math. I've never taken your course.

*I.M.:*   Oh! Well, in that case—you have taken *some* math, haven't you? You know the proof of the fundamental theorem of calculus—or the fundamental theorem of algebra?

*Student:*   I've seen arguments in geometry and algebra and calculus that were called proofs. What I'm asking you for isn't *examples* of proof, it's a definition of proof. Otherwise, how can I tell what examples are correct?

*I.M.:*   Well, this whole thing was cleared up by the logician Tarski, I guess, and some others, maybe Russell or Peano. Anyhow, what you do is, you write down the axioms of your theory in a formal language with a given list of symbols or alphabet. Then you write down the hypothesis of your theorem in the same symbolism. Then you show that you can transform the hypothesis step by step, using the rules of logic, till you get the conclusion. That's a proof.

*Student:*   Really? That's amazing! I've taken elementary

|          | and advanced calculus, basic algebra, and topology, and I've never seen that done. |
| :------- | :--- |
| *I.M.:* | Oh, of course no one ever really *does* it. It would take forever! You just show that you could do it, that's sufficient. |
| *Student:* | But even that doesn't sound like what was done in my courses and textbooks. So mathematicians don't really do proofs, after all. |
| *I.M.:* | Of course we do! If a theorem isn't proved, it's nothing. |
| *Student:* | Then what is a proof? If it's this thing with a formal language and transforming formulas, nobody ever proves anything. Do you have to know all about formal languages and formal logic before you can do a mathematical proof? |
| *I.M.:* | Of course not! The less you know, the better. That stuff is all abstract nonsense anyway. |
| *Student:* | Then really what *is* a proof? |
| *I.M.:* | Well, it's an argument that convinces someone who knows the subject. |
| *Student:* | Someone who knows the subject? Then the definition of proof is subjective; it depends on particular persons. Before I can decide if something is a proof, I have to decide who the experts are. What does that have to do with proving things? |
| *I.M.:* | No, no. There's nothing subjective about it! Everybody knows what a proof is. Just read some books, take courses from a competent mathematician, and you'll catch on. |
| *Student:* | Are you sure? |
| *I.M.:* | Well—it is possible that you won't, if you don't have any aptitude for it. That can happen, too. |
| *Student:* | Then *you* decide what a proof is, and if I don't learn to decide in the same way, you decide I don't have any aptitude. |
| *I.M.:* | If not me, then who? |

Then the ideal mathematician met a positivist philosopher.

*P.P.:*  This Platonism of yours is rather incredible. The silliest undergraduate knows enough not to multiply entities, and here you've got not just a handful, you've got them in uncountable infinities! And nobody knows about them but you and your pals! Who do you think you're kidding?

*I.M.:*  I'm not interested in philosophy, I'm a mathematician.

*P.P.:*  You're as bad as that character in Molière who didn't know he was talking prose! You've been committing philosophical nonsense with your "rigorous proofs of existence." Don't you know that what exists has to be observed, or at least observable?

*I.M.:*  Look, I don't have time to get into philosphical controversies. Frankly, I doubt that you people know what you're talking about; otherwise you could state it in a precise form so that I could understand it and check your argument. As far as my being a Platonist, that's just a handy figure of speech. I never thought hypersquares existed. When I say they do, all I mean is that the axioms for a hypersquare possess a model. In other words, no formal contradiction can be deduced from them, and so, in the normal mathematical fashion, we are free to postulate their existence. The whole thing doesn't really mean anything, it's just a game, like chess, that we play with axioms and rules of inference.

*P.P.:*  Well, I didn't mean to be too hard on you. I'm sure it helps you in your research to imagine you're talking about something real.

*I.M.:*  I'm not a philosopher, philosophy bores me. You argue, argue and never get anywhere. My job is to prove theorems, not to worry about what they mean.

The ideal mathematician feels prepared, if the occasion should arise, to meet an extragalactic intelligence. His first effort to communicate would be to write down (or otherwise transmit) the first few hundred digits in the binary expansion of pi. He regards it as obvious that any intelligence capable of intergalactic communication would be mathematical and that it makes sense to talk about mathematical intelligence apart from the thoughts and actions of human beings. Moreover, he regards it as obvious that binary representation and the real number pi are both part of the intrinsic order of the universe.

He will admit that neither of them is a natural object, but he will insist that they are discovered, not invented. Their discovery, in something like the form in which we know them, is inevitable if one rises far enough above the primordial slime to communicate with other galaxies (or even with other solar systems).

The following dialogue once took place between the ideal mathematician and a skeptical classicist.

*S.C.:* You believe in your numbers and curves just as Christian missionaries believed in their crucifixes. If a missionary had gone to the moon in 1500, he would have been waving his crucifix to show the moon-men that he was a Christian, and expecting them to have their own symbol to wave back.* You're even more arrogant about your expansion of pi.

*I.M.:* Arrogant? It's been checked and rechecked, to 100,000 places!

*S.C.:* I've seen how little you have to say even to an American mathematician who doesn't know your game with hypersquares. You don't get to first base trying to communicate with a theoretical physicist; you can't read his papers any more than he can

---

* Cf. the description of Coronado's expedition to Cibola, in 1540: ". . . there were about eighty horsemen in the vanguard besides twenty-five or thirty foot and a large number of Indian allies. In the party went all the priests, since none of them wished to remain behind with the army. It was their part to deal with the friendly Indians whom they might encounter, and they especially were bearers of the Cross, a symbol which . . . had already come to exert an influence over the natives on the way" (H. E. Bolton, *Coronado,* University of New Mexico Press, 1949).

read yours. The research papers in your own
field written before 1910 are as dead to you as
Tutankhamen's will. What reason in the world is
there to think that you could communicate with
an extragalactic intelligence?

*I.M.:*   If not me, then who else?

*S.C.:*   Anybody else! Wouldn't life and death, love and
hate, joy and despair be messages more likely to
be universal than a dry pedantic formula that nobody but you and a few hundred of your type will
know from a hen-scratch in a farmyard?

*I.M.:*   The reason that my formulas are appropriate for
intergalactic communication is the same reason
they are not very suitable for terrestrial communication. Their content is not earthbound. It is
free of the specifically human.

*S.C.:*   I don't suppose the missionary would have said
quite that about his crucifix, but probably something rather close, and certainly no less absurd
and pretentious.

The foregoing sketches are not meant to be malicious;
indeed, they would apply to the present authors. But it is a
too obvious and therefore easily forgotten fact that mathematical work, which, no doubt as a result of long familiarity, the mathematician takes for granted, is a mysterious,
almost inexplicable phenomenon from the point of view of
the outsider. In this case, the outsider could be a layman, a
fellow academic, or even a scientist who uses mathematics
in his own work.

The mathematician usually assumes that his own view of
himself is the only one that need be considered. Would we
allow the same claim to any other esoteric fraternity? Or
would a dispassionate description of its activities by an ob-
servant, informed outsider be more reliable than that of a
participant who may be incapable of noticing, not to say
questioning, the beliefs of his coterie?

Mathematicians know that they are studying an objective
reality. To an outsider, they seem to be engaged in an esoteric communion with themselves and a small clique of

friends. How could we as mathematicians prove to a skeptical outsider that our theorems have meaning in the world outside our own fraternity?

If such a person accepts our discipline, and goes through two or three years of graduate study in mathematics, he absorbs our way of thinking, and is no longer the critical outsider he once was. In the same way, a critic of Scientology who underwent several years of "study" under "recognized authorities" in Scientology might well emerge a believer instead of a critic.

If the student is unable to absorb our way of thinking, we flunk him out, of course. If he gets through our obstacle course and then decides that our arguments are unclear or incorrect, we dismiss him as a crank, crackpot, or misfit.

Of course, none of this proves that we are not correct in our self-perception that we have a reliable method for discovering objective truths. But we must pause to realize that, outside our coterie, much of what we do is incomprehensible. There is no way we could convince a self-confident skeptic that the things we are talking about make sense, let alone "exist."

# A Physicist Looks at Mathematics

HOW DO PHYSICISTS view mathematics? Instead of answering this question by summarizing the writings of many physicists, we interviewed one physicist whose scientific feelings were judged to be representative. Since the summary which follows cannot represent his full and precise views, his name has been changed.

Professor William F. Taylor is an international authority in Engineering Science. He is actively engaged in teaching and research, and maintains extensive scientific connections. In August, 1977, the writer interviewed Professor Taylor in Wilmington, Vermont where he and his wife were on vacation enjoying tennis and the Marlboro Concerts. In the interview, an attempt was made not to confront the interviewee with opposing views and not to engage in argumentation.

Professor Taylor says that his professional field lies at the intersection of physics, chemistry, and materials science. He does not care to describe this combination by a single word. Although he uses mathematics extensively, he says he is definitely not an applied mathematician. He thinks, though, that many of his views would be held by applied mathematicians.

Taylor makes frequent computations. When asked whether he thought of himself as a creator or a consumer of mathematics, he answered that he was a consumer. He added that most of the mathematics he uses is of a nineteenth century variety. With respect to contemporary mathematical research he says that he feels drawn to it intellectually. It appears to unify a wide variety of complex structures. He is not, however, sufficiently motivated to learn any of it because he feels it has little applicability to his work. He thinks that much of the recently developed mathematics has gone beyond what is useful.

He seemed to be aware of the broad outline of the newly developed "nonstandard" analysis. He said,

> That subject looks very interesting to me, and I wish I could take out the time to master it. There are numerous places in my field where one is confronted with things that are going on simultaneously at totally different size scales. They are very difficult to deal with by conventional methods. Perhaps nonstandard analysis with its infinitesimals might provide a handle for this sort of thing.

Taylor asserts that only seldom in his professional work does he think along philosophic lines. He has done a small amount of reading in the philosophy of science and the philosophy of physics, principally in the area of quantum physics. He finds questions as to how and to what extent processes are affected by the mode of observation particularly interesting. He says that such questions have affected his professional work and outlook somewhat although he has not written anything of a formal nature about it.

Although his personal familiarity with the philosophy of science may be said to be slight, he believes it to be an important line of inquiry, and he welcomed the present interview and framed his answers thoughtfully and with gusto.

Taylor is unaware of the main classical issues of mathematical philosophy. In response to the question of whether there were or had been any crises in mathematics, he answered that he had heard of Russell's Paradox, but it seemed to be quite remote from anything he was interested in. "It was nothing I should worry about," he said.

Taylor's approach to science, to mathematics, and to a variety of related philosophic issues can be summed up by saying that he is a strong and eloquent spokesman for the model theory or approach. This holds that physical theories are provisional models of reality. He uses the word "model" frequently and brings around his arguments to this approach. Mathematics itself is a model, he says. Questions as to the truth or the indubitability of mathematics are not important to him because all scientific work of every kind is of a provisional nature. The question should be not how true it is but how good it is. In the interview, he elaborated at length on what he meant by "good" and this was done from the vantage point of models.

As part of his elaboration, he answered along the following lines. There are many situations in physics that are very messy. They may contain too many mutually interacting phenomena of equal degrees of importance. In such a situation there is no hope whatever of setting up something which can be asserted to be the "real thing." The best one can hope for is a model which is a partial truth. It is a tentative thing and one hopes the best for it. All physical theories are models. A model should be able at the very least to describe certain phenomena fairly accurately. Even at this level one runs into trouble in constructing models. The models that one constructs are of course dependent upon one's state of knowledge. Ideally, a model should have predictive value. Therefore it is no good to construct

46

a model which is too complex to support reason. Whether it is or it is not too complex may depend upon the current state of the mathematical or computational art. But one has to be in a position to derive mathematical and hence physical consequences from the model, and if this is found to be impossible—and it may be so for a variety of reasons—then the model has little significance.

Professor Taylor was asked to comment on the contemporary view that the scientific method can be summed up by the sequence: induction, deduction, verification, iterated as often as necessary. He replied that he went along with it in its broad outlines. But he wanted to elaborate.

> Induction is related to my awareness of the observations of others and of existing theories. Deduction is related to the construction of a model and of physical conclusions drawn from it by means of mathematical derivations. Verification is related to predictions of phenomena not yet observed and to the hope that the experimentalist will look for new phenomena.
>
> The experimentalist and the theoretician need one another. The experimentalist needs a model to help him lay out his experiments. Otherwise he doesn't know where to look. He would be working in the dark. The theoretician needs the experimentalist to tell him what is going on in the real world. Otherwise his theorizing would be empty. There must be adequate communication between the two and, in fact, I think there is.

When asked why the profession splits into two types—experimentalists and theoreticians—he said that apart from a general tendency to specialize, it was probably a matter of temperament. "But the gap is always bridged—usually by the theoretician."

Professor Taylor was asked how he felt about the often quoted remark of a certain theoretician that he would rather his theories be beautiful than be right.

> This cuts close to the bone. It really does. But as I see it, mere aesthetics doesn't pay dividends. In my experience, I should be inclined to replace the word "beautiful" by the word "analyzable." I should like my models to be beautiful,

47

effective, and predictive. But the real goal is the understanding of a situation. Therefore the models must be ana-
lyzable because understanding can come only through analyzability. If one has all of these things, then this is a great
and rare achievement, but I should say that my immediate
goal is analyzability.

What were his views on mathematical proof? Professor
Taylor said that his papers rarely contain formal proofs of
a sort that would satisfy a mathematician. To him, proofs
were relatively uninteresting and they were largely unnecessary in his personal work. Yet, he felt that his work contained elements that could be described as mathematical
reasoning or deduction. Truth in mathematics, he said, is
reasoning that leads to correct physical relationships. Empirical demonstrations are possible. True reasoning should
be capable of being put into the format of a mathematical
proof. It is nice to have this done ultimately. Proof is for
cosmetic purposes and also to reduce somewhat the edge
of insecurity on which one always lives. However, for him
to engage in mathematical proof would seriously take him
away from his main interests and methodology.

In view of Professor Taylor's familiarity with computational procedures, he was asked to comment on the current opinion that the object of numerical science or numerical physics is to replace experimentation. He thought
a while and then replied,

> I think one has to distinguish here between the require-
> ments of technology and those of pure science. To the for-
> mer, I would reply a limited "yes"; to the latter "no." Con-
> sider a problem in technology. One has a pressure vessel
> which is subject to many many cycles of heating and cool-
> ing. How many cycles can it stand? Now, if one really knew
> the process that leads to failure (which is not yet the case)
> one could say that in a specific instance it might be much
> more effective to make a computer experiment than an ac-
> tual experiment. Here one is dealing with something like a
> "production" situation.
>
> On the other hand, in pure science, the elimination of
> experimentation is a contradiction in terms. The way one
> finds out what is going on in the universe is through ex-

perimentation. This is where new experiences, new facts come from.

There is no point to run experiments on bodies falling in a vacuum. Newtonian mechanics is known to be an adequate model. But if one goes, say, to cosmology, where it isn't known whether existing models are adequate or are not adequate, then numerical computation is insufficient.

Asked whether it would be possible to imagine a kind of theoretical physics without mathematics, Professor Taylor answered that it would not be possible.

Asked the same question for technology, he answered again that it would not be possible.

He added that the mathematics of technology was perhaps more elementary and more completely studied than that of modern physics, but it was mathematics, nonetheless. The role of mathematics in physics or in technology is that of a powerful reasoning tool in complex situations.

He was then asked why mathematics was so effective in physics and technology. The interviewer underlined that the word "effective" was one used by Professor Eugene Wigner in a famous article, "The Unreasonable Effectiveness of Mathematics in the Natural Sciences." "This has to do," he answered,

> with our current convention or system of beliefs as to what constitutes understanding. In these fields we mean by 'understanding' precisely those things which are explainable or predictable by mathematics. You may think this is going around in circles, and so it may be. The question of course is fundamentally unanswerable, and this is the way I care to frame my answer. Understanding means understanding through mathematics.

"Do you rule out other types of understanding?"

> There is what might be called humanistic or cross-cultural understanding. I have been reading Jacques Barzun and Theodore Roszak recently. What is the great concern with numbers and decimal points, they seem to be asking. One sees it in the old poem of Walt Whitman called "The Astronomer." Whitman had heard a lecture in astronomy in Cooper Union Hall. After the lecture he went outside,

looked up at the heavens, and felt a certain release at being freed from theories and symbols. He felt the exhilaration of being confronted by naked experience, if you will.

Now this may be a valid point of view, but it leads to a different end result. Quantitative science—that is, science with mathematics—has proved effective in altering and controlling nature. The majority of society backs it up for this reason. At the present moment, they want nature altered and controlled—to the extent, of course that we can do it and the results are felicitous. The humanist point of view is a minority point of view. But it is influential—one sees this among young people. It seems to have a defensive nature to it, a chip on its shoulder, but because it is a minority point of view, it poses only a minor threat to quantitative science.

"With regard to the conflict of the 'Two Worlds,' which of the two, the scientist and the humanist, knows more about the other man's business?"

The scientist very definitely knows more about the humanities than the other way around. The scientist—well, many that I know anyway—are forever reading novels, essays, criticism, etc., go to concerts, theatres, to art shows. The humanists very seldom read anything about science other than what they find in the newspaper. Part of the reason for this lies in the fact that the locus of the humanities is to be found in sound, vision, and common language. The language of science with its substantial sublanguage of mathematics poses a formidable barrier to the humanist.

The goals of society may change, of course. If they do, then the goals of quantitative science may be weakened. Science and mathematics might be pursued only by a small but interested minority. It might not be possible to make a living at it. We saw a very slight indication of this in the late sixties and early seventies.

"Can there be knowledge without words, without symbols?"

Knowledge, as I understand it in the technical sense, implies that it can be expressed in symbols. Moving towards humanistic questions, one might say that a skillful writer evokes a mood by his use of words. Or when a Mozart score

50

is played, it evokes a kind of conscious state. The symbolic words and the music are a model for the state.

"Does a cat have knowledge?"

"A cat knows certain things. But this is knowledge of a different kind. We are not dealing here with theoretical knowledge."

"When a flower brings forth a blossom with six-fold symmetry, is it doing mathematics?"

"It is not."

"Would you care to comment on the old Greek saying that God is a Mathematician?"

"This conveys nothing to me. It is not a useful concept."

"What is scientific or mathematical intuition?"

"Intuition is an expression of experience. Stored experience. There is an inequality in people with respect to it. Some people gain intuition more rapidly than others."

"To what extent can one be deceived by intuition?"

"This occurs not infrequently. It is a large part of my own work. I say to myself, this model seems to be sufficient, but it just doesn't sit right. Or, I ask myself, is my model a better one than their model? And I probably have to answer on the basis of intuition."

The final question put to Taylor was whether he is a mathematical Platonist in the sense that he believes that mathematical concepts exist in the world apart from the people that do mathematics. He replied that he was, but in a limited sense. Certainly not in a "theological" sense. He believed that certain concepts turn out to be so far superior to others that it is only a matter of time before these concepts prevail and are universally adopted. This is something like a Darwinian process, a survival of the fittest ideas, models, constructs. The evolution of mathematics and theoretical physics is something like the evolution of biosystems.

# I. R. Shafarevitch and the New Neoplatonism

ONE OF THE WORLD'S leading researchers in algebraic geometry is also one of the leading figures among the "dissenters" in the Soviet Union. I. R. Shafarevitch was mentioned in a survey article in the *New York Times* as a representative of that ideological tendency in Russia which sees Orthodox Christianity as a central and essential element in the life and character of the Russian people.

Shafarevitch discussed his views on the relation between mathematics and religion in a lecture he gave on the occasion of his receiving a prize from the Academy of Science at Göttingen, West Germany. We quote from his lecture.

"A superficial glance at mathematics may give an impression that it is a result of separate individual efforts of many scientists scattered about in continents and in ages. However, the inner logic of its development reminds one much more of the work of a single intellect, developing its thought systematically and consistently using the variety of human individualities only as a means. It resembles an orchestra performing a symphony composed by someone. A theme passes from one instrument to another, and when one of the participants is bound to drop his part, it is taken up by another and performed with irreproachable precision.

"This is by no means a figure of speech. The history of mathematics has known many cases when a discovery made by one scientist remains unknown until it is later reproduced by another with striking precision. In the letter written on the eve of his fatal duel, Galois made several assertions of paramount importance concerning integrals of algebraic functions. More than twenty years later Riemann, who undoubtedly knew nothing about the letter of Galois, found anew and proved exactly the same assertions. Another example: after Lobachevski and Bolyai laid the foundation of non-Euclidean geometry independently of one another, it became known that two other men, Gauss and Schweikart, also working independently, had both come to the same results ten years before. One is overwhelmed by a curious feeling when one sees the same designs as if drawn by a single hand in the work done by four scientists quite independently of one another.

"One is struck by the idea that such a wonderfully puzzling and mysterious activity of mankind, an activity that has continued for thousands of years, cannot be a mere chance—it must have some goal. Having recognized this we inevitably are faced by the question: *What is this goal?*

*I. R. Shafarevitch*

"Any activity devoid of a goal, by this very fact loses its sense. If we compare mathematics to a living organism, mathematics does not resemble conscious and purposeful activity. It is more like instinctive actions which are repeated stereotypically, directed by an external or internal stimulus.

"Without a definite goal, mathematics cannot develop any idea of its own form. The only thing left to it, as an ideal, is uncontrolled growth, or more precisely, expansion in all directions. Using another simile, one can say that the development of mathematics is different from the growth of a living organism which preserves its form and defines its own border as it grows. This development is much more akin to the growth of crystals or the diffusion of gas which will expand freely until it meets some outside obstacle.

"More than two thousand years of history have convinced us that mathematics cannot formulate for itself this final goal that can direct its progress. Hence it must take it from outside. It goes without saying that I am far from attempting to point out a solution of this problem, which is not only the inner problem of mathematics but the problem of mankind at large. I want only to indicate the main directions of the search for this solution.

Apparently there are two possible directions. In the first place one may try to extract the goal of mathematics from its practical applications. But it is hard to believe that a superior (spiritual) activity will find its justification in the inferior (material) activity. In the "Gospel according to Thomas" discovered in 1945,* Jesus says ironically:

> If the flesh came for the sake of the spirit, it
> is a miracle. But if the spirit for the sake
> of the flesh—it is a miracle of miracles.

All the history of mathematics is a convincing proof that such a "miracle of miracles" is impossible. If we look upon the decisive moment in the development of mathematics, the moment when it took its first step and when the ground on which it is based came into being—I have in mind *logical proof*—we shall see that this was done with material that actually excluded the very possibility of practical applications. The first theorems of Thales of Miletus proved statements evident to every sensible man—for instance that a diameter divides the circle into two equal parts. Genius was needed not to be convinced of the justice of these statements, but to understand that they need proofs. Obviously the practical value of such discoveries is nought.

In ending, I want to express a hope that . . . mathematics may serve now as a model for the solution of the main problem of our epoch: to reveal a supreme religious goal and to fathom the meaning of the spiritual activity of mankind."

Thus, Shafarevitch—a surprising statement to come from the lips of any contemporary mathematician in or out of Russia. But it is hardly a new statement. The Greek philosophers thought of mathematics as a bridge between theology and the perceptible, physical world, and this view was stressed and developed by the Neoplatonists. The quadrivium: arithmetic, music, geometry, astronomy, already known to Protagoras (d. 411 B.C.), was thought to

---

* (Footnote added by P.J.D.) The Gospel of Thomas is probably the most significant of the books discovered in the 1940s at Nag Hammadi in Egypt. It is a compilation of the "sayings of Jesus," placed in a Gnostic context. Gnosticism asserts that there is a secret knowledge (gnosis) through which salvation can be achieved and that this knowledge is superior to ordinary faith. (See R. M. Grant, "Gnosticism, Marcion, Origen" in "The Crucible of Christianity," A. Toynbee, ed., London: Thames and Hudson, 1969.)

lead the mind upward through mathematics to the heavenly sphere where the eternal movements were the perceptible form of the world soul.

### Further Readings. See Bibliography

P. Merlan; I. R. Shafarevitch

# Unorthodoxies

OST MATHEMATICIANS have had the following experience and those whose activities are somewhat more public have had it often: an unsolicited letter arrives from an unknown individual and contained in the letter is a piece of mathematics of a very sensational nature. The writer claims that he has solved one of the great unsolved mathematical problems or that he has refuted one of the standard mathematical assertions. In times gone by, circle squaring was a favorite activity; in fact, this activity is so old that Aristophanes parodies the circle squarers of the world. In more recent times, proofs of Fermat's "Last Theorem" have been very popular. The writer of such a letter is usually an amateur, with very little training in mathematics. Very often he has a poor understanding of the nature of the problem he is dealing with, and an imperfect notion of just what a mathematical proof is and how it operates. The writer is usually male, frequently a retired person with leisure to pursue his mathematics, often he has achieved considerable professional status in the larger community and he exhibits his status symbols within the mathematical work itself.

Very often the correspondent not only "succeeds" in solving one of the great mathematical unsolvables, but has also found a way to construct an antigravity shield, to interpret the mysteries of the Great Pyramid and of Stone-

55

henge, and is well on his way to producing the Philosophers' Stone. This is no exaggeration.

If the recipient of such a letter answers it, he will generally find himself entangled with a person with whom he cannot communicate scientifically and who exhibits many symptoms of paranoia. One gets to recognize such correspondents on sight, and to leave their letters unanswered, thus unfortunately increasing the paranoia.

I have on my desk as I write a paper of just this sort which was passed on to me by the editor of one of the leading mathematical journals in the United States. For self-protection I shall change the personal details, retaining the flavor as best I can. The paper is nicely and expensively printed on glossy stock and comes from the Philippines. It is written in Spanish and purports to be a demonstration of Fermat's Last Theorem. There is a photograph of the author, a fine-looking gentleman in his eighties, who had been a general in the Philippine army. Along with the mathematics there is a lengthy autobiography of the author. It would appear that the author's ancestors were French aristocrats, that after the French Revolution the cadet branch was sent to the East, whence the family made its way to the Philippines, etc. There are also included in this paper on Fermat's Last Theorem, nice engravings of the last three reigning Louis of France and a long plea for the restoration of the Bourbon dynasty. After page one, the mathematics rapidly wanders into incomprehensibility. I spent ten minutes with this paper; your average editor would spend less. Why? The Fermat "Last Theorem" is at the time of this writing a great unsolved problem. Perhaps the man from the Philippines has solved it. Why did I not examine his work carefully?

There are many types of anomalous or idiosyncratic writing in mathematics. How does the community strain out what it wants? How does one recognize brilliance, genius, crankiness, madness? Anyone can make an honest error. Shortly after World War II, Professor Hans Rademacher of the University of Pennsylvania, one of the leading number theoreticians in the world, thought he had proved the famous Riemann Hypothesis. (See page 363 for

56

a statement of this conjecture.) The media got wind of this news and an account was published in *Time* magazine. It is not often that a mathematical discovery makes the popular press. But shortly thereafter, an error was found in Rademacher's work. The problem is still open as these words are being written.

This is an example of incorrect mathematics produced within the bounds of mathematical orthodoxy—and detected there as well. This happens to the best of us every day of the week. When the error is pointed out, one recognizes it as an error and acknowledges it. This kind of situation is dealt with routinely.

At the opposite pole, there is the type whose psychopathology has just been described above. This type of writing is usually dismissed at sight. The probability that it contains something of interest is extremely small and it is a risk that the mathematical community is willing to take. But it is not always easy to draw the line between the crank and the genius.

An obscure and poor young man from a little-known place in India writes a letter around 1913 to G. H. Hardy, the leading English mathematician of the day. The letter betrays signs of inadequate training, it is intuitive and disorganized, but Hardy recognizes in it brilliant pearls of mathematics. The Indian's name was Srinivasa Ramanujan. If Hardy had not arranged for a fellowship for Ramanujan, some very interesting mathematics might have been lost forever.

*Srinivasa Ramanujan*
*1887–1920*

Then there was the case of Hermann Grassman (1809–1877). In 1844 Grassman published a book called *Die lineale Ausdehnungslehre*. This work is today recognized as a work of genius. It was an anticipation of what would be subsequently worked out as vector and tensor analysis and associative algebras (quaternions). But because Grassman's exposition was obscure, mystical, and unusually abstract for its period, this work repelled the mathematical community and was ignored for many years.

Less known than either Grassman or Ramanujan is the story of Jozef Maria Wronski (1776–1853), whose personality and work combined elements from pretentious na-

57

iveté to genius near madness. Today Wronski is chiefly remembered for a certain determinant $W[u_1, u_2, \ldots, u_n] =$

$$
\begin{vmatrix}
u_1 & u_2 & \cdots & u_n \\
u_1' & u_2' & \cdots & u_n' \\
\cdot & & & \\
\cdot & & & \\
\cdot & & & \\
u_1^{(n-1)} & u_2^{(n-1)} & \cdots & u_n^{(n-1)}
\end{vmatrix}
$$

formed from $n$ functions $u_1, \ldots, u_n$.

This determinant is related to theories of linear independence and is of importance in the theory of linear differential equations. Every student of differential equations has heard of the Wronskian.

Wronski was a Pole who fought with Kosciuszko for Polish independence, yet, dedicated his book "Introduction à la Philosophie des Mathématiques et Technie de l'Algorithmie" to His Majesty, Alexander I, Autocrat of all the Russias. A political realist, one would think.

On the 15th of August 1803, Wronski experienced a revelation which enabled him to conceive of "the absolute." His subsequent mathematical and philosophical work was motivated by a drive to expound the absolute and its laws of unification. In addition to his mathematics and philosophy, Wronski pursued theosophy, political and cultural messianism (he wrote five books on this topic), promoted the ideas of arithmosophism, mathematical vitalism, and something which he called "séchelianisme" (from the Hebrew; sechel: reason). This latter purported to change Christianity from a revealed religion to a proved religion. Wronski distinguished three forces which control history: providence, fatality (destiny), and reason. He constructed almost all of his system around the negation of the principle of inertia. Inasmuch as the material has no inertia it does not compete with the spiritual. The scientific ideal would be a kind of panmathematism which unites the knowledge of the formation of mathematical systems with the laws of living beings.

Wronski's philosophy is, apparently, not uninteresting and ties in with the later writings of Bergson.

What do we find, mathematically speaking, when we open up the first volume of his *Oeuvres Mathématiques?*

It appears, at a quick glance, to be mixture of the theory of infinite series, difference equations, differential equations, and complex variables. It is long, rambling, polemical, tedious, obscure, egocentric, and full of philosophical interpolations giving unifying schemata. The "Grand Law of the Generation of Quantities," which contains the Key to the Universe, appears as equation (7). Wronski sold it to a wealthy banker. The banker did not pay up and Wronski aired his complaints publicly. Here is the Grand Law:

*Wronski's key to the universe; placed in a cartouche, sanctified by the zodiac, guarded by a sphinx, and printed on all his works.*

$$\text{``}Fx = A_0\Omega_0 + A_1\Omega_1 + A_2\Omega_2 + A_3\Omega_3 + A_4\Omega_4 + \text{ etc. à l'infini.''}$$

What does it mean? It appears to be a general scheme for the expansion of functions as linear combinations of other functions; a kind of generalized Taylor expansion which contains all expansions of the past and all future expansions.

It is not possible for me to grasp the essential spirit of Wronski's work; and it would take a profound student of eighteenth century mathematics to tell what, if anything, is new or useful in the four volumes. I am only too willing to accept the judgment of history that Wronski deserves to be remembered only for the Wronskian. The doors of the mathematical past are often rusted. If an inner chamber is difficult of access, it does not necessarily mean that treasure is to be found therein.

There is work, then, which is wrong, is acknowledged to be wrong and which, at some later date may be set to rights. There is work which is dismissed without examination. There is work which is so obscure that it is difficult to interpret and is perforce ignored. Some of it may emerge later. There is work which may be of great importance—such as Cantor's set theory—which is heterodox, and as a result, is ignored or boycotted. There is also work, perhaps the bulk of the mathematical output, which is admittedly correct, but which in the long run is ignored, for lack of interest, or because the main streams of mathematics did not choose to pass that way. In the final analysis, there can be no formalization of what is right and how we know it is right, what is accepted, and what the mechanism for acceptance is. As Hermann Weyl has written, "Mathematizing may well be a creative activity of man . . . whose historical decisions defy complete objective rationalization."

**Further Readings. See Bibliography**

J. M. Wronski

# The Individual and the Culture

THE RELATIONSHIP BETWEEN the individual and society has never been of greater concern than it is today. The opposing tendencies of amalgamation versus fragmentation, of nationalism versus regionalism, of the freedom of the individual as opposed to the security within a larger group are acting out a drama on history's stage which may settle a direction for civilization for the next several centuries. Running perpendicularly to these struggles is the conflict between the "Two Cultures": the humanistic and the technological.

Mathematics, being a human activity, possesses all four components. It profits greatly from individual genius, but thrives only with the tacit approval of the wider community. As a great art form, it is humanistic; it is scientific-technological in its applications.

To understand just where and how mathematics fits into the human condition, it is important that we pay heed to all four of these components.

There are two extreme positions on the history of discovery. The first position holds that individual genius is the wellspring of discovery. The second position is that social and economic forces bring forth discovery. Most people do not hold with the one or with the other in a pure form, but try to find a mixture which is compatible with their own experiences.

60

The doctrine of the individual is the more familiar of the two, the easier of the two, and we are rather more comfortable with it. As teachers, we try our best to concentrate on the individual student; we do not attempt to teach people in their multitude. Methods of teaching en masse, through media of some sort, all postulate an individual at the receiving end. On the contrary, the word "indoctrination," which implies a kind of group phenomenon, worries us.

We study mathematical didactics and strategies of discovery as in Pólya's books (See Chapter 6) and try to transfer some of the insights of a great mathematician to our students. We read biographies of great geniuses and study their works carefully.

One of the most striking statements of the doctrine of the individual in mathematics was put forward in an article by Alfred Adler. The author is a professional mathematician and his article is as eloquent as it is dramatic. The article is also a very personal statement; its views are romanticized, manic-depressive, and apocalyptic.

Adler begins by putting the case for an extreme form of élitism:

> Each generation has its few great mathematicians, and mathematics would not even notice the absence of the others. They are useful as teachers, and their research harms no one, but it is of no importance at all. A mathematician is great or he is nothing.

This is accompanied by the statement of "The Happy Few."

> But there is never any doubt about who is and who is not a creative mathematician, so all that is required is to keep track of the activities of these few men.

"The Few"—or at least five of them—are then identified (as of 1972).

It is noted that the creation of mathematics appears to be a young man's business:

> The mathematical life of a mathematician is short. Work rarely improves after the age of twenty-five or thirty. If little has been accomplished by then, little will ever be ac-

61

complished. If greatness has been attained, good work may continue to appear, but the level of accomplishment will fall with each decade.

Adler records the intense joy of the artist:

> A new mathematical result, entirely new, never before conjectured or understood by anyone, nursed from the first tentative hypothesis through labyrinths of false attempted proofs, wrong approaches, unpromising directions, and months or years of difficult and delicate work— there is nothing, or almost nothing, in the world that can bring a joy and a sense of power and tranquillity to equal those of its creator. And a great new mathematical edifice is a triumph that whispers of immortality.

He winds up with a mathematical Götterdämmerung:

> There is a constant awareness of time, of the certainty that mathematical creativity ends early in life, so that important work must begin early and proceed quickly if it is to be completed. There is the focus on problems of great difficulty, because the discipline is unforgiving in its contempt for the solution of easy problems and in its indifference to the solution of almost any problems but the most profound and difficult ones.
>
> What is more, mathematics generates a momentum, so that any significant result points automatically to another new result, or perhaps to two or three new results. And so it goes—goes, until the momentum all at once dissipates. Then the mathematical career is, essentially, over; the frustrations remain, but the satisfactions have vanished.

And so we leave our ageing hero as he knocks tentatively on the gates of a Valhalla which itself may be illusory.

Lest any reader be deterred from a mathematical career by this dismal picture, we must report that there are many instances of mathematicians continuing to do first-class research past the age of fifty; for example, Paul Lévy, one of the creators of modern probability theory was close to forty when he wrote his first paper in this area; he continued doing profound, original work into his sixties.

When we speak of the culture as being the main source

of discovery, we are on grounds that are far more tenuous, far less well understood. This is the doctrine of "The Many." This is Hegel's Zeitgeist, the spirit of the age: the ideas, the attitudes, the conceptions, the needs, the modes of self-expression that are common to a time and to a place. These are the things that are "in the air." Read Tolstoy's retrospective final chapter of *War and Peace* and see how he comes to the conclusion that the trends initiated in Europe by the French Revolution would have worked themselves through with or without Napoleon. There is a tendency on the part of theoretical Marxists to favor the doctrine of the culture. So, for example, one might read how the British scientist and Marxist J. D. Bernal works it out in the area of the natural sciences.

We know in our bones that culture makes a difference. We know that there are cultures in which symphonic music has flourished and those in which it has not. But the explication by culture does not come easily. The record of a single man is easier to read than the traces of a whole civilization. Why did the small country of Hungary in the years since 1900 produce such a large number of first-rate mathematicians? Why have governments since 1940 supported mathematical research while prior to that date they did not? Why did the Early Christians find Christ and Euclid incompatible, while a thousand years later, Newton was able to embrace them both?

For contemporary history, where the facts are available or fresh in mind and where the principal actors might yet be alive, it would be possible to write easily and convincingly of the cultural reasons for this or that. So, for example, it might be possible—and very worthwhile—to spell out the extramathematical, extratechnological reasons which have led in one short generation to the development of the electronic computing machine. (See the book of H. Goldstine.) It would be rather harder to explain the rise of function algebras along the same lines. When it comes to the deep past, one puts it together by inference or by statistics as best as one can. A whole new subject, cliometrics— mathematical treatment of historical records—has just been born; but what comes out is as often as not romanticized fabrications, oversimplifications and misinterpretations.

The doctrine of the culture is buttressed, strangely, by the platonic view of mathematics. If, after all, $e^{\pi i} = -1$ is a fact of the universe, an immutable truth, existing for all time, then surely Euler's discovery of this fact was mere accident. He was merely the medium through which the fact was vented. Sooner or later, so the argument goes, it would have—of necessity it had to have—been discovered by any one of a hundred other mathematicians.

Neither of the extreme views presented is adequate. Why did mathematics go to sleep for at least 800 years from about 300 to 1100? Presumably the genes of mathematical genius were present in the Mediterranean populace of the year 600 as they were in the days of Archimedes. Or take Tolstoy's philosophy of history. Despite his relegation of Napoleon to historical nonnecessity, everything that is of interest in *War and Peace* derives from the perception of individuals in their uniqueness. Despite the penchant of Marxists for cultural explanations, the relevance of V. I. Lenin to the Russian Revolution is not for them a subject of silent contemplation.

In the final analysis, the dichotomy between the doctrine of the individual and the doctrine of the culture is a false dichotomy, something like the argument of mind over matter or of the spirit and the flesh. Attempts have been made to reconcile the extreme views in a variety of ways. There is the reconciliation by means of time scale. This opinion holds that in the short run (say less than 500 years) the individual is important. In the long run (say more than 500 years), the individual is no longer important, but the culture is.

An intermediate view of great appeal was put forward by the American psychologist and philosopher William James. In his essay "Great Men and Their Environment," James wrote,

> The community stagnates without the impulse of the individual; the impulse dies away without the sympathy of the community.

Now this is a very simple and undramatic formulation stating what must be apparent to most observers, that both elements are necessary. I was brought up in a textile town and have my own private formulation of the Jamesian synthesis. Woven cloth consists of two perpendicular sets of interlaced threads: the warp and the woof. Neither holds without the other. Similarly, the warp of society requires the woof of the individual.

Having now summarized James' view of the matter in this brief quotation, we can now pose a major question.

*Is it possible to write a history of mathematics along the lines suggested by this quotation?*

It would be nice to think so, but it has not been done and it is not at all certain it can be done.

## Further Readings. See Bibliography

A. Adler; J. D. Bernal; S. Bochner [1966]; P. J. Davis [1976]; B. Hessen. For a rebuttal see G. N. Clark; W. James [1917], [1961]; M. Kline [1972]; T. S. Kuhn; R. L. Wilder [1978].

The relation between society and the physical sciences has been rather more intensively explored than with mathematics. Here are some books in that direction:

A. H. Dupree; G. Basalla; L. M. Marsak; J. Ziman.

# 3

# OUTER ISSUES

# Why Mathematics Works: A Conventionalist Answer

**E**VERYONE KNOWS THAT if you want to do physics or engineering, you had better be good at mathematics. More and more people are finding out that if you want to work in certain areas of economics or biology, you had better brush up on your mathematics. Mathematics has penetrated sociology, psychology, medicine, and linguistics. Under the name of cliometry, it has been infiltrating the field of history, much to the shock of old-timers. Why is this so? What gives mathematics its power? What makes it work?

One very popular answer has been that God is a Mathematician. If, like Laplace, you don't think that deity is a necessary hypothesis, you can put it this way: the universe expresses itself naturally in the language of mathematics. The force of gravity diminishes as the second power of the distance; the planets go around the sun in ellipses; light travels in a straight line, or so it was thought before Einstein. Mathematics, in this view, has evolved precisely as a symbolic counterpart of the universe. It is no wonder, then, that mathematics works; that is exactly its reason for existence. The universe has imposed mathematics upon humanity.

This view of mathematics goes well with what is often called the Platonic view. Mathematical Platonism is the view that mathematics exists independently of human

*Pierre Simon Laplace*
*1749–1827*

68

*The astronomer reaches for truth. He is depicted as breaking through the shell of appearances to arrive at an understanding of the fundamental mechanism that lies behind appearances. (Woodcut from Camille Flammarion,* L'Atmosphère Metéorologie Populaire, *1888.)*

*Courtesy: Deutsches Museum, Munich*

beings. It is "out there somewhere," floating around eternally in an all-pervasive world of Platonic ideas. Pi is in the sky. For example, if one contemplated communicating with the creatures on Galaxy X–9, one should do it in the language of mathematics. There would be no point in asking our extragalactic correspondent about his family, or his job, or his government, or his graphic arts, for these objects of existence might have no meaning for him. On the other hand, stimulate him with the digits of pi (3, 1, 4, 1, 5, . . .) and, so the argument goes, he will be sure to respond. The universe will have imposed essentially the same mathematics upon Galaxy X-9 as upon terrestrial men. It is universal.

In this view, the job of the theorist is to listen to the universe sing and to record the tune.

But there is another view of the matter. This opinion holds that applications of mathematics come about by fiat. We create a variety of mathematical patterns or structures. We are then so delighted with what we have wrought, that we deliberately force various physical and social aspects of the universe into these patterns as best we can. If the slipper fits, as it did with Cinderella, then we have a beautiful theory; if not—and the world of hard facts is more like the ugly sister; the slipper always pinches—back to the drawing board of theory.

This view is related to the opinion that theories of applied mathematics are merely "mathematical models." The utility of a model is precisely its success in mimicking or predicting the behavior of the universe. If a model is inadequate in some respect, one looks around for a better model or an improved version. There is no philosophical truth in either the statement, the "earth goes around the sun" or in the statement, "the sun goes around the earth." Both are models, and which one we operate with is determined by such things as simplicity, fruitfulness, etc. Both were derived from prior mathematical experiences of a simple nature.

This philosophical view has become increasingly popular. Courses in increasing number are being taught under the name "Mathematical Modelling." What would have been taught in a previous generation as "the theory of such and such"; now is known merely as "model for such and such." Truth has abdicated and expediency reigns.

## Some Simple Instances of Mathematics by Fiat

Of course, hardly any scientists live by a consistent creed. Scientists believe simultaneously both in theories and in models, in truth and in expediency.

As far as the "average thinking man" is concerned, I would guess that he is a Platonist. In fact, I would guess that he is so much of one that he finds it difficult to conceive how mathematical structures can be imposed upon the world. I should like to explain this, using as an example something that everyone is familiar with: the mathematical operation of addition.

After the recitation of the integers one, two, three, . . ., and an intuitive recognition of serial order, addition is the very first operation that one learns. One can distinguish

70

three aspects of addition. The first is the algorithmic aspect. This refers to the rules of manipulation by which you (or your hand computer) are able to work sums. The second (which was unduly stressed by the "new mathematics") relates to the formal laws that sums obey, e.g., $a + b = b + a$, or $(a + b) + c = a + (b + c)$, $a + 1 > a$. The third is the applications of addition: Under what circumstances does one add?

The first two are easy. The third is hard, and the fun begins there. These are the "word problems" of grade school. There are many children who know how to add, but do not know when to add. Do you think the adult knows when to add? We shall see.

Why is there any problem about when to add? Two apples and three apples are five apples and where's the mystery? Now I shall put forward for discussion a list of word problems that ostensibly call for addition.

Problem 1.   One can of tuna fish costs $1.05. How much do two cans of tuna fish cost?

Problem 2.   A billion barrels of oil costs $x$ dollars. How much does a trillion barrels of oil cost?

Problem 3.   A bank in computing a credit rating allows two points if you own your house, adds one point if your salary is over $20,000, adds one point if you have not moved in the last five years, subtracts one point if you have a criminal record, subtracts one point if you are under 25, etc. What does this sum mean?

Problem 4.   An intelligence test adds one point if you can answer correctly a question about George Washington, one point if you answer about polar bears, one point if you know about Daylight Saving Time, etc. What does the final sum represent?

Problem 5.   A cup of milk is added to a cup of popcorn. How many cups of the mixture will result?

Problem 6.   One man can paint a room in one day. A second man who can paint a room in two days is added to the work force. How many days will it take both men working together?

Problem 7.   A rock weighs one pound. A second rock weighs two pounds. How much will both rocks together weigh?

Now for some comments on these problems.

**Problem 1.**   My market sells a can of tuna fish for $1.05 and two cans for $2.00. Well, you might say that the "real" price is $2.10 and the grocer has not charged you the "real" price. I say that the "real" price is what the grocer charges, and if he finds that simple addition does not adequately suit his business, then he exhibits no qualms about modifying it. Discounts are so widespread that we all understand the inadequacies of addition in this context.

If we buy a can of tuna fish at $1.05 and a can of peaches at 60¢ and add them to arrive at a bill of $1.65, then this reflects a reduction of all goods to a common value system. This reduction is then followed by addition of the individual prices, and is one of the great fiats of the economic world. There have been times, e.g., during periods of rationing, when a pound of meat cost 40¢ plus one red token and a pound of sugar cost 30¢ plus one blue token. We have here an example of "vector" pricing where the price comprises several different components and the "vector" addition exhibits the arbitrary nature of the process.

**Problem 2.**   Here we have the same problem but in reverse: What price shall be charged for a diminishing resource? Surely a penalty and not a discount is called for, so that ordinary addition is inappropriate. Formulating an absurd but not unrelated question one might ask: If the Mona Lisa painting is valued at $10,000,000, what would be the value of two Mona Lisa paintings?

**Problem 3.**   The bank has arrived at what might be called a figure of merit for its potential customer. Does it really make sense to say that a criminal record is counterbalanced by a salary of over $20,000? Perhaps it does.

Figures of merit, such as this one, are widely employed. One state has a system of demerits for traffic offenses.

Such ratings—in other fields—might be the basis of automatized ethics or computer-dispensed justice, or computer-dispensed medicine. The ad hoc nature of the scheme seems fairly apparent.

One is reminded of the story of the man with a cup who sat in Times Square begging. He had pinned on him this sign

| | |
|---------|---|
| Wars | 2 |
| Legs | 1 |
| Wives | 2 |
| Children | 4 |
| Wounds | 2 |
| Total | 11 |

**Problem 4.**   Most tests add up the results of the individual parts. This is commonly accepted. If one gives a mathematics test in college, and the test is not of the multiple choice variety, then students scream for partial credit in the individual parts. Teachers know that such credit can be given only subjectively. The whole business of addition of points is a widely accepted, but nonetheless an ad hoc affair. We bypass the difficult question now raging of just what an individual question tests in any case.

**Problem 5.**   A cup of popcorn will very nearly absorb a whole cup of milk without spillage. The point here is that the word "add" in a specific physical or even popular sense does not necessarily correspond to "add" in the mathematical sense.

**Problem 6.**   Similarly, by a confusion of language, we allow the popular "add" to imply the mathematical "add." One sees this clearly in this problem which comes straight out of the high school algebra books.

**Problem 7.**   It is possible to discuss physical measurements meaningfully only within the context of a theory—say Newtonian mechanics. Weight is proportional to mass and mass is additive. That is, by definition, the mass of the union of two bodies is the sum of the masses of separate bodies. If two rocks are weighed on a spring balance, then if the rocks are sufficiently heavy, the response of the spring may be seen to be nonlinear (and be compensated for by suitable calibration) because of the already accepted additive definition. Simple addition of spring displacements may not be appropriate.

The upshot of this discussion is: *There is and there can be no comprehensive systematization of all the situations in which it is appropriate to add.* Conversely, *any systematic application of addition to a wide class of problems is done by fiat.* We simply say: Go ahead and add, hoping that past and future experience will bear out the act as a reasonable one.

If this is true for addition, it is much more so for the other more complex operations and theories of mathematics. This, in part, explains the difficulty people have with "word" problems and, at a higher level, the grave difficulties that confront the theoretical scientist.

**One final example.**  A bake shop does a thriving business. The owner, in order to establish peace and quiet among his customers, has put in a system of numbers. Many shops have such a system. How should he do it? Well, you say, just work it out so that the customers are served in order of arrival. But this is only one possible criterion. The universe is not crying out for this criterion, nor would it vanish in a thunderclap if another criterion were put into operation. Perhaps the waiting lines are long and the owner decides to add zest to the process by inserting lucky numbers which take the holder immediately to the head of the line. Mathematics is capable of providing this. Perhaps, if your number is even, you get gas while someone whose number is odd does not. Strange? But something like this was in effect during the recent gasoline shortage.

The imposition of mathematics is by fiat, but once established, it carries with it many social consequences. The mathematics of income tax is by fiat, the mathematics of welfare is by fiat; and each has huge computer backups to facilitate the operation. Once such a system has been set in motion, it is not easy to "pull out the plug" without risking social disruption. And it is not, in my view, an accident that an increasing amount of social mathematizing has occurred precisely at a time when there is an increased belief in a philosophy of science which confers on equations only the status of a model.

74

## Fiat in the Physical Sciences?

How can man, who is a mere speck in the universe, impose his mathematical will on the great cosmic processes? Here the argument is harder to understand but can be made along the following lines.

We shall consider two theories of the motion of the planets, the first given by Claudius Ptolemy (second century) and the second by Isaac Newton (1642–1727). In the Ptolemaic system, the earth is fixed in position while the sun moves, and all the planets revolve about it. Fixing our attention, say, on Mars, one assumes that Mars circulates about the earth in a certain eccentric circle and with a certain fixed period. Compare this theory now with the observation. It fits, but only partially. There are times when the orbit of Mars exhibits a retrograde movement which is unexplainable by simple circular motion.

*Claudius Ptolemy*
*c.145* A.D.

To overcome this limitation, Ptolemy appends to the basic circular motion a second eccentric circular motion with its own smaller radius and its own frequency. This scheme can now exhibit retrograde motion, and by the careful adjustment of the radii and the eccentricities and periods, we can fit the motion of Mars quite well. If still more precision is wanted, then a third circle of smaller radius still and with yet a different period may be added. In this way, Ptolemy was able to achieve very good agreement between theory and observation. This is one of the earliest examples in science of *curve fitting*—not unlike harmonic analysis—but no deeper explanation of the process, no unification from planet to planet was found possible.

Fifteen hundred years later God said, according to Pope, "'Let Newton be,' and all was light." The Newtonian theory of planetary motion provided a model with a modern flavor and of immense theoretical and historical importance. Here the organic basis lies much deeper. Here new elements enter the picture: masses, accelerations, the Law of Motion $F = mA$, the inverse square law of gravitation. These physical laws find mathematical expression as differential equations. The laws are postulated to be of universal validity, applying not only to the sun and the earth,

75

but to Mars and Venus and all the other planets, comets and satellites. Whereas the Ptolemaic scheme appears static and ad hoc—mere curve fitting—divorced from reality, the Newtonian scheme appears by contrast richly dynamic, grounded in the reality of matter, force, acceleration. The resulting differential equation came to appear closer to the ultimate truth as to how the universe is governed.

But is the matter really so simple? Take the differential equation for Mars and solve it. It predicts that Mars goes around the sun in an ellipse. Check this out against observations. It doesn't check exactly. There are discrepancies. What are they due to? Well, we've got the force slightly wrong. In addition to the force of the sun, there is the force of Jupiter, a massive planet, which perhaps we should take account of. Well, put in Jupiter. It still doesn't work precisely. There must be other forces to account for. How many other forces are there? It's hard to know; there are an unlimited number of possible forces and some may be of importance. *But there is no systematic way of telling a priori what forces exist and should be taken into account.* It goes without saying that historical modifications of Newton such as relativistic mechanics cannot be anticipated. The criterion of success is still in operation, and an accurate prediction based upon up to date celestial mechanics emerges, like Ptolemy's, as a patchwork job—a theory by fiat. We are still curve fitting, but are doing it on the basis of a more versatile vocabulary of the solutions of differential equations rather than on a vocabulary of "ready made" simple curves such as circles.

### Further Readings. See Bibliography

E. Wigner
**On Problem 6:**
F. P. Brooks, Jr. for interesting statistics on the productivity rates of programming groups

# Mathematical Models

WHAT IS A MODEL? Before generalizing, let us consider some concrete examples. As just mentioned, Newton's theory of planetary motion was one of the first of the modern models.

Under the simplifying assumption of a sun and one planet, Newton was able to deduce mathematically that the planet will describe an orbit in accordance with the three laws that Kepler had inferred from examination of a considerable quantity of astronomical observations. This conclusion was a great triumph of physical and mathematical analysis and gave the Newtonian fiat its completely compelling force.

If there are three, four, five . . . bodies interacting, then the system of differential equations becomes increasingly complicated. Even with just three bodies we may not have "closed-form" solutions à la Kepler. There is often a gap between what we would like our theory to do and what we are able to have it do. This may control the course of the subsequent methodology. If we want to know where Jupiter will be so as to plan properly the Jupiter shot, then we may proceed in one mathematical direction. If we are interested in whether the solar system is dynamically stable or unstable, we will have to proceed in another.

In view of the inherent difficulties of the mathematics, the art of modelling is that of adopting the proper strategy. Take, as a less familiar example, the chemical engineering problem of a stirred tank reaction (see R. Aris, pp. 152–164).

A cylindrical tank is provided with incoming pipes and an outgoing pipe. The incoming pipes bring in reactants, and the outgoing pipe takes away a mixture of products and what is left over of the reactants. The tank is surrounded by a cylindrical water jacket to cool it, and both the reactor tank and the jacket are stirred to achieve perfect mixing.

Now, apart from the geometrical hypotheses which may be only approximately true, one is confronted at the very least with eleven laws or assumptions $H_0, H_1, \ldots, H_{10}$ on which to formulate a mathematical model. $H_0$ asserts the laws of conservation of matter, energy and Fourier's law of heat conduction. $H_1$ asserts that the tank and jacket volumes are constant, as are the flow rates and the feed temperatures. The mixing is perfect so that the concentrations and the reaction temperatures are independent of position. Thus, hypothesis after hypothesis is proposed. $H_9$ asserts that the response of the cooling jacket is instantaneous. $H_{10}$ asserts that the reaction is of the first-order and is irreversible with respect to the key species.

Now, on this basis, six principal models can be proposed, using various assumptions. The most general assumes only $H_0, \ldots, H_4$ and leads to six simultaneous equations, while the simplest assumes $H_0, \ldots, H_{10}$ and leads to two equations.

"A mathematical model," says Aris, is "any complete and consistent set of mathematical equations which are designed to correspond to some other entity, its prototype. The prototype may be a physical, biological, social, psychological or conceptual entity, perhaps even another mathematical model." For "equations" one might substitute the word "structure," for one does not always work with a numerical model.

Some of the purposes for which models are constructed are (1) to obtain answers about what will happen in the physical world (2) to influence further experimentation or observation (3) to foster conceptual progress and understanding (4) to assist the axiomatization of the physical situation (5) to foster mathematics and the art of making mathematical models.

The realization that physical theories may change or may be modified (Newtonian mechanics vs. Einsteinian mechanics, for example), that there may be competing theories, that the available mathematics may be inadequate to deal with a theory in the fullest sense, all this has led to a pragmatic acceptance of a model as a "sometime thing," a convenient approximation to a state of affairs rather than an expression of eternal truth. A model may be considered good or bad, simplistic or sophisticated, aesthetic or ugly, useful or useless, but one is less inclined to label it as "true" or "false." Contemporary concentration on models as opposed to theories has led to the study of model-making as an art in its own right with a corresponding diminution of interest in the specific physical situation that is being modelled.

### Further Readings. See Bibliography

R. Aris; P. Duhem; H. Freudenthal [1961]; L. Iliev.

# Utility

### 1. Varieties of Mathematical Uses

For a thing to be useful is for it to have the capacity for satisfying a human need. Mathematics is commonly said to be useful, but as the variety of its uses is large, it will pay us to see what different meanings can be found for this word. A pedagogue—particularly of the classical variety—might tell us that mathematics is useful in that it teaches us how to think and reason with precision. An architect or sculptor—again of a classical sort—might tell us that mathematics is useful because it leads to the perception and creation of visual beauty. A philosopher might tell us that mathematics is useful insofar as it enables him to escape from the realities of day-to-day living. A teacher might say that mathematics is useful because it provides him with bread and butter. A book publisher knows that mathematics is useful for it enables him to sell many textbooks. An astronomer or a physicist will say that mathematics is useful to him because mathematics is the language of science. A civil engineer will assert that mathematics enables him to build a bridge expeditiously. A mathematician will say that within mathematics itself, a body of mathematics is useful when it can be applied to another body of mathematics.

Thus, the meanings of the expression "mathematical utility" embrace aesthetic, philosophical, historic, psychological, pedagogical, commercial, scientific, technological, and mathematical elements. Even this does not include all possible meanings. I have the following story from Professor Roger Tanner of Sydney, Australia. Two students walked into a colleague's office and told him that they would like to take his advanced course in applied mathematics. The professor, delighted, gave his prospective students a big sales talk on his course: what the syllabus was, how it connected with other subjects, etc., etc. But the two students interrupted: "No, no. You don't understand. We are Trotskyites. We want to take your course because it is completely useless. If we take it, 'they' can't turn us to counterrevolutionary purposes." Thus, even uselessness is useful.

We shall concentrate here on mathematical utility that occurs within scientific or technological activity. One can distinguish between utility within the field itself and utility to other fields. Even with these subdivisions, the notion of utility is exceedingly slippery.

## 2. On the Utility of Mathematics to Mathematics.

What does it mean when it is said that a piece of mathematics is used or applied to mathematics itself? For example, it may be said that the theory of ideals is useful in the theory of numbers. This means that some results of the theory of ideals were used to prove the impossibility of special instances of Fermat's Last Problem. It means that in order for you to understand the proof of this impossibility, you had better understand such and such theorems of the theory of ideals. (Historically, the facts are turned around: the theory of ideals evolved as part of an attempt to establish Fermat's Last Theorem.) In this sense, then, one can speak of the application of tensor analysis to elasticity theory, of complex variable theory to number theory, of nonstandard analysis to Hilbert space theory, or of fixed-point theory to differential equations.

Application of theory A to theory B within mathematics means, then, that the materials, the structure, the techniques, the insights of A are used to cast light or to derive inferences with regard to the materials and the structures of B. If a piece of mathematics is used or connected up with another piece of mathematics, then this aspect is often called "pure." Thus, if the algebraic theory of ideals is used in a discussion of Fermat's Last Theorem, then we are talking about a pure aspect or a pure application. If, on the other hand, ideal theory finds application to telephone switching theory (I don't know whether it has been used this way), then such a use would be called applied.

Now methods and proofs are not unique; theorems may be proved in different ways. Therefore a certain application of something in A may be inessential as far as establishing the truth of something in B. It may be preferable for historical or other reasons to establish B by means of C or of D. In fact, it may even be part of the game to do so. Thus, for many years, the prime number theorem (see Chapter 5) was proved via the theory of functions of a complex variable. Since the concept of a prime number is simpler than that of a complex number, it was considered a worthy goal to establish this theory without depending on the use of complex numbers. When this goal was finally achieved, then the utility of complex variable in number theory had changed.

Time may bring about changes in utility in the reverse way. Thus, when the first proof of the fundamental theorem of algebra was given, topology was still in its infancy, and the topological aspects of the proof were thought to be obvious or unimportant. One hundred and fifty years later, with a ripe topology at hand, the topological aspects of the problem are considered crucial and a fine application of the notion of winding number.

We may distinguish between a useful theorem, i.e. one for which an application has been found, a very useful theorem, i.e., one for which many applications have been found and a useless theorem, i.e., one for which no application has thus far been found. Of course, one can always splice something onto a given theorem T to arrive at a theorem T′ so as to give T an application. But such tricks go against common standards of mathematical aesthetics and exposition. The mathematical literature contains millions of theorems and very likely most of them are useless. They are dead ends.

It is true also that there is a tendency to thread one's thinking—and later on one's exposition—through well-known standard or famous theorems such as the mean value theorem or the fixed point theorem or the Hahn-Banach theorem. To some extent this is arbitrary in the way that the Chicago Airport is an arbitrary transfer point for an air passenger from Providence, R.I., to Albuquerque, N.M. But reasons for doing it are not hard to find.

Great store and reputation is set by theorems which are very useful. This is somewhat paradoxical, for if a theorem is the fruit or a goal of a mathematical activity, then this goal, as an aesthetic object, should be valuable whether it is itself the progenitor of other goals.

This high regard for "useful" results combined with confusion as to the meaning of utility, is at the basis of acrimonious discussion as to what is useful or fruitful and what is not. Judgments on this issue affect all aspects of mathematics from teaching to research, and sometimes lead to unstable trendy enthusiasms.

This regard, also, lies at the base of an overemphasis on the *process* of mathematizing at the expense of the *results* of mathematizing. All too many mathematical textbooks today have a nervous, breathless quality in which a fixed goal is systematically and inexorably pursued. The goal having been attained, one is left not with a feeling of exhilaration but of anticlimax. Nowhere in such books is any appreciation to be found of why or how the goal is important, other, possibly, than the statement that the goal may now be used as the starting point for reaching other, deeper goals, which considerations of space, alas, prevent the author from pursuing. Blame it on Euclid, if you want, for the tendency was already in his exposition.

### 3. On the Utility of Mathematics to Other Scientific or Technological Fields.

The activity in which mathematics finds application outside its own interests is commonly called *applied mathematics*. Applied mathematics is automatically cross-disciplinary, and ideally should probably be pursued by someone whose primary interests are not mathematics. If the cross-discipline is, say, physics, it may be hard to know what to classify as applied mathematics and what as theoretical physics.

The application of mathematics to areas outside itself raises issues of another kind. Let us suppose we have an application, say, of the theory of partial differential equations to the mathematical theory of elasticity. We may now inquire whether elasticity theory has an application outside itself. Suppose it has in theoretical engineering. We may inquire now whether that theory is of interest to the practical engineer. Suppose it is; it enables him to make a stress analysis of an automobile door. Again we raise the question, asking how this might affect the man in the street. Suppose the stress analysis shows that a newly designed door satisfies minimal strength requirements set by law. In this way, we can trace the application of mathematics from the most abstract level down to the consumer level. Of course, we don't have to stop there. We can inquire whether the automobile is useful for something. For commuting. Is commuting useful? . . . etc., etc.

Let us agree to call the utility that extends all the way to the man in the street *common utility*. (This assumes that we know what the man in the street is really interested in, which again is a questionable assumption.) We do not suggest that the criterion of the street should be set up as the sole criterion for judging mathematical utility. It would be disastrous to do so. But as life proceeds to a large measure by the activities of making and consuming, buying, selling and exchanging, one should have as firm a grasp as possible as to where one's subject stands with respect to these basic activities.

Which applications of mathematics have common utility? The answer to this question obviously has great implications for education, for the preparation of texts and for research. Yet the answer is shrouded in myth, ignorance, misinformation, and wishful thinking. Some instances of common utility are as plain as day. When the clerk in a supermarket tots up a bag of groceries, or when a price is arrived at in an architect's office, we have a clear application of mathematics at the level of common utility. These computations may be trivial and may be performable by mathematically unsophisticated people; nonetheless, they are mathematics, and the computations that refer to counting, measuring, and pricing constitute the bulk of all mathematical operations at the level of common utility.

When one moves to higher mathematics, such applications are harder to observe and to verify. It would be of enormous importance to the profession if some lively and knowledgeable investigator would devote several years to this task, and by visiting a number of businesses, laboratories, plants, etc. document just where this occurs.*

An organization may employ people well trained in mathematics, it may have a sophisticated computer system, because the theoretical aspects of its business may be cast in mathematical terms. All this does not yet mean that the mathematics being done reaches down to the level of common utility. The emergence at the level of common utility of potentially applicable mathematics may be blocked or frustrated for dozens of different reasons. It may be too

---

* This is not so easy.

A story is told of how, some years ago, a group of specialists were asked by Agency A to evaluate its sponsorship of mathematical research. What mathematical work, sponsored by Agency A, led directly to applications at the level of common utility in which Agency A was interested, and such that those applications would have been impossible without it? After several days of pondering, the experts decided they could not identify any such work, but the sponsorship of mathematical research should be justified on other grounds. For example, it should be justified, in that it kept a pool of researchers in training against "future needs."

difficult, too expensive, or too inaccurate to compute the stresses on an automobile door via a mathematical model. It might be faster, cheaper, and more reliable to test the door out in a testing machine or in a crash. Or a mathematical model may call for the knowledge of many parameters and these parameters may simply not be available.

In a typical book on applied mathematics, one finds, for example, a discussion of the Laplace problem for a two-dimensional region. This has important applications, says the author, in electrodynamics and in hydrodynamics. So it may be, but one should like to see the application pinpointed at the level of common utility rather than of pious potentiality.

## 4. Pure vs. Applied Mathematics

There is a widely held principle that mind stands higher than matter, the spirit higher than the flesh, that the mental universe stands higher than the physical universe. This principle might have its origin in human physiology and in the feeling which identifies "self" with "mind" and locates the mind in the brain. Replace an organ such as a leg or an eye with an artificial or a transplanted organ, and this does not appear to alter or to threaten the self. But if one imagines a brain transplant or the dumping of the contents of someone else's brain into one's own, then the self seems to shriek bloody murder, it is being destroyed.

*Godfrey Harold Hardy
1877–1947*

The reputed superiority of mind over matter finds mathematical expression in the claim that mathematics is at once the noblest and purest form of thought, that it derives from pure mind with little or no assistance from the outer world, and that it need not give anything back to the outer world.

Current terminology distinguishes between "pure" and "applied" mathematics and there is a pervasive unspoken sentiment that there is something ugly about applications. One of the strongest avowals of purity comes from the pen of G. H. Hardy (1877–1947), who wrote,

> I have never done anything "useful." No discovery of mine has made, or is likely to make, directly or indirectly,

for good or ill, the least difference to the amenity of the world. I have helped to train other mathematicians, but mathematicians of the same kind as myself, and their work has been, so far at any rate as I have helped them to it, as useless as my own. Judged by all practical standards, the value of my mathematical life is nil; and outside mathematics it is trivial anyhow. I have just one chance of escaping a verdict of complete triviality, that I may be judged to have created something worth creating. And that I have created something is undeniable: the question is about its value.

The case for my life, then, or for that of any one else who has been a mathematician in the same sense in which I have been one, is this: that I have added something to knowledge, and helped others to add more; and that these somethings have a value which differs in degree only, and not in kind, from that of the creations of the great mathematicians, or of any of the other artists, great or small, who have left some kind of memorial behind them.

Hardy's statement is extreme, yet it expresses an attitude that is central to the dominant ethos of twentieth-century mathematics—that the highest aspiration in mathematics is the aspiration to achieve a lasting work of art. If, on occasion, a beautiful piece of pure mathematics turns out to be useful, so much the better. But utility as a goal is inferior to elegance and profundity.

In the last few years, there has been a noticeable shift in the attitudes predominant among American mathematicians. Applied mathematics is becoming stylish. This trend is certainly not unrelated to changes in the academic job market. There are not enough jobs to go around for Ph.D. mathematicians in American universities. Of the jobs one sees advertised, many call for competence in statistics, in computing, in numerical analysis or in applied mathematics. As a consequence, there is a visible attempt by many mathematicians to find a link between their own specialty and some area of application. It is not clear whether this shift in attitude is transitory or permanent. There is little evidence of a change in the basic value system among mathematicians, which makes the goal of utility an inferior goal.

The assertion of the superiority of mind over matter casts its shadow over the writing of the history of mathematics. By far the bulk of the standard writings on the subject have to do with inner developments or issues, that is, the relationship of mathematics to its own self. Despite the vast amount of raw material available on outer issues, this material remains unevaluated, undervalued, or misrepresented. For example, the role of positional astronomy in the development of the theory of functions of a complex variable is ignored. It is known that a good deal of the motivation for this theory came from the desire to solve Kepler's positional equation for planetary motion.

Quite apart from issues of superiority, one can assert firmly that in a number of respects it is harder to work in applications than in pure mathematics. The stage is wider, the facts are more numerous and are more vague. The precision and aesthetic balance which is so often the soul of pure mathematics may be an impossibility.

### 5. From Hardyism to Mathematical Maoism

Hardyism is the doctrine that one ought only to pursue useless mathematics. This doctrine is given as a purely personal credo in Hardy's *A Mathematician's Apology*.

Mathematical Maoism by contrast is the doctrine that one ought to pursue only those aspects of mathematics which are socially useful. "What we demand," wrote Chairman Mao Tse-Tung, "is the unity of politics and art."

At some point during the Mao regime, a moratorium was declared on scientific research work. During this time review committees were supposed to assess the importance of fields and subfields, keeping in mind the criterion that research should be directed towards practical problems and that teaching should be based upon concrete applications. Pressure was put on researchers to get out of some areas, e.g., topology. The "open-door" policy of scientific research was to be stressed wherein "scientific research should serve proletarian politics, serve the workers, peasants and the soldiers and be integrated with productive labor." The research workers were supposed to get out of their ivory towers and take jobs in factories or communes;

reciprocally, peasants and workers were to be marched to institutions to propose research. Research was supposed to combine the efforts of the administrators, the researchers and the workers, the old, the middle-aged, and the young. This is known as the "three-in-one" principle.

In 1976 a delegation of distinguished U.S. mathematicians visited the Peoples Republic of China. During this visit, the delegates gave lectures, listened to lectures, and had an opportunity to meet informally with Chinese mathematicians. They issued a report "Pure and Applied Mathematics in the Peoples Republic of China." Here is one of the more interesting dialogues. (Kohn is Professor J. J. Kohn of Princeton University.)

*Dialogue on the Beauty of Mathematics:* from a discussion at the Shanghai Hua-Tung University

*Kohn:* Should you not present beauty of mathematics? Couldn't it inspire students? Is there room for the beauty of science?

*Answer:* The first demand is production.

*Kohn:* That is no answer.

*Answer:* Geometry was developed for practice. The evolution of geometry could not satisfy science and technology; in the seventeenth century, Descartes discovered analytical geometry. He analyzed pistons and lathes and also the principles of analytical geometry. Newton's work came out of the development of industry. Newton said, "The basis of any theory is social practice." There is no theory of beauty that people agree on. Some people think one thing is beautiful, some another. Socialist construction is a beautiful thing and stimulates people here. Before the Cultural Revolution some of us believed in the beauty of mathematics but failed to solve practical problems; now we deal with water and gas pipes, cables, and rolling mills. We do it for the country and the workers appreciate it. It is a beautiful feeling.

What one wants to achieve, of course, in mathematics as well as in life in general, is a balance. Does a proper balance exist in any country? No one knows. After Mao's death, the imbalance of Mathematical Maoism became obvious and corrective measures were taken. It is my impression gained from talking to Chinese mathematicians visiting the U.S. in the spring of 1979 that research is now pursued in China pretty much as it is everywhere else.

**Further Readings. See Bibliography**

D. Bernstein; Garrett Birkhoff; R. Burrington; A. Fitzgerald and S. Mac-Lane; G. H. Hardy; J. von Neumann; K. Popper and J. Eccles; J. Weissglass

# Underneath the Fig Leaf

A NUMBER OF ASPECTS of mathematics are not much talked about in contemporary histories of mathematics. We have in mind business and commerce, war, number mysticism, astrology, and religion. In some instances the basic information has not yet been assembled; in other instances, writers, hoping to assert for mathematics a noble parentage and a pure scientific existence, have turned away their eyes. Histories have been eager to put the case for science, but the Handmaiden of the Sciences has lived a far more raffish and interesting life than her historians allow.

The areas just mentioned have provided and some still provide stages on which great mathematical ideas have played. There is much generative power underneath the fig leaf.

## 1. Mathematics in the Marketplace

The activities of trading, pricing, coinage, borrowing, and lending have obviously been a strong source of concept formation in mathematics. Despite the contemporary

*Luca Pacioli*
*1445–1514*

*Simon Stevin*
*1548–1620*

conspiracy of silence, a remarkable amount is known about the interplay between business and mathematics. The main features of the development of arithmetic in the middle ages are clear, and there are books on the history of bookkeeping. In the medieval period and early Renaissance, some great mathematicians concerned themselves with bookkeeping. For example, in 1202 Fibonacci in his *Liber Abaci* introduced accounts with parallel roman and arabic numerals. In 1494 Luca Pacioli devoted three chapters of his *Summa de Arithmetica, Geometria, Proportioni et Proportionalita* to trade, bookkeeping, money, and exchange. In later centuries the Flemish mathematician Simon Stevin (1548–1620) and the English mathematician Augustus de Morgan (1806–1871) paid some attention to bookkeeping. In our own century, electronic computers have become indispensible in business; the development of these machines engaged some of the most brilliant minds in mathematics and physics. This story is told in the detailed history by Herman Goldstine. In the ancient world as now, *trade has been the principal consumer of mathematical operations measured in terms of the sheer number of operations performed.*

In trade, we find the four arithmetic operations: addition to find a total, subtraction to strike a balance, multiplication for replication, division for equal partition. Logically prior to these operations, though not chronologically, are a number of more primitive notions. There is exchange or equivalence: two sheep for a goat. There is assignment of abstract measures of value: everything has a price. In this way, equivalence classes of value are set up. The abstract representatives of the equivalence classes, coins, are originally perceived to have intrinsic value, but gradually this value tends to become symbolic as one moves toward paper money, checks, credit lines, bits in a computer memory.

There is the idea that all symbolic values are intermiscible and operable upon by the laws of arithmetic. If one goat = 2 sheep, and one cow = 3 goats, then one can compute that one cow = $3 \times (2 \text{ sheep}) = 6 \text{ sheep}$.

There is comparison: the concept of "greater than" and the institutionalization of the arithmetic laws of inequality:

1. $a < b$ or $a = b$ or $a > b$
   (everything has comparative value)
2. $a < b$ and $b < c$ implies $a < c$
   (the value system is transitive)

The notion of the discrete as opposed to the continuous is emphasized by coinage, which goes by standardized units. If the coins are perceived as being too valuable, they may be broken. If the coins are not sufficiently large, the articles in the exchange may be subdivided. This leads to the idea of fractions ("breakings").

As one moves from the ancient world to more modern times, one can point to a variety of operations and notions which come into mathematics directly from the experience of money or are reinforced through these means. The algorithms of arithmetic have been formed under the impact of business and are in constant flux. The currently taught algorithms of elementary school are hardly a hundred years old. Who knows how the children of the next generation will do their sums, what with hand-held computers or better at their disposal. The idea of interest, compound interest, discount, have analogies with and applications to calculus and thence to a variety of theories of growth.

The theory of probability entered mathematics through gambling—a financial transaction of great antiquity—and now finds application in the most elevated positions of theoretical science. The notion of a coin, repeatedly tossed, has become one of the fundamental schemata of mathematical experience, the paradigm of randomness, independence, and equiprobability.

The probabilistic notions of expectation and risk also came from gambling, and later became essential in life insurance, as part of the science of statistics. Derived from these classical theories are the modern mathematical theories of queueing, traffic, and optimization.

When, together with this, we consider theories of modern mathematical economics, we find a rich palette of higher mathematics in use. The principal tool is the theory of differential equations and other functional equations.

From: L. C. Anderson and K. Carlson "St. Louis Model Revisited," in "Econometric Model Performance," Klein and Burmeister, Univ. of Penna. Press, 1976.

## THE SAINT LOUIS MODEL

### Estimated Equations of the St. Louis Model

I. Total Spending Equation

A. Sample period: I/1953 − IV/1968

$$\Delta Y_t = \underset{(2.69)}{2.30} + \underset{(6.69)}{5.35\ \Delta M_{t-i}} + \underset{(.15)}{.05\ \Delta E_{t-i}}$$

II. Price Equation

A. Sample period: I/1955 − IV/1968

$$\Delta P_t = \underset{(6.60)}{2.95} + \underset{(9.18)}{.09\ D_{t-i}} + \underset{(5.01)}{.73\ \Delta P_t^A}$$

III. Unemployment Rate Equation

A. Sample period: I/1955 − IV/1968

$$U_t = \underset{(67.42)}{3.94} + \underset{(1.33)}{.06\ G_t} + \underset{(6.15)}{.26\ G_{t-i}}$$

IV. Long-Term Interest Rate

A. Sample period: I/1955 − IV/1968

$$R_t^L = \underset{(4.63)}{1.28} - \underset{(-2.40)}{.05\ \dot{M}_t} + \underset{(8.22)}{1.39\ Z_t} + \underset{(2.55)}{.20\ \dot{X}_{t-i}} + \underset{(11.96)}{.97\ \dot{P}/(U/4)_{t-i}}$$

V. Short-Term Interest Rate Equation

A. Sample period: I/1955 − IV/1968

$$R_t^S = \underset{(-2.43)}{-.84} - \underset{(-3.72)}{.11\ M_t} + \underset{(2.78)}{.50\ Z} + \underset{(9.28)}{.75\ \dot{X}_{t-i}} + \underset{(12.24)}{1.06\ \dot{P}/(U/4)_{t-i}}$$

Symbols are defined as:

$\Delta Y$ = dollar change in total spending (GNP in current prices)
$\Delta M$ = dollar change in money stock
$\Delta E$ = dollar change in high-employment federal expenditures
$\Delta P$ = dollar change in total spending (GNP in current prices) due to price change
$D = Y - (X^F - X)$
$X^F$ = potential output
$X$ = output (GNP in 1958 prices)
$\Delta P^A$ = anticipated price change (scaled in dollar units)
$U$ = unemployment as a percent of labor force
$G = ((X^F - X)/X^F) \cdot 100$
$R^L$ = Moody's seasoned corporate AAA bond rate
$\dot{M}$ = annual rate of change in money stock
$Z$ = dummy variable (0 for I/1955 − IV/1960) and (1 for I/1961 − end of regression period)
$\dot{X}$ = annual rate of change in output (GNP in 1958 prices)
$\dot{P}$ = annual rate of change in GNP price deflator (1958 = 100)
$U/4$ = index of unemployment as a percent of labor force (base = 4.0)
$R^S$ = four- to six-month prime commercial paper rate.

*Mathematical economics predicts the course of the economy by such models as this.*

Fixed-point theory for existence of equilibria is also important. The theory of business cycles has analogies within mathematical physics. There is hardly an area of modern mathematics which might not be called upon for contributions to economics. In very recent times, the theory of nonstandard analysis is being applied, wherein an analogy is found between small individual firms and infinitesimals. While the material just cited represents contributions to economics from mathematics, it also goes the other way: there are contributions from economics to mathematics. Thus, Brownian motion enters into the mathematical literature first via the motion of the Bourse in the early work of L. Bachelier.

On the mechanical (or electronic) side, the demands of big business and big government have led to a wide variety of computing machines (think of the "B" in "IBM"). This, on the one hand, has fostered a new branch of mathematical learning known as computer science, a science which has logical, linguistic, combinatorial, and numerical features. On the other hand, the existence of the machines has fed back into and changed the traditional arrangements and attitudes of business itself. (Think of credit cards.) Thus, there is a very strong interplay between mathematics and the marketplace, and if the business of the country is business, as Calvin Coolidge proclaimed, we should expect this strong reciprocal feedback to increase.

At still a deeper level, one may raise the question of the relations between the socioeconomic condition and the whole of science, technology, and mathematics. This is what Joseph Needham calls the Great Debate of the history of science. We look next at a prominent example of such a relation.

### Further Readings. See Bibliography

H. Goldstine, A. Littleton and B. Yamey

### 2. Mathematics and War

Legend has it that Archimedes put his science at the service of warfare. He is reputed to have devised compound pulleys to launch galleys, to have invented a variety of catapults and military engines and most spectacularly to have focused the sun's rays on besieging ships by means of a paraboloidal mirror. All this for King Hieron of Syracuse who was under attack by the Romans. Now Archimedes was the most brilliant scientist and mathematician of his age, but the achievements just listed, although they can be explained by the mathematical theories of mechanics and optics, do not appear to have involved mathematics at the basic level of application.

What is the relationship between mathematics and war? In the beginning, the contribution was meager. A few mathematical scribes to take the census and to arrange for induction into the army. A few bookkeepers to keep track of ordnance and quartermaster. Perhaps a bit of surveying and a bit of navigation. In their capacity as astrologers, the principal contribution of the ancient mathematicians was probably to consult the stars and to tell the kings what the future held in store. In other words, military intelligence.

Modern warfare is considered by some authorities to have begun with Napoleon, and with Napoleon one begins to see an intensification of the mathematical involvement. The French Revolution found France supplied with a brilliant corps of mathematicians, perhaps the most brilliant in its history: Lagrange, Condorcet, Monge, Laplace, Legendre, Lazare Carnot. Condorcet was Minister of the Navy in 1792; Monge published a book on the manufacture of cannons. Under Napoleon, mathematicians continued to bloom. It is reported that Napoleon himself was fond of mathematics. Monge and Fourier accompanied Napoleon on his Italian and Egyptian campaigns, and if these men did not do anything directly mathematical during these army hitches (Monge supervised booty while Fourier wrote the *Description of Egypt*), one is left with the feeling that Napoleon thought that mathematicians were useful fellows to have around.

Arriving at World War II, one finds mathematical and scientific talent in widespread use in the Army, Navy, and Air Force, in government research laboratories, in war industries, in governmental, social and business agencies. A brief list of the variety of things that mathematicians did would include aerodynamics, hydrodynamics, ballistics, development of radar and sonar, development of the atomic bomb, cryptography and intelligence, aerial photography, meteorology, operations research, development of computing machines, econometrics, rocketry, development of theories of feedback and control. Many professors of mathematics were directly involved in these things, as were many of their students. This writer was employed as a mathematician-physicist at NACA (later NASA), Langley Field, Virginia, with only a bachelor's degree to his credit, and many of his contemporaries at Langley Field subsequently occupied chairs of mathematics throughout the country.

With the explosion of the atomic bomb over Japan and the subsequent development of more powerful bombs, atomic physicists who had hitherto lived ivory-tower academic existences experienced a sense of sin. This sense of sin spread simultaneously over the mathematical community. Individual mathematicians asked themselves in what way they, personally, had unleashed monsters on the world, and if they had, how they could reconcile it with whatever philosophic views of life they held. Mathematics, which had previously been conceived as a remote and Olympian doctrine, emerged suddenly as something capable of doing physical, social, and psychological damage. Some mathematicians began to compartmentalize their subject into a good part and a bad part. The good part: pure mathematics, the more abstract the better. The bad part: applied mathematics of all kinds. Some mathematicians and a rising generation of students left applications forever. Norbert Wiener, who had been engaged in developing theories of prediction and feedback control, renounced government support of his work and devoted the remainder of his life to doing "good works" in biophysics and to propagandizing against the nonhuman use of human beings.

World War II was followed by the Cold War, during which the shock of Sputnik occurred. The intensified space activities employed many thousands of mathematicians, as did the development, practically *ex nihilo,* of the whole computer industry.

During the protest against the Vietnam War, there were direct physical attacks against mathematical institutions. Two of the main centers of research in applied mathematics are at New York University and the University of Wisconsin. At NYU there is a large computing center, which is sponsored by the Energy Research and Development Authority—formerly the Atomic Energy Commission. At Wisconsin, there is a large building which houses the Mathematics Research Center—formerly the Army Mathematics Research Center. In 1968, a bomb was exploded at the center in Madison, killing a graduate student who happened to be in the building working late at night. At NYU, the Computing Center was captured and held for ransom, and an unsuccessful attempt was made to blow it up.

Many opponents of the war considered it immoral to work in military-supported institutions. It no longer mattered if one was working on military or nonmilitary problems; the whole institution was regarded as contaminated by evil.

One began to hear it said that World War I was the chemists' war, World War II was the physicists' war, World War III (may it never come) will be the mathematicians' war. With this, there entered into the general consciousness the full realization that mathematics is inevitably bound up in the general fabric of life, that mathematics is good or bad as people make it so, and that no activity of the human mind can be free from moral issues.

### Further Readings. See Bibliography

N. P. Davis; H. Goldstine [1972]; J. Needham, vol. III, p. 167.

### 3. Number Mysticism

"We who are heirs to three centuries of science," writes Sir Kenneth Clark in his marvellous *Landscape into Art,* "can hardly imagine a state of mind in which all material objects were regarded as symbols of spiritual truths or episodes in sacred history. Yet, unless we make this effort of imagination, mediaeval art is largely incomprehensible." We who are heirs to three recent centuries of scientific development can hardly imagine a state of mind in which many mathematical objects were regarded as symbols of spiritual truths or episodes in sacred history. Yet, unless we make this effort of imagination, a fraction of the history of mathematics is incomprehensible.

Read how Plutarch (40 A.D.–120 A.D.), in describing the Isis cult of Egypt, blends sacred history and mathematical theorems.

> The Egyptians relate that the death of Osiris occurred on the seventeenth (of the month), when the full moon is most obviously waning. Therefore the Pythagoreans call this day the "barricading" and they entirely abominate this number. For the number seventeen, intervening between the square number sixteen and the rectangular number eighteen, two numbers which alone of plane numbers have their perimeters equal to the areas enclosed by them,* bars, discretes, and separates them from one another, being divided into unequal parts in the ratio of nine to eight. The number of twenty-eight years is said by some to have been the extent of the life of Osiris, by others of his reign; for such is the number of the moon's illuminations and in so many days does it revolve through its own cycle. When they cut the wood in the so-called burials of Osiris, they prepare a crescent-shaped chest because the moon, whenever it approaches the sun, becomes crescent-shaped and suffers eclipse. The dismemberment of Osiris into fourteen parts is interpreted in relation to the days in which the planet wanes after the full moon until a new moon occurs.

"All is number," said Pythagoras, and number mysticism takes this dictum fairly literally. The universe in all its aspects is governed by number and by the idiosyncrasies of number. Three is the trinity, and six is the perfect number, and 137 was the fine-structure constant of Sir Arthur Eddington, who was a number mystic and a distinguished physicist.

In the year 1240, the most triumphal year in the reign of Frederick II of Sicily, western Europe was beset by rumors

---

* This is a nice theorem. Prove it.

of a great king in the far East who ruled over a vast kingdom and who was making his way slowly and relentlessly
westward. One Islamic kingdom after another had fallen
to his sword. Some Christians interpreted the news as pre-
saging the arrival of the legendary Prester John who would
unite with the kings of the West in Jerusalem and seal the
doom of the Islamic religion. The Jews of Europe, for reasons to be explained shortly, held this Eastern monarch to
be King Messiah, the scion of David, and proposed going
forth to meet him in joy and celebration. Other Christians,
while agreeing with the messianic interpretation, held that
Frederick himself, *stupor et dominus mundi,* the marvel and
master of the world, one of the most remarkable intellects
ever to sit on a royal throne, was the promised Messiah.

Now on what basis was it concluded that the Messiah was
arriving? Simply that the year 1240 in the Christian calendar corresponded to the year 5000 in the Jewish calendar
and that, according to some theories, the Messiah was to
appear at the beginning of the sixth millenium. Here we
have a piece of number mysticism of a sort which is incredible to the modern mind. (At this point, we should inform
our curious reader that the eastern king was neither Prester John nor the Messiah but Batu, son of Genghis Khan
and the founder of the Golden Horde, who slaughtered
his way up to Liegnitz in Silesia.)

But the sacred merges imperceptibly with the practical.
Mathematics, asserted Henry Cornelius Agrippa, a popular philosophical magician of the sixteenth century, is absolutely necessary for magic, "for everything which is done
through natural virtue is governed by number, weight and
measure. When a magician follows natural philosophy and
mathematics and knows the middle sciences which come
from them—arithmetic, music, geometry, optics, astronomy, mechanics—he can do marvelous things."* One of
the ways in which number mysticism works itself out is
through the art of gematria (the word itself is derived
from "geometry"). Gematria is based on the fact that the
classic alphabets of Latin, Greek, and Hebrew normally

* Frances Yates, *Giordano Bruno and the Hermetic Tradition.*

have numerical equivalents. In its simpler form, gematria equates words with equivalent numbers and interprets the verbal equivalents.

Here is an example from the period of Frederick II. The name "Innocentius Papa" (Pope Innocent IV) has the numerical equivalent 666. This is the "Number of the Beast" of Revelations 13:18 and hence Innocent equals the Antichrist. (Frederick was violently antipope.)

What sort of nonsense and intellectual trash is this, one wonders, particularly when one realizes that public policy may have been based upon such reasoning. One hopes that contemporary political reasoning is based on firmer stuff. Yet this kind of reasoning, this riding rough-shod with numbers, may have fostered number skills and interests that far outweighed the damage done.

To the medieval mind, a number, particularly if it were a sacred number, was a manifestation of divine and spiritual order. It could be turned into an aesthetic principle. As an example, we mention a recent analysis by Horn of a master plan for a monastic settlement drawn up in Aachen in 816. This is the so-called "Plan of Saint Gall." Horn finds that the designing architect kept the sacred numbers three, four, seven, ten, twelve, and forty in his mind and worked with them repeatedly. We shall bypass the credentials or the certification of the holiness of these specific numbers and pass to the architectural detail.

In the plan, there are three major areas—east, central, and west. There are three building sites, three cloisters, three bake and brew houses, three bathhouses, three medical installations, three walled gardens, three poultry pens, and three milling installations.

There are four circular structures, four altars in the transept and four in each aisle, and four pieces of liturgical furniture in the nave. There are four rows of plantings. Four also plays a role in the basic modules of the layout.

Seven buildings form the core of the monastic settlement. There are seven steps which raise the presbytery above the crossing, seven desks for the scribes in the scriptorium. There are seventy-seven beds in the monks' dormitory. There are seven liturgical stations in the axis of the

*Giordano Bruno*
*1548–1600*

*Hermetic Figures.*

From Giordano Bruno, *Articuli centum et sexaginta adversus huius tempestatis mathematicos atque philosophos*, Prague, 1588 (p. 313 ff.).

church. There are seventeen (10 + 7) altars in the whole of the church and twenty-one (3 × 7) altars in the whole of the plan.

We shall not pursue Horn's extensive catalogue of sacred numbers through forty. If the skeptic finds that these occurrences are accidental, or merely evidence that all numbers can be decomposed by sums or products into a list of sacred numbers as in the prime decomposition theorem, or Goldbach's conjecture (see Chapter 5), let him work with the master plan of his local Howard Johnson's motel and discern what sacred plan it conforms to.

### Further Readings. See Bibliography

J. Griffiths; W. Horn; E. Kantorowicz; F. Yates

### 4. Hermetic Geometry

The dream of philosophic magic (to be distinguished from conjurer's magic) has persisted for millennia. One of the fundamental assumptions of this magic is that spiritual forces in the universe may be induced to enter in and influence the material forces. The spiritual is celestial, the material is earthly. Earthly forms are often represented as geometrical figures and as such are thought to be aspects of the pure celestial forms. By proper representation and arrangement, the material figure induces a kind of sympathetic resonance with its celestial counterpart and, as a result, the figure is endowed with the potency of a talisman. This potency is then applied to strictly practical ends—the curing of illness, the achievement of business success, destruction of one's enemies, practical erotics, and many others.

The accompanying illustration shows three pieces of hermetic geometric art dating from 1588. They have been taken from the book *Articuli . . . adversus . . . mathematicos atque philosophos* written by Giordano Bruno. Bruno was an ex-Dominician, a brilliant philosopher, and a philosophic magician.

These figures strike us today as pleasant designs which would make agreeable tile work. One supposes that the specific arrangements are not arbitrary, but have been

100

created according to some principle. Many have thought (and think) they have discovered the keys to the universe. One should not hold a key in contempt because it opens up only a minor chamber at the periphery.

A fourth hermetic figure is the magic hieroglyph of John Dee (1564). It may remind one of the peace symbol of the early 1970s. The mathematical and magical properties of this symbol are set forth in a book called *Monas Hieroglyphica* which explains its construction and its interpretation through a sequence of "theorems." The reader should compare the theorematic material here with that of Euclid given in Chapter 5.

*Dee's hieroglyph*

### Further Readings. See Bibliography

G. Bruno; J. Dee; C. Josten; F. Yates

### 5. Astrology

The role of astrology in the development of mathematics, physics, technology, and medicine has been both misrepresented and downplayed; contemporary scholarship has been restoring proper perspective to this activity. We are dealing here with a prescience and a failed science. It can be called a false or a pseudoscience only insofar as it is practiced with conscious deception.

The roots of astrology can be found in Babylonia of the fourth century B.C., if not earlier. Astrology and divination were widespread in the East and can be found today as an integral part of life in various portions of the East. In the West, one sees remnants of astrology as a kind of pop culture in the numerology of the newspaper, computer horoscopes, and zodiac books.

Astrology has waxed and waned and waxed again. At times it has been denounced as superstition—pagan superstition—naturally. It has played an official role in governments and religious establishments, it has been reviled and proscribed, it has been thought immoral and iniquitous; it was considered heretical, for does it not contradict free will? It has been tolerated and winked at. It has been used as a teaser, as when the famous mathematician and physician Cardano cast a horoscope of Jesus. When in the

(*Continued on p. 103*)

*Reproduction and Trans-
lation by C. H. Josten,
Ambix, vol. 12 (1964).*

# MONAS HIEROGLYPHICA:
## IOANNIS DEE, LONDINENSIS,

*Mathematicè, Magicè, Cabalisticè, Anagogicéque,*
*explicata: Ad*

SAPIENTISSIMVM,

ROMANORVM, BOHEMIAE, ET HVNGARIAE,

REGEM,

MAXIMILIANVM.

### *THEOREMA* I.

PEr Lineam rectam, Circulumque, Prima, Sim-
plicissimaque suit Rerum, tum, non existentiū,
tum in Naturæ latentium Inuolucris, in Lu-
cem Productio, representatioque.

### *THEOREMA* II.

AT nec sine Recta, Circulus; nec sine Puncto, Recta artifi-
ciose fieri potest. Puncti proinde, Monadisque ratione,
Res, & esse cœperūt primò: Et quæ peripheria sunt affectæ,
(quantæcūque fuerint) Centralis Puncti
nullo modo carere possunt Ministerio.

### *THEOREMA* III.

MOnadis, Igitur, HIEROGLYPHICAE Conspicuū
Centrale Punctum, TERRAM refert; circa quam, tum
SOL tum LVNA, reliquiǵue Planetæ
suos conficiunt Cursus. Et in hoc mune-
re, quia dignitatem SOL obtinet sum-
mam, Ipsum, (per excellentiam,) Circulo
notamus Integro, Centroque Visibili.

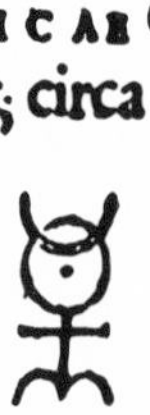

MONAS
HIERO-
GLYPHI-
CA.

*THEO:*

The Hieroglyphic Monad
of John Dee, of London

mathematically, magically, cabbalistically, and anagogically explained, [and addressed] to the most wise Maximilian, King of the Romans, of Bohemia, and of Hungary.

### Theorem I.

The first and most simple manifestation and representation of things, non-existent as well as latent in the folds of Nature, happened by means of straight line and circle.

### Theorem II.

Yet the circle cannot be artificially produced without the straight line, or the straight line without the point. Hence, things first began to be by way of a point, and a monad. And things related to the periphery (however big they may be) can in no way exist without the aid of the central point.

### Theorem III.

Thus the central point to be seen in the centre of the hieroglyphic monad represents the earth, around which the Sun as well as the Moon and the other planets complete their courses. And since in that function the Sun occupies the highest dignity, we represent it (on account of its superiority) by a full circle, with a visible centre.

---

*(Continued from page 101)*

1500s it was pursued honestly as a science, it could not help but influence the course of scientific discovery.

Astrology proceeds from the belief that the celestial bodies affect the affairs of men. The position of the man on the moon, and of the planets as they make their way across the backdrop of the zodiacal constellations, are said to influence the fate of individuals, of kings and rulers, and of nations in a vital way. That there is something to this, no one can deny. Does not the sun energize our life? Do not changes in its radiation pattern affect the weather, and the radio reception? Do not the sun and the moon together control the tides? Does not, according to a modern cosmologist, the existence of matter at the furthest reaches of the universe create the gravity that keeps our tennis balls from flying away? What a grand conception of the universe this

## *THEOREMA IIII.*

LVnæ Hemicyclium, licet hic, Solari sit Circulo quasi Superius Priusque: Tamen S o l e m tanquam Dominum, Regemeque suum obseruat: eiusdem Forma ac vicinitate adeo gaudere videtur, vt & illum in Semidiametri æmuletur Magnitudine, (Vulgaribus apparente hominibus,) & ad eundem, semper suum conuertat Lumen : S o l a r i- ı v s'q v ɛ ita tandem imbui Radijs appetat, vt in eundem quasi Transformata, totu'disparcat Cælo : donec aliquot post Diebus, omnino hac qua depinximus, appareat corniculata figura.

## *THEOR. V.*

ET Lunari certè Semicirculo ad Solare complementum perducto: Factum est Vespere & Mane Dies vnus. Sit ergo Primus, quo L v x est facta Philosophorum.

## *THEOR. VI.*

SO l ɛ m, L v n a m'q v ɛ, Rectilineæ Cruci, inniti, hic videmus. Que, tum T ɛ ʀ n a ʀ ı v m, tum Q v a t ɛ ʀ n a- ʀ ı v m, apposite satis, ratione significare Hieroglyphica, potest. T ɛ ʀ n a ʀ ı v m quidem: ex duabus Rectis, & Communi vtrisque, quasi Copulatiuo Puncto. Q v a t ɛ ʀ n a ʀ ı v m vero: ex 4. Rectis, includentibus 4. Angulos rectos. Singulis, bis, (ad hoc) repetitis; (Sicque, ibidem, secretissime, etiam O c t o n a ʀ ı v s, sese offert; quem, dubito an nostri Prædecessores, Magi, vnquam conspexerint: Notabisque maxime.) Primorū Patrum, & Sophorum T ɛ ʀ- n a ʀ ı v s, Magitus, C ɒ ʀ ᴘ ᴏ ʀ ɛ, s ᴘ ı ʀ ı t v, & a n ı- m a, constabat. Vnde, Manifestum bro Primariū habemus S ɛ ᴘ t ɛ n a ʀ ı v m. Ex duabus numiuin, Rectis, ɛt Communi Puncto: Deinde ex 4. Rectis, a ɒ Vno Puncto, sese, Separantibus.

*THEOR.*

### Theorem IIII.

Although the half-circle of the Moon appears here to be, as it were, above the solar circle, and more important than it, she respects the Sun all the same as her master and King. She seems to find so much delight in his shape and his vicinity that she emulates the size of [his] radius (as it appears to the vulgar) and always turns her light towards him. And so much, in fine, does she long to be imbued with solar rays, that, when she has been, as it were, transformed into him, she disappears from the sky altogether until, after a few days, she appears in horned shape, exactly as we have depicted her.

### Theorem V.

And, surely, one day was made out of evening and morning[38] by joining the lunar half-circle to its solar complement. Be it accordingly the first [day] on which the light of the philosophers was made.

### Theorem VI.

We see Sun and Moon here resting upon a rectilinear cross which, by way of hieroglyphic interpretation, may rather fittingly signify the ternary as well as the quaternary: the ternary, [in so far as it consists] of two straight lines and one point which they have in common and which, as it were, connects them; the quaternary [in so far as it consists] of four straight lines including four right angles, each [line being] (for this purpose) twice repeated[39]. (And so here also the octonary offers itself in a most secret manner, of which I doubt whether our predecessors [among] the *magi* ever beheld it, and which you will especially note.) The magical ternary of the first [of our] forefathers and wise men consisted of body, spirit, and soul. Thence we see here manifested a remarkable septenary, [consisting] to be sure of two straight lines and a point which they have in common, and of four straight lines separating themselves from one point.

---

## Etc. Etc. Altogether: Twenty Three "Theorems."

[38] *Cf.* Genesis 1, 5.

[39] This passage is somewhat obscure. It means that, when the cross is considered as formed from four right angles, four pairs of lines containing these angles coincide in one line each.

is, uniting the immense with the miniscule, the near with the far! This is the Grand Plan and the problem now becomes, how can we read this plan for our own purposes?

Among the classic forms of astrological practice we find genethlialogy, catarchic astrology, and interrogatory astrology, all three interrelated. Genethlialogy asserts that the celestial omens at the moment of one's birth affect the course of one's life. To predict this course one needs to know the exact moment and place of birth. One must calculate where the planets were and calculate certain relations between them such as conjunctions and oppositions.

Catarchic astrology asserts that any act is influenced by the celestial omens at the moment of the inception of the act. Therefore, from a knowledge of the future position of the planets, one can forecast auspicious dates for the occurrence of important events.

Interrogatory astrology invites questions of all sorts. Where did I lose my wallet? Should I marry such a one? It asserts that the moment of interrogation affects the correct answer. Thus does astrology give answers in a world full of bitter problems, where advice is rare and of doubtful quality.

To practise astrology in its intensive forms, one had to know astronomy, mathematics, medicine and much besides, for its methods (algorithms) were elaborate. When a patient came in with a complaint, the first thing to do was to cast his nativity. This would be based on information as exact as one had about the time of his birth. Almanacs were available which would tell the practitioner about the state of the heavens at the moment of birth. How did one know the moment of birth? This, remember, was in a period in which the average person could neither read nor write nor reckon, when a knowledge of the number of miles from London to Canterbury was given as one of the great reasons for learning arithmetic. But given a nativity of even an approximate sort, the astrologer-mathematician-physician would proceed to examine the patient's symptoms (in particular the color of his urine), and then come up with a prescription. This was the ten-dollar job.

Naturally, when the patient was a member of the nobility

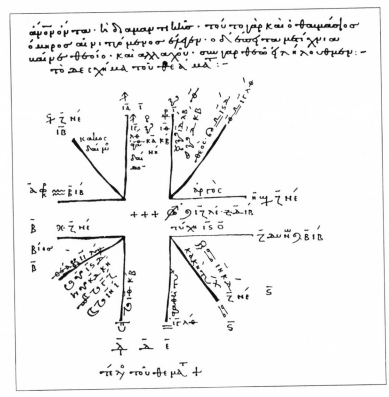

*Greek horoscope dated October 28, 497.*

*From Neugebauer and van Hoesen*

or the clergy or when the patient was the astrologer himself, first-order accuracy was thought insufficient. An exact horoscope was necessary and this, of course, would cost a hundred dollars. This created the necessity for accurate tables of planetary positions, instruments whose accuracy exceeded that of the run-of-the-mill astrolabe or transit, accurate clocks, convenient and accurate modes of mathematical computation. Thus, to be a physician in, say, the thirteenth century, one had, ideally, also to be a herbalist, an alchemist, a mathematician, an astronomer, and a maker of scientific instruments. The pursuit of accurate knowledge of time and position leads through Brahe, Kepler, Galileo, and Newton to contemporary physics and mathematics.

Astrologers must have chalked up quite a few successes. Even tossing a coin is often a useful device for instituting

107

policy and by the law of averages should be right every now and again. So one hears of the spectacular job done by John Dee for young Queen Elizabeth. Dee, asked to provide an auspicious date for the coronation of Elizabeth I, worked his science, and came up with the date that led to England's dominance of the world for more than three hundred years.

But in the long run, though astrology was a prescience, and though many of its practitioners pursued it in the spirit of modern scientific inquiry, it was a failure, one of many failed theories having a mathematical core. It provided a wrong model of reality and this led to its decline and to its intellectual trivialization.

**Further Readings. See Bibliography.**

D. Pingree; L. White, Jr.

### 6. Religion

Mathematics, in an earlier view, is the science of space and quantity; in a later view, it is the science of pattern and deductive structure. Since the Greeks, mathematics is also the science of the infinite. Hermann Weyl speculates that the presence of the infinite in mathematics runs parallel to religious intuition:

*Hermann Weyl*
*1885–1955*

> . . . purely mathematical inquiry in itself, according to the conviction of many great thinkers, by its special character, its certainty and stringency, lifts the human mind into closer proximity with the divine than is attainable through any other medium. *Mathematics is the science of the infinite,* its goal the symbolic comprehension of the infinite with human, that is finite, means. It is the great achievement of the Greeks to have made the contrast between the finite and the infinite fruitful for the cognition of reality. Coming from the Orient, the religious intuition of the infinite, the ἄπειρον, takes hold of the Greek soul . . .
> This tension between the finite and the infinite and its conciliation now become the driving motive of Greek investigation.

108

Like mathematics, religions express relationships between man and the universe. Each religion seeks an ideal framework for man's life and lays down practices aimed at achieving this ideal. It elaborates a theology which declares the nature of God and the relationship between God and man. Insofar as mathematics pursues ideal knowledge and studies the relationship between this ideal and the world as we find it, it has something in common with religion. If the objects of mathematics are conceptual objects whose reality lies in the common consciousness of human minds, then these shared mathematical concepts may constitute the dogma of mathematical belief.

It is the writer's impression that most contemporary mathematicians and scientists are agnostics, or if they profess to a religious belief, they keep their science and their religion in two separate boxes. What might be described as the "conventional scientific" view considers mathematics the foremost example of a field where reason is supreme, and where emotion does not enter; where we know with certainty, and know that we know; where truths of today are truths forever. This view considers religion, by contrast, a realm of pure belief unaffected by reason. In this view, all religions are equal because all are equally incapable of verification or justification.

However, this perceived dichotomy between mathematics and religion, though now widespread, is not universal; and over the centuries, the interplay between mathematics and religion has taken various fruitful forms.

Religious considerations have, for instance, spurred some kinds of mathematical creation and practice. Scholars like A. Seidenberg have sought the origins of counting and geometry in ancient rituals. The development of the calendar is another example. To what extent was the development of the calendar influenced and fostered by a desire to standardize periodic ritual events?

We have also observed in regard to number symbolism and number mysticism, how religious practice may be affected by mathematics.

At a somewhat deeper level of cultural influence we can

109

see how notions of mathematical proof have contributed to the development of theology. The medieval schoolmen looked for rational proofs of theological theorems, so that points of dogma might be established Q.E.D.

Nicolas of Cusa (1450) believed that the true love of God is *amor Dei intellectualis* and that the intellectual act through which the divine is revealed is mathematics. (The divine is to be reached through many paths, e.g., through cleanliness, or through uncleanliness as in the case of the Desert Fathers. Nicolas asserts that the divine is to be reached through "thinkliness.")

The German romantic epigrammatist Novalis (Friedrich von Hardenberg, d. 1801) said that "Pure mathematics is religion," because, as later explained by the choreographer Oskar Schlemmer in 1925, "It is the ultimate, the most refined and the most delicate." Novalis, who had read quite a bit of contemporary mathematics, also wrote "Das Leben der Götter ist Mathematik" and "Zur Mathematik gelangt Man nur durch eine Theophanie."

As further instances of this tendency consider Spinoza's treatment of ethics, *more geometrico,* and John Locke's statement in "An Essay on Human Understanding":

> Upon this ground it is that I am bold to think that morality is capable of demonstration, as well as mathematics: since the precise real essence of the things moral words stand for may be perfectly known, and so the congruity and incongruity of the things themselves be certainly discussed; in which consists perfect knowledge.

Conversely, religious views of the world have posited mathematics as a paradigm of Divine thought. The nun-playwright Hrosvita of Gandersheim (980), in her play "Sapientia," after a rather long and sophisticated discussion of certain facts in the theory of numbers, has Sapientia say that

> this discussion would be unprofitable if it did not lead us to appreciate the wisdom of our Creator, and the wondrous knowledge of the Author of the world, Who in the beginning created the world out of nothing, and set everything in number, measure and weight, and then, in time and the

age of man, formulated a science which reveals fresh wonders the more we study it.

Or listen to Kepler in *Harmonia Mundi* (1619):

> I thank thee, O Lord, our Creator, that thou hast permitted me to look at the beauty in thy work of creation; I exult in the works of thy hands. See, I have completed the work to which I felt called; I have earned interest from the talent that thou hast given me. I have proclaimed the glory of thy works to the people who will read these demonstrations, to the extent that the limitations of my spirit would allow.

*Johann Kepler*
*1571–1630*

These, of course, are instances of the Platonic idea that mathematical law and the harmony of nature are aspects of the divine mind-soul. In this frame of reference, the Euclid myth discussed in Chapter 7 appears as an essential and congruous element.

Belief in a nonmaterial reality removes the paradox from the problem of mathematical existence, whether in the mind of God or in some more abstract and less personalized mode. If there is a realm of nonmaterial reality, then there is no difficulty in accepting the reality of mathematical objects which are simply one particular kind of nonmaterial object.

So far, we have discussed the interaction between the discipline of mathematics and established religions. We might also ask to what extent does mathematics itself function as a religion. Insofar as the "laws of mathematics" are properties possessed by certain shared concepts, they resemble doctrines of an established church. An intelligent observer seeing mathematicians at work and listening to them talk, if he himself does not study or learn mathematics, might conclude that they are devotees of exotic sects, pursuers of esoteric keys to the universe.

Nonetheless, there is remarkable agreement among mathematicians. While theologians notoriously differ in their assumptions about God, still more in the inferences they draw from these assumptions, mathematics seems to be a totally coherent unity with complete agreement on all important questions; especially with the notion of *proof*, a procedure by which a proposition about the unseen reality

111

**God wields the compass.** →

*William Blake,* The Ancient of Days, *Whitworth Art Gallery, University of Manchester*

**Archimedes wields the compass.**

**Dr. Dee wields the compass.**

*From French,* John Dee.

**The compass wields itself.**
*The Mystical Compass, From Robert Fludd,* Utriusque cosmi—historia, *II* (I), *p. 28 (p. 407).*

*From: Yales "Giordano Bruno"*

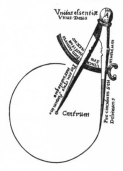

## Who wields the compass? God? Archimedes? John Dee? Or does the compass wield itself?

can be established with finality and accepted by all adherents. It can be observed that if a mathematical question has a definite answer, then different mathematicians, using different methods, working in different centuries, will find the same answers.

Can we conclude that mathematics is a form of religion, and in fact the true religion?

### Further Readings. See Bibliography

Dyck; F. von Hardenberg; H. Weyl; A. Seidenberg

# Abstraction and
# Scholastic Theology

ABSTRACTION IS the life's blood of mathematics, and conversely, as P. Dirac points out, "Mathematics is the tool specially suited for dealing with abstract concepts of any kind. There is no limit to its power in this field." But abstraction is ubiquitous. It is almost characteristic or synonymous with intelligence itself. Among the many fruits of abstraction of a mathematical type can be listed systematic scholastic theology. In the view of Bertrand Russell (*History of Western Philosophy*, p. 37), systematic scholastic theology derives directly from mathematics. It is especially interesting to trace this in the writings of Sa'id ibn Yusuf (882–942).

Sa'id ibn Yusuf (Saadia Gaon), philosopher, theologian, prominent leader of Babylonian Jewry, was born in the Faiyum district of Egypt. In 922 he moved to Babylonia and was appointed head of the Pumbidita Academy. His major philosophical work, *Kitab al-Amanat wa-al Itiqadat* (The Book of beliefs and opinions), makes ample references to Biblical and Talmudic authority, but in addition draws on medicine, anatomy, mathematics, astronomy and music. In the views of contemporary mathematicians, Saadia (we now use this more common spelling) had thoroughly mastered the mathematical sciences, and it is this aspect of the Kitab that we shall examine.

Saadia is fascinating because in him can be seen not only the mathematics of his day, but in his systematic theology there were already present the methods, the drives, the processes of thought which characterize nineteenth and twentieth-century mathematics.

The mathematics of the tenth century is present. Thus Saadia says (p. 93),

I do not demand of Reuben 100 drachmas, but I demand of him the square root of ten thousand.

113

This is a turn of phrase which would probably not have occurred to the man on the street in Pumbedita, but it is surely not the most exciting thought that tenth-century mathematics could have dreamed up. But there it is, in a religious context.

There is a discussion of time which might remind one of a reversed paradox of Achilles and the tortoise, not put to a destructive purpose as with Zeno but to the positive purpose of proving The Creation. If the world were uncreated, says Saadia, then time would be infinite. But infinite time cannot be traversed. Hence, the present moment couldn't have come to be. But the present moment clearly exists. Hence, the world had a beginning.

In Treatise II: concerning the belief that the Creator of all things . . . is one, Saadia begins his Exordium by saying (p. 88)

> the data with which the sciences start out are concrete, whereas the objectives they strive for are abstract.

This is certainly spoken like a modern scientist, and one wonders whether this spirit is the imposition of a modern translator who renders "big" as "concrete" and "fine" as "abstract." But I think not, for in the example which Saadia then gives, it is clear that what is "fine" is an explanation which is less specific, more general, and therefore an explanation which is capable of dealing with groups of collateral phenomena, i.e., an abstract theory. Further on he says (p. 90)

> the last rung in the ladder of knowledge is the most abstract and subtle of all.

This is all prelude to his insistence that God must be understood, in fact can only be understood, through the process of abstraction. Saadia's Deity is accordingly highly abstract, highly intellectualized. One of the major programs of modern mathematics is the abstract program. This may come as a surprise to the nonmathematician for whom such things as numbers, points, lines, equations are already sufficiently abstract. But to the mathematician whose pro-

114

fession has been dealing with these objects for three thousand years, they have become quite concrete, and he has found it essential to impose additional levels of abstraction in order to explain adequately certain common features of these more prosaic things. Thus, there have arisen over the past hundred years such abstract structures as "groups," "spaces," and "categories" which are generalizations of fairly common and simple mathematical ideas.

In his role of abstractor, the mathematician must continually pose the questions "What is the heart of the matter?," "What makes this process tick?," "What gives it its characteristic aspect?" Once he has discovered the answer to these questions, he can look at the crucial parts in isolation, blinding himself to the whole.

Saadia arrives at his concept of God in very much the same way. He has inherited a backlog of thousands of years of theological experience, and these he proceeds to abstract:

> the idea of the Creator . . . must of necessity be subtler than the subtlest, more recondite than the recondite, and more abstract than the most abstract.

Though there is in the corporeal something of God, God is not corporeal. Though there is in motion, in the accidents of space and time, in emotions or in qualities something of God, God is not identical with these. Though these attributes may pertain to Him, for He is (p. 134) eternal, living, omnipotent, omniscient, the Creator, just, not wasteful, etc., he has been abstracted by Saadia out of these attributes. The Deity emerges as a set of relationships between things some of which are material, some spiritual, these relationships being subject to certain axiomatic requirements. When Saadia seeks to know God through the process of abstraction, he finds a very mathematical God.

Having gone through this program of abstracting the Deity, Saadia asks (p. 131),

> How is it possible to establish this concept in our minds when none of our senses have ever perceived Him?

115

He answers this by saying,

> It is done in the same way in which our minds recognize the impossibility of things being existent and nonexistent at the same time, although such a situation has never been observed by the senses.

That is, we recognize that "A" and "not A" cannot coexist despite the fact that we may not ever have experienced either "A" or "not A."

One might amplify Saadia's answer by pointing out that it can be done by the process of abstraction just as an abstract graph is not a labyrinth, nor a simple arithmetic or geometric representation of a labyrinthine situation, but the abstracted essence of the properties of traversing and joining. Conversely, a labyrinth is a concrete manifestation of an abstract graph. (See Chapter 4, Abstraction as Extraction.)

With respect to the current trend of extreme abstraction, the mathematical world finds itself divided. Some say that while abstraction is very useful, indeed necessary, too much of it may be debilitating. An extremely abstract theory soon becomes incomprehensible, uninteresting (in itself), and may not have the power of regeneration. Motivation in mathematics has, by and large, come from the "coarse" and not from the "fine." Researchers carrying out an ultra-abstract program frequently devote the bulk of their effort to straightening out difficulties in the terminology they have had to introduce, and the remainder of their effort to reestablishing in camouflaged form what has already been established more brilliantly, if more modestly. Programs of extreme abstraction are frequently accompanied by attitudes of complete hauteur on the part of their promulgators, and can be rejected on emotional grounds as being cold and aloof.

The same limitations are present in Saadia's conception of the Deity. By its very nature impossible to conceptualize, it appeals to the intellect and not to the emotions. Even the intellect has difficulty in dealing with it. There is a story about a professor of mathematics whose lectures were always extremely abstract. In the middle of such a lecture—

he was proving a certain proposition—he got stuck. So he went to a corner of the blackboard and very sheepishly drew a couple of geometric figures which gave him a concrete representation of what he was talking about. This clarified the matter, and he proceeded merrily on his way —*in abstracto*. Saadia's concept of the Deity suffers from the same defect. It requires bolstering from below. As part of religious practice, it must be supplemented emotionally by metaphors. Saadia himself seems to have been aware of this and so spends much time discussing the various anthropomorphisms associated with God. He then makes a statement that all proponents of ultra-abstract programs should remember! (p. 118)

> Were we, in our effort to give an account of God, to make use only of expressions which are literally true . . . there would be nothing left for us to affirm except the fact of His existence.

Saadia also speaks (p. 95) of

> . . . a proof of God's uniqueness . . .

The whole development in this section has a surprisingly mathematical flavor. One of the standard mathematical activities is the proving of what are called "existence and uniqueness theorems." An existence theorem is one which asserts that, subject to certain restrictions set down a priori, there will be a solution to such and such a problem. This is never taken for granted in mathematics, for many problems are posed which demonstrably do not possess solutions. The restrictions under which the problem was to have been solved may have been too severe, the conditions may have been inherently self-contradictory. Thus, the mathematician requires existence theorems which guarantee to him that the problem he is talking about can, indeed, be solved. This kind of theorem is frequently very difficult to establish.

If Saadia had been a theologian with the background of a modern mathematician, he would surely have begun his treatise with a proof of the existence of God. Even Maimonides (1135–1204) does this (to a certain extent). Thus

117

in *Mishneh Torah* Book I, Chapter I, he says,

> The basic principle is that there is a First Being who brought every existing thing into being, for if it be supposed that he did not exist, then nothing else could possibly exist . . .

To the mathematical ear, this sounds like proof by contradiction (a much-used device); the fact that the mathematician might be inclined to label Maimonides' syllogism a nonsequitur is irrelevant here.

But Saadia does not, as far as I can see, proceed in this way. The existence of God is given, i.e. is postulated. His uniqueness is then proved, and later, the properties which characterize him are inferred through a curious combination of abstraction and biblical syllogisms. Here the method of the Greeks is fused with Jewish tradition.

This brings us now to the question of "uniqueness theorems." Just as an existence theorem asserts that under such and such conditions a problem has a solution, a uniqueness theorem asserts that under such and such conditions a problem can have no more than one solution. The expression "one and only one solution" is one which is frequently heard in mathematics. Much effort is devoted to proving uniqueness theorems, for they are as important as they are hard to prove. In fact, one might say that there is a basic drive on the part of mathematicians to prove them.

Uniqueness implies a well-determined situation, wholly predictable. Nonuniqueness implies ambiguity, confusion. The mathematical sense of aesthetics loves the former and shuns the latter. Yet there are many situations in which uniqueness is, strictly, not possible. But the craving for uniqueness is so strong that mathematicians have devised ways of suppressing the ambiguities by the abstract process of identifying those entities which partake of common properties and creating out of them a superentity which then becomes unique. This is no mere verbalism, for the ambiguities are far better understood by this seemingly artificial device of suppressing them. The drive toward deistic uniqueness might be explained in much the same terms.

Browsing still further in Saadia, consider the following

quotation which occurs as part of his uniqueness proof,

> For if He were more than one, there would apply to him the category of number, and he would fall under the laws governing bodies.

And later,

> I say that the concept of quantity calls for two things neither of which can be applied to the Creator.

It appears, then, that God cannot be quantized. Yet God can be reasoned about, can be the subject matter of a syllogism. This may strike one as analogous to the fact—which is less than 150 years old—that mathematics can deal with concepts which do not directly involve numbers or spatial relations.

In sum, in Saadia's chapter on God, one finds the process of abstraction, the use of the syllogism including some interesting logical devices as "proof by contradiction." There are also certain logical concepts which have become standard since Russell and Whitehead such as the formation of the unit class consisting of a sole element. Furthermore, there is the realization of the central position that existence and uniqueness theorems must play within a theory.

### Further Readings. See Bibliography

Saadia Gaon, J. Friedman.

# 4
# INNER ISSUES

# Symbols

T HE SPECIAL SIGNS that constitute part of the written mathematical record form a numerous and colorful addition to the signs of the natural languages. The child in grade school soon learns the ten digits 0, 1, 2, 3, . . . , 9 and ways of concatenating, decimalizing, and exponentiating them. He also learns the operative signs $+$, $-$, $\times$ (or $\cdot$), $\div$ (or $/$), and $\sqrt{\phantom{x}}$, $\sqrt[3]{\phantom{x}}$. He learns signs for special mathematical numbers such as $\pi$ ($= 3.14159$ . . .), or special interpretations such as the degree sign in 30° or 45°. He learns grouping signs such as ( ), { }. He learns relational signs such as $=$, $>$, $<$. These signs have the effect of imputing to a page of arithmetic a secret, mystical quality—so much so that when crackpots invent their own private mathematics, they often take great care and pride in the invention of their own vocabulary of eccentric signs.

Further immersion in mathematics takes the student to algebra where common letters now reappear in an altogether surprising and miraculous context: as unknowns or variables.

Calculus brings in further symbols: $\frac{d}{dx}$, $\int$, $\iint$, $dx$, $\Sigma$, $\infty$, $\lim$, $\frac{\partial}{\partial x}$, etc. The mathematical font of special symbols cur-

122

rently in common use comprises several hundred symbols with new symbols created every year. Among the more visually interesting, one might cite

$$a_{ijk}, \cong, \triangle, \triangledown, \square, \uparrow, \#, \oint, \otimes, \oplus, \ulcorner, \urcorner, \cap, \cup, \exists, \sim, \forall, \pi, \infty, \wp$$

Computer science embraces several varieties of mathematical disciplines but has its own symbols: →, END, DECLARE, IF, WHILE, ÷ , +, *, etc.

The creation of some symbols can be ascribed to specific authors. Thus, the notation *n!* for the repeated product $1 \cdot 2 \cdot 3 \cdots n$ is due to Christian Kramp in 1808. The letter *e* to designate 2.71828 . . . is due to Euler (1727). But the inventors of the digits 0, 1, 2, . . . , 9 or of their primitive forms are lost in the mists of time. Some symbols are abbreviated forms of words: + is the medieval contraction of the word "et"; $\pi$ denotes the initial letter of "periphery"; ∫ is the mediaeval long *s*, the initial letter of "summa," sum. Others are pictorial or ideographic: △ for triangle, ○ for circle. Still others seem perfectly arbitrary: ÷, ∴.

There is undoubtedly a law of the survival of the fittest among symbols; Cajori's monumental study of mathematical symbols is, in part, a graveyard for dead symbols. The reader will find in it many obsolete symbols, some of them so complicated visually as to be almost comic. One damper to the free creation of new mathematical symbols is that if a manuscript is to be reproduced in some sort of printed form, a new typeface must be created. This has always been an expensive process. Today authors often restrict themselves to the symbols that are found on a standard typewriter. This can have disadvantages. It leads to the overuse of certain symbols (e.g., *). Certain forms of computerized printing offer a potentially infinite number of symbols, specifiable by the printer-programmer, but the practice over the past few years has been fairly conservative. It may be easy to create a new symbol, but the creator cannot guarantee widespread acceptance, without which the symbol becomes useless.

The principal functions of a symbol in mathematics are to designate with precision and clarity and to abbreviate. The reward is that, as Alfred North Whitehead put it, "by

123

relieving the brain of all unnecessary work, a good notation sets it free to concentrate on more advanced problems, and, in effect, increases the mental power of the race." In point of fact, without the process of abbreviation, mathematical discourse is hardly possible.

Consider, for an example, a list of abbreviations in formal logic. The following is adapted from *Mathematical Logic* by W. V. O. Quine.

D1. $\sim\phi$ for $\phi \downarrow \phi$

D2. $\phi \cdot \psi$ for $\sim\phi \downarrow \sim\psi$

D3. $\phi \vee \psi$ for $\sim(\phi \downarrow \psi)$

D4. $\phi \supset \psi$ for $(\sim\phi \vee \psi)$

D5. $\phi \cdot \psi \cdot \chi$ for $(\phi \cdot \psi) \cdot \chi$, etc

D6. $\phi \vee \psi \vee \chi$ for $(\phi \vee \psi) \vee \chi$, etc

D7. $\exists\, \alpha$ for $\sim(\alpha)\sim$

The complete list contains 48 such definitions. Any statement in formal logic such as

$$(x)(y)\ y \in x \equiv (\exists z)(y \in z \cdot z = x)$$

can, in principle, be expanded back into primitive atomic form. In practise this cannot be carried out, because the symbol strings quickly become so long that errors in reading and processing become unavoidable.

The demands of precision require that the meaning of each symbol or each symbol string be razor sharp and unambiguous. The symbol 5 is perceived in a way which distinguishes it from all other symbols, say, 0, 16, +, $\sqrt{\ }$, or *, and the meaning of the symbol is to be agreed upon, universally. Whether or not symbols can, in fact, achieve absolute sharpness, fidelity, and unambiguousness is a question which is by no means easy to answer. At the end of the nineteenth century various committees were set up for the purpose of symbol standardization. They achieved only limited success. It would seem that mathematical symbols share with natural languages an organic growth and change that cannot be controlled by the ukase of a committee.

124

What do we do with symbols? How do we act or react upon seeing them? We respond in one way to a road sign on a highway, in another way to an advertising sign offering a hamburger, in still other ways to good-luck symbols or religious icons. We act on mathematical symbols in two very different ways: we calculate with them, and we interpret them.

In a calculation, a string of mathematical symbols is processed according to a standardized set of agreements and converted into another string of symbols. This may be done by a machine; if it is done by hand, it should in principle be verifiable by a machine.

Interpreting a symbol is to associate it with some concept or mental image, to assimilate it to human consciousness. The rules for calculating should be as precise as the operation of a computing machine; the rules for interpretation cannot be any more precise than the communication of ideas among humans.

The process of representing mathematical ideas in symbolic form always entails an alteration in the ideas; a gain in precision and a loss in fidelity or applicability to its problem of origin.

Yet it seems at times that symbols return more than was put into them, that they are wiser than their creators. There are felicitous or powerful mathematical symbols that seem to have a kind of hermetic power, that carry within themselves seeds of innovation or creative development. For example, there is the Newtonian notation for the derivatives: $\dot{f}, \ddot{f}$, etc. There is also the Leibnitzian notation: $Df, D^2f$. The Leibnitzian notation displays an integer for the number of successive differentiations, and this suggests the possibility of fruitful interpretation of $D^\alpha f$ for negative and fractional $\alpha$. The whole of the operational calculus derives from this extension, which contributed powerfully to the development of abstract algebra in the mid-nineteenth century.

## Further Readings. See Bibliography

F. Cajori [1928–29]; H. Freudenthal [1968]; C. J. Jung

# Abstraction

I T IS COMMONLY HELD that mathematics began when the perception of three apples was freed from apples and became the integer three. This is an instance of the process of abstraction, but as this word is used in several different but related senses in mathematics, it is important to explain them.

### (a) Abstraction as Idealization

A carpenter, using a metal ruler, draws a pencil line across a board to use as a guide in cutting. The line he has drawn is a physical thing; it is a deposit of graphite on the surface of a physical board. It has width and thickness of varying amounts, and in following the edge of the ruler, the tip of the pencil reacts to the inequalities in the surface of the board and produces a line which has warp and corrugation.

Alongside this real, concrete instance of a straight line, there exists the mental idea of the mathematical abstraction of an ideal straight line. In the idealized version, all the accidentals and imperfections of the concrete instance have been miraculously eliminated. There exists an idealized, cosmeticized, verbal description of a straight line: "That which lies evenly with the points on itself" (Euclid: Definition 4) or that curve every part of which is the shortest distance between two of its points. The line is conceived of as potentially extending to infinity on both sides. Perhaps experience in pulling string between the fingers is at the bottom of these descriptions, or experience with the propagation of rays of light, or with folding pieces of paper. But no matter how, idealization has occurred as the end result of a process of perfection. In the end, we may lose or abandon all notions of what a straight line "really is," and merely replace it (if we are at the level of axiomatics) with statements of how the straight line acts or combines. Thus, for example: a straight line is a set of points,

containing at least two points. Two distinct points are contained in one and only one line, etc. (See The Stretched
String, this chapter.)

Along with the straight line, we arrive at many idealizations and perfections: planes, squares, polygons, circles,
cubes, polyhedra, spheres. Some of these, in particular
geometrical expositions, will be undefined, e.g. points,
lines, planes. Others may be defined in terms of simpler
concepts, e.g., the cube. It is well understood that any concrete instance of a cube, say a cubic crystal, will exhibit imperfections, and that any property of the cube inferred
mathematically can be verified only approximately by real-
world approximations. It is proved in plane geometry that
in any triangle the three angle bisectors intersect in a single
point; but real-world experience teaches us that no matter
how carefully a draftsman draws this figure the bisectors
will be only approximately concurrent; the figure will embody a degree of blur or fuzz which the eye will perceive
but which the mind will obligingly overlook.

The idealizations just mentioned have proceeded from
the world of spatial experience to the mathematical world.
Aristotle has described this process by saying (*Metaphysics*,
1060a, 28–1061b, 31) that the mathematician strips away
everything that is sensible, for example, weight, hardness,
heat, and leaves only quantity and spatial continuity. The
making of contemporary mathematical models exhibits
updated versions of Aristotle's process. As an example, we
quote from a popular book on differential equations (S. L.
Ross, Blaisdell, N.Y., 1964, pp. 525–6):

> We now make certain assumptions concerning the string,
> its vibrations, and its surroundings. To begin with, we as-
> sume that the string is perfectly flexible, is of constant lin-
> ear density $\rho$, and is of constant tension $T$ at all times. Con-
> cerning the vibrations, we assume that the motion is
> confined to the $xy$ plane and that each point on the string
> moves on a straight line perpendicular to the $x$ axis as the
> string vibrates. Further, we assume that the displacement $y$
> at each point of the string is small compared to the length $L$
> and that the angle between the string and the $x$ axis at each

*The abstraction
process*

127

point is also sufficiently small. Finally, we assume that no external forces (such as damping forces, for example) act upon the string.

Although these assumptions are not actually valid in any physical problem, nevertheless they are approximately satisfied in many cases. They are made in order to make the resulting mathematical problem more tractable. With these assumptions, then, the problem is to find the displacement $y$ as a function of $x$ and $t$.

*The mathematical idealization.*

Under the assumptions stated it can be shown that the displacement $y$ satisfies the *partial differential equation,*

$$\alpha^2 \frac{\partial^2 y}{\partial x^2} = \frac{\partial^2 y}{\partial t^2}, \qquad (14.20)$$

where $\alpha^2 = T/\rho$. This is the one-dimensional wave equation.

Since the ends of the string are fixed at $x = 0$ and $x = L$ for all time $t$, the displacement $y$ must satisfy the *boundary conditions*

$$\begin{aligned} y(0,t) = 0, \qquad & 0 \le t < \infty; \qquad (14.21) \\ y(L,t) = 0, \qquad & 0 \le t < \infty. \end{aligned}$$

At $t = 0$ the string is released from the initial position defined by $f(x)$, $0 \le x \le L$ with initial velocity given by $g(x)$, $0 \le x \le L$. Thus the displacement $y$ must also satisfy the *initial conditions.*

$$y(x, 0) = f(x), \qquad 0 \le x \le L; \qquad (14.22)$$

$$\frac{\partial y(x, 0)}{\partial t} = g(x), \qquad 0 \le x \le L.$$

This, then, is our problem. We must find a function $y$ of $x$ and $t$ which satisfies the partial differential equation (14.20), the boundary conditions (14.21), and the initial conditions (14.22).

The system of mathematical equations is an idealization of an exceedingly complex set of physical conditions.

The relationship between the real and the ideal is illustrated by the accompanying diagram (though, strictly speaking, we cannot draw the ideal objects on the right side of the diagram).

Intimately related to mathematical idealization is Plato's

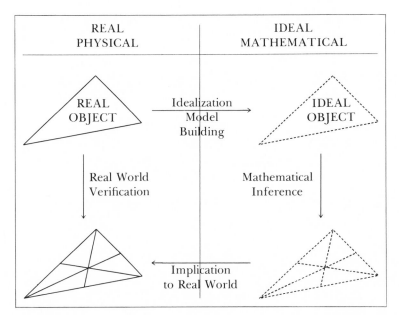

notion of the world of idealized objects. The so-called real world of experience, says Plato, is not real at all. We are like dwellers in a cave, who perceive the shadows of the external world and mistake the shadow for the true thing (*Republic* VII.514–517). The objects of mathematics are all abstract and the Platonic world is the dwelling place of the true circle, and the true square. It is the dwelling place of the true forms, the true perfections, and for this world the language of mathematics is said to provide true descriptions. "Without doubt," wrote Kepler in 1611, "the authentic type of these figures exists in the mind of God the Creator and shares His eternity."

### (b) Abstraction as Extraction

Four birds are eating bread crumbs in my back yard. There are four oranges on my kitchen table. The very use of the word "four" implies the existence of a process of abstraction wherein a common feature of the birds and of the oranges has been separated out. For each bird there is an orange. For each orange there is a bird, and in this way, there is a one-to-one correspondence between birds and

129

oranges. Here, in one setting, are objects. There, in another, are abstract numbers, existing, apparently, in isolation from birds or oranges.

"Arithmetic," says Plato (*Republic* VII.525), "has a very great and elevating effect, compelling the soul to reason about abstract number, and rebelling against the introduction of visible or tangible objects into the argument."

Today mathematics largely leaves aside the interesting psychohistorical problem of how abstractions come about and concentrates on a set-theoretic description of abstraction formation. The abstract notion of four is, according to Russell and Whitehead (*Principia Mathematica,* vol. I), the set of all sets that can be put into a one-to-one correspondence with the four birds on my lawn.

It will be illuminating to illustrate the process of mathematical abstraction by another example which is at once simple and modern. It is drawn from the theory of abstract graphs. Take a look at the two figures here and ask, what do these figures have in common? At the first glance it may seem that they have nothing in common. The first figure seems to be a series of boxes within boxes, while the second might represent a simplified version of a pearl necklace. There is no doubt, however, that the second is a much simpler figure than the first. Yet there is one very important respect in which these figures are completely identical. Think of the first figure as the plan of a maze or a labyrinth. Starting from the outside, we try to find our way into the innermost chamber. We walk down the corridors more or less at random trying to find a door which will take us one more layer further in, and hoping that we do not

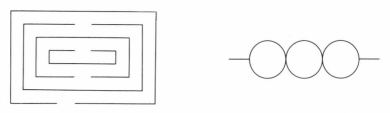

Figure 1. Figure 2.

130

come back to a spot where we've already been. Once we have made a complete investigation of the labyrinth, we can describe it completely. We may even do it verbally. Suppose that the maze is labeled as in the third figure. Then a description might go as follows:

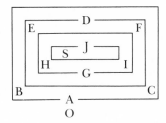

*Figure 3.*

From the outside O go to the door at A. The door at A leads to two halls, B and C, both of which lead to a door D. The door D leads to two halls, E and F, both of which lead to a door G. The door G leads to two halls, H and I, both of which lead to the door J. The door J leads to the innermost chamber S.

Now suppose that we mark Figure 2 as follows.

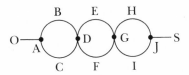

Then you can easily see that the verbal description which applies to the first maze applies also to the above figure as we traverse it from left to right. These two figures are therefore identical in this respect, and the second is a lot simpler to work with conceptually. Naturally, the second figure does not contain as much information about the labyrinth as the first, which may have been an exact floor plan. But if we are interested in the problem of *traversing* the maze, then the second diagram is adequate.

131

An object such as Figure 2 wherein we are interested only in all possible traversings is known as a *graph*.

Additional examples of graphs are drawn here.

Notice that certain aspects of the geometrical configuration of Figure 2 are inessential. Thus, it makes no difference whether we draw

As regards the traversing process, we can adopt the one which is simplest visually.

The process of abstraction can be continued further, and can depart from geometry completely. In the above graphs, I have marked by small dots the places where alternatives occur. Suppose that I label these decision points with letters, thus:

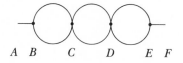

The lines which connect the decision points are now merely symbolic.

It would serve our purpose as well to write "AB, 1; BC, 2; CD, 2; DE, 2; EF, 1" by way of indicating that one hall connects A with B, two separate halls connect B with C, etc., and there are no other connections. One can even put all this information in a tableau as follows.

132

|   | A | B | C | D | E | F |
|---|---|---|---|---|---|---|
| A | 0 | 1 | 0 | 0 | 0 | 0 |
| B | 1 | 0 | 2 | 0 | 0 | 0 |
| C | 0 | 2 | 0 | 2 | 0 | 0 |
| D | 0 | 0 | 2 | 0 | 2 | 0 |
| E | 0 | 0 | 0 | 2 | 0 | 1 |
| F | 0 | 0 | 0 | 0 | 1 | 0 |

This is a so-called "incidence matrix" of Poincaré. The labyrinth of Figure 1 has been completely described in an arithmetic fashion.

If we are interested in setting up a theory of graphs which describes only the properties of traversings and nothing else, we need nothing more than the information in this matrix.

With this as our raw material, we proceed to make deductions. Although the original inspiration may have been geometric, we have stripped away all the geometric accoutrements, and the theory of abstract graphs is purely combinatorial. The abstract graph emerges as a set of things (nodes) together with a set of relationships between these nodes (paths) which satisfy certain axiomatic requirements. More than this is not necessary.

### Further Readings. See Bibliography

J. Weinberg

# Generalization

"**M**AN MUSS IMMER generalisieren," (one should always generalize) wrote Jacobi in the 1840s. The words "generalization" and "abstraction" are often used interchangeably, but there are several particular meanings of the former which should be elucidated. Let us suppose that at some fictitious time in antiquity mathematician Alpha announced, "If ABC is an equilateral triangle, then the angle at A equals the angle at B." Suppose that some time later, mathematician Beta said to himself that while what Alpha said is perfectly true, it is not necessary that ABC be equilateral for the conclusions to hold: it will suffice if side BC = side AC. He then could have announced, "In an isosceles triangle, the base angles are equal." The second statement is a generalization of the first. The hypotheses of the first imply those of the second, but not vice versa, while the conclusion is the same. We have the strong impression that we are getting more for our money in the second version. The second version is an improvement, a strengthening, or a generalization of the first.

More examples come to mind easily:

Statement: Every number that ends in 0 is divisible by 2

Generalization: Every number that ends in 0, 2, 4, 6 or 8 is divisible by 2.

Generalization is not always accompanied by identical conclusions:

Statement: In a right triangle $c^2 = a^2 + b^2$.

Generalization: In any triangle $c^2 = a^2 + b^2 - 2ab \cos C$.

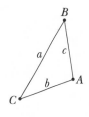

Here we obtain an altered conclusion corresponding to a generalized assumption, but the former may be recovered from the latter by the insertion of $C = 90°$ for which $\cos C = 0$.

134

Generalization may also come about by a radical change in the environment.

Statement: If a three-dimensional box has edges $x_1, x_2, x_3$, then its diagonal $d$ is given by $d = \sqrt{x_1^2 + x_2^2 + x_3^2}$.

Generalization: If an $n$-dimensional box has edges $x_1$, $x_2, \ldots, x_n$, then its diagonal $d$ is given by $d = \sqrt{x_1^2 + x_2^2 + \ldots + x_n^2}$.

Encouraged by the "coincidence" that in two dimensions one has $d = \sqrt{x_1^2 + x_2^2}$, and in three dimensions $d = \sqrt{x_1^2 + x_2^2 + x_3^2}$, we generalize by looking for a mathematical environment in which we can assert $d = \sqrt{x_1^2 + x_2^2 + \ldots + x_n^2}$. This is found in the theory of $n$-dimensional Euclidean spaces.

It should be noticed carefully that while the general includes some aspects of the particular, it cannot include all aspects, for the very particularity grants additional privileges. Thus the theory of equilateral triangles is not contained in the theory of isosceles triangles. Numbers that end in 0 are divisible both by 2 and 5 whereas those that end in 2, 4, 6 or 8 are not. The general theory of continuous functions contains only a small amount of interesting information about the particular continuous function $y = e^x$. Thus, in moving from the particular to the general, there may occur a dramatic refocusing of interest and a reorientation as to what is significant.

One benefit of generalization is a consolidation of information. Several closely related facts are wrapped up neatly and economically in a single package.

Statements:     If a number ends in 0 it is divisible by 2.
                     If a number ends in 2 it is divisible by 2.

Consolidation:  If a number ends in an even digit it is divisible by 2.

Statements:     The Legendre polynomials satisfy a three-term recurrence.
                     The Tchebyscheff polynomials satisfy a three-term recurrence.

Consolidation:   Any set of orthogonal polynomials satisfy
a three-term recurrence.

Insofar as generalization expands the stage on which the
action occurs, it results in an expansion of material. Consolidation may not be possible if the original narrow stage
has vital peculiarities.

# Formalization

FORMALIZATION IS THE process by which
mathematics is adapted for mechanical processing. A computer program is an example of a
formalized text. To program a computer to balance your checkbook, you must know the computer's vocabulary. You must know the rules of grammar in the computer's systems program. Mathematical texts of the usual
sort are never completely formalized. They are written in
English or other natural languages, because they are intended to be read by human beings. Nevertheless, it is believed that every mathematical text *can* be formalized. Indeed, it is believed that every mathematical text can be
formalized within a single formal language. This language
is the language of formal set theory.

Every text on mathematical logic explains the rules of
syntax for this language. The picture on page 138 shows
the axioms of Zermelo-Fraenkel-Skolem, the most often
used axiom system for set theory. The axioms are shown
written in the formal language; below each formal statement is a translation into English.

Four of the symbols are specifically related to set theory.
These are the symbol for "union," $\cup$; the symbol for "is a
subset of," $\subseteq$; the symbol for "is a member," $\in$; and the
symbol for "the empty set," $\varnothing$. The other symbols are symbols of logic; they would be used in any formalized mathematical theory. For example, one could write down the
axioms of plane geometry in a formal language. Instead of

*Abraham A. Fraenkel*
*1891–1965*

136

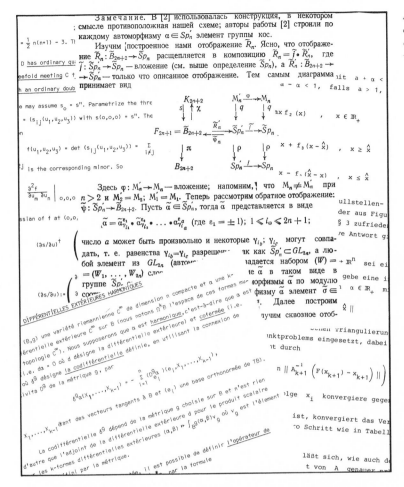

*Mathematical discourse is carried out in a mixture of formalized and natural languages*

$\cup$, $\subseteq$, $\in$, $\varnothing$ one would introduce symbols for "point," "line," "intersects," and "parallel," and write formulas which would be translations into the formal language of the ordinary English statements of the axioms.

The motives for using formal languages have undergone a significant evolution. Formal languages were first introduced by Peano and Frege in the late nineteenth century with the intention of making mathematical proof more rigorous—that is, of increasing the certainty of the conclusion of a mathematical argument. However, this purpose was not served so long as the argument was addressed to a human reader. The *Principia Mathematica* of

*Giuseppe Peano*
*1858–1932*

137

| | | |
|---|---|---|
| $\forall$ FOR ALL | $\leftrightarrow$ IF AND ONLY IF | $\in$ IS A MEMBER (ELEMENT) OF |
| $\exists$ THERE EXISTS | V OR | = EQUALS |
| $\exists!$ THERE EXISTS UNIQUELY | & AND | $\neq$ DOES NOT EQUAL |
| U UNION | $\sim$ NOT | $\phi$ THE EMPTY SET |
| $\to$ IMPLIES | $\subseteq$ IS A SUBSET OF | |

### 1. AXIOM OF EXTENSIONALITY

$$\forall x, y \ (\forall z \ (z \in x \to z \in y) \to x = y).$$

Two sets are equal if and only if they have the same members.

### 2. AXIOM OF THE NULL SET

$$\exists x \ \forall y \ (\sim y \in x).$$

There exists a set with no members (the empty set).

### 3. AXIOM OF UNORDERED PAIRS

$$\forall x, y \ \exists z \ \forall w \ (w \in z \leftrightarrow w = x \lor w = y).$$

If x and y are sets, then the (unordered) pair $\{x,y\}$ is a set.

### 4. AXIOM OF THE SUM SET OR UNION

$$\forall x \ \exists y \ \forall z \ (z \in y \leftrightarrow \exists t \ (z \in t \ \& \ t \in x) \ ).$$

If x is a set of sets, the union of all its members is a set. (For example, if $x = \{ {a,b,c \atop a,c,d,e} \}$, then the union of the (two) elements of x is the set {a,b,c,d,e}.)

### 5. AXIOM OF INFINITY

$$\exists x \ (\phi \in x \ \& \ \forall y \ (y \in x \to y \ U \ \{y\} \in x).$$

There exists a set x that contains the empty set, and that is such that if y belongs to x, then the union of y and $\{y\}$ is also in x. The distinction between the element y and the singleton set $\{y\}$ is basic. This axiom guarantees the existence of infinite sets.

### $6_n$. AXIOM OF REPLACEMENT

$$\forall t_1,...,t_n \ (\forall x \ \exists! \ y \ A_n(x,y;t_1,...,t_n) \to \forall u \ \exists v \ B(u,v) \ ) \quad \text{where} \quad B(u,v) \ \Xi \ \forall r \ (r \in v \leftrightarrow \exists s \ (s \in u \ \& \ A_n(s,r;t_1,...,t_n) \ ) \ ).$$

This axiom is difficult to restate in English. It is called $6_n$ rather than 6 because it is really a whole family of axioms. We suppose that all the formulas expressible in our system have been enumerated; the nth is called $A_n$. Then the axiom of replacement says that if for fixed $t_1,...,t_n$, $A_n(x,y;t_n)$ defines y uniquely as a function of x, say $y = \varphi(x)$, then for each u the range of $\varphi$ on u is a set. This means, roughly, that any ("reasonable") property that can be stated in the formal language of the theory can be used to define a set (the set of things having the stated property).

### 7. AXIOM OF THE POWER SET

$$\forall x \ \exists y \ \forall z \ (z \in y \leftrightarrow z \subseteq x).$$

This axiom says that there exists for each x the set y of all subsets of x. Although y is thus defined by a property, it is not covered by the replacement axiom because it is not given as the range of any function. Indeed, the cardinality of y will be greater than that of x, so that this axiom allows us to construct higher cardinals.

### 8. AXIOM OF CHOICE

If $a \to A_a \neq \phi$ is a function defined for all $a \in x$, then there exists another function f(a) for $a \in x$, and $f(a) \in A_a$.

This is the well-known axiom of choice, which allows us to do an infinite amount of "choosing" even though we have no property that would define the choice function and thus enable us to use $6_n$ instead.

### 9. AXIOM OF REGULARITY

$$\forall x \ \exists y \ (x = \phi \ \lor \ (y \in x \ \& \ \forall z \ (z \in x \to \sim z \in y)) \ ).$$

This axiom explicitly prohibits x ∈ x, for example.

Russell and Whitehead was the great attempt actually to carry out the formalization of mathematics. It has been accepted as the outstanding example of an unreadable masterpiece.

Whereas human readers have an insuperable aversion to formal languages, computers thrive on them. With the appearance of electronic computers shortly after World War II, formal languages became a growth industry. Under the name of "software," texts written in a formal language have become one of the characteristic artifacts of our culture.

A formalized text is a string of symbols. When it is manipulated, by a mathematician or a machine, it is transformed into another string of symbols. Such symbol manipulations can themselves be the subject of a mathematical theory. When the manipulation is thought of as being done by a machine, the theory is called "automata theory" by computer scientists or "recursion theory" by logicians. When the manipulation is thought of as being done by a mathematician, the theory is called "proof theory."

If we imagine that mathematicians work in a formal language, we can construct a mathematical theory about mathematics. For the purpose of logical analysis of mathematics, it is necessary to conceive of mathematics as expressed in a formal language. On this supposition, logicians have been able to create imposing theories about the properties of mathematical systems. However, actual mathematical work, including the work done by mathematical logicians, continues to be done in natural languages augmented by special mathematical notations. The question of how the findings of mathematical logic relate to the actual practice of living mathematicians is a difficult one, because it is a philosophical question, not a mathematical one.

An ordinary page of mathematical exposition may occasionally consist entirely of mathematical symbols. To a casual eye, it may seem that there is little difference between such a page of ordinary mathematical text and a text in a formal language. But there is a crucial difference which becomes unmistakable when one reads the text. Any steps which are purely mechanical may be omitted from an ordinary mathematical text. It is sufficient to give the starting point and the final result. The steps that are included in such a text are those that are *not* purely mechanical—that involve some constructive idea, the introduction of some new element into the calculation. To read a mathematical text with understanding, one must supply the new idea which justifies the steps that are written down.

One could almost say that the rules for writing mathematics for human consumption are the opposite of the rules for writing mathematics for machine consumption (i.e., formalized texts). For the machine, nothing must be left unstated, nothing must be left "to the imagination." For the human reader, nothing should be included which is so obvious, so "mechanical" that it would distract from the ideas being communicated.

### Further Readings. See Bibliography

P. Cohen and R. Hersh; K. Hrbacek and T. Jech; G. Takeuti and W. M. Zaring

# Mathematical Objects and Structures; Existence

INFORMAL MATHEMATICAL discourse, as part of natural discourse, is composed of nouns, verbs, adjectives, etc. The nouns denote mathematical objects: for example, the number 3, the number $e = 2.178 \ldots$, the set of primes, the matrix $\left(\begin{smallmatrix}100\\012\end{smallmatrix}\right)$, the Riemann zeta function $\zeta(z)$. Mathematical structures, for example the real number system or the cyclic group of order 12, are somewhat more complex nouns and consist of mathematical objects linked together by certain relationships or laws of combination. The symbols of combination or of relation, such as "equal," "is greater than," addition, differentiation, play a role similar to that of verbs. Mathematical adjectives are restrictors or qualifiers, for example, group vs. cyclic group.

Without pushing this gramatical analogy too far, let us come back to the objects and structures. In contemporary usage a mathematical structure consists of a set of objects $S$, which can be thought of as the carrier of the structure, a set of operations or relations, which are defined on the carrier, and a set of distinguished elements in the carrier, say 0, 1, etc. These basic ingredients are said to constitute the *"signature"* of the structure, and they are often displayed in $n$-tuple form. For example: $<R, +, \cdot, 0, 1>$ means the set of real numbers combined by addition and multiplication, with two distinguished elements, 0 and 1. When a signature has been subjected to a set of axioms that lay down requirements on its elements, then a mathematical structure emerges. Thus, a semigroup is $<S, \circ>$ where $\circ$ is a binary associative operation. Expanding this a bit, a semigroup is a set of objects $S$ any two elements of which can be combined to produce a third element: if $a, b \in S$ then $a \circ b$ is defined and is in $S$; for any three elements $a, b, c \in S$, $a \circ (b \circ c) = (a \circ b) \circ c$. Similarly, a monoid is a signature $<S, \circ, 1>$ where $\circ$ is a binary associative operation and where 1 is a two-sided identity for $\circ$.*

The distinction between a mathematical object and a mathematical structure is vague. It seems to be time- and

---

* All language when carried to extreme becomes mannerism and jargon. Hans Freudenthal parodies this mode of talking when he describes a "mathematical theory of meetings" which begins by setting up a "model" of a meeting.

"A meeting is an ordered set $<M, P, c, s, C_1, C_2, b, i_1, i_2, S i_3>$ consisting of

a bounded part $M$ of Euclidean space;

a finite set $P$, that of the participants;

two elements $c$ and $s$ of $P$ called chairman and secretary;

a finite set $C_1$, called the chairs;

a finite set $C_2$, called the cups of coffee;

an element $b$, called bell;

an injection $i_1$ of $P$ into $C_1$;

a mapping $i_2$ of $C_2$ into $P$;

an ordered set $S$, the speeches;

a mapping $i_3$ of $S$ into $P$ with the property that $c$ belongs

to the image of $i_3$.

If $i_3$ is a surjection, it is usual to say that everybody has had the floor."

utility-dependent. If a mathematical structure is used frequently over a long period of time and a body of experience and intuition is built upon it, then it might be regarded as an object. Thus the real numbers $R$, a structure, may be thought of as an object when one takes the direct product $R \times R$ to form pairs of real numbers.

Standardized mathematical objects, structures, problems, . . . are built into mathematical tables, programs, articles, books. Thus, the more expensive hand-held computers contain real-world realizations of decimalized subsets of the rational numbers together with numerous special functions. Books contain lists of special functions, their principal properties, and, of more contemporary research interest, lists of special function-spaces and their properties.

It is important to realize that as we move back in time what is now regarded as a simple mathematical object, say a circle, or an equilateral triangle or a regular polyhedron, might have carried the psychological impact of a whole structure and might well have influenced scientific methodology (e.g., astronomy). An individual number, say three, was regarded as a whole structure, with mystic implications derived therefrom. A mathematical object considered in isolation has no meaning. It derives meaning from a structure and it plays out its role within a structure.

The term "mathematical object" implies that the object in question has some kind of existence. One might think that the notion of existence is clearcut, but in fact there are severe logical and psychological difficulties associated with it. A conception of small integers such as 1, 2, 3, etc. may come about as an act of abstraction. But what shall we say to the number 68,405,399,900,001,453,072? Since it is extremely unlikely that anyone has seen or dealt with an assemblage of that number of items, and perceived its unique numerical flavor, it is clear that the existence of this large number as a mathematical object is based on other considerations. We have, in fact, written it down. We can, if we like, manipulate it; for example, we can easily double it. We can answer certain questions about it: is it even or odd? does it exceed 237,098? In this way, despite the fact that the multiplicity of the number is not experienced directly, one asserts confidently the existence of the number in a different sense.

Working within finitary mathematics and with just a few symbols, one may lay down definitions which lead to integers of such enormous size that the mind is baffled in visualizing even the decimal representation.

One of the most amusing of such constructions comes from the Polish mathematician Hugo Steinhaus and the Canadian mathematician Leo Moser. Here is Steinhaus' economical definition and notation.

Let $\triangle{a} = a^a$, e.g. $\triangle{2} = 2^2 = 4$; let $\boxed{b} = b$ with $b$ $\triangle$'s around it, e.g. $\boxed{2} = \triangle{\triangle{2}} = \triangle{4} = 4^4 = 256$; let $\bigcirc{c} = c$ with $c$ $\square$'s around it. Now

a *mega* is defined to be $= \bigcirc{2} = \boxed{2} = \boxed{256} = 256$ with 256 triangles around it

$= 256^{256}$ with 255 triangles around it

$= (256^{256})^{256^{256}}$ with 254 triangles around it, etc.

Not content to let large enough alone, Moser continued the pattern with hexagons, heptagons, etc.; an $n$-gon containing the number $d$ is defined as the number $d$ with $d$ $(n-1)$-gons around it. The *moser* is defined as 2 inside a mega-gon.

The existence of the moser poses no existential problems for conventional mathematics; yet what else other than the fact that it is a huge power of two, can one really say about it?

The existence of more complicated mathematical objects may be likewise experienced in terms of how one interacts with the objects. The set $N$ of positive integers 1, 2, 3, 4, . . . is the basic infinite set in mathematics. It contains as just one member the gigantic number just written. $N$ cannot be experienced completely through the sensation of multiplicity. Yet a mathematician works with it constantly, manipulates it, answers questions about it; for example, is every number in $N$ either even or odd, or is there a number $x$ in $N$ such that $x = x + 1$? Some people have psychological difficulties with this set and have even de-

143

clared that its existence is nonsense. If so, they would differ from the mathematicians of the world and in this rejection could cut themselves off from most of mathematics.

We go beyond. Assuming that we can accept the idea of a simple infinity, we can easily then accept the idea of some specific infinite sequence of the digits 0, 1, . . . , 9. For example, the sequence 1,2,3,1,2,3,1,2,3,1,2,3, . . . where the digits 1, 2, 3 repeat cyclically. Or, a more complicated instance: the decimal expansion of the number $\pi$: 3.14159 . . . . Now let us imagine the set of all possible sequences of digits 0, . . . , 9. This, essentially, is the real number system, and is the fundamental arena for the subject of mathematical analysis, i.e., calculus and its generalizations. This system has so many elements that as Cantor proved, it is impossible to arrange its elements in a list, even an infinite list. This system is not part of the experience of the average person, but the professional is able to accumulate a rich intuition of it. The real numbers form the underpinning for analysis, and the average student of calculus (and the average teacher) swiftly accepts the existence of this set and confines his attention to the formal manipulative aspects of the calculus.

We go beyond. We construct what is called the Fréchet ultrafilter, a concept which is of great use in many parts of point set topology and which is fundamental to the concept of the system of hyperreal numbers, which is itself related to nonstandard analysis. (See Chapter 5). To this end, consider all the infinite subsets of the integers. For example, (1, 2, 3, . . .) or (2, 4, 6, 8, 10, . . .) or (1, 1000, 1003, 1004, 20678, . . .).

In the last subset one isn't even thinking of a particular rule to generate the members of the subset. We now wish to restrict the subsets under consideration.

Let $X$ be a set and let $F$ be a family of non-empty subsets of $X$ such that (1) If two sets $A$ and $B$ belong to the family $F$, then so does their intersection.

(2) If $A$ belongs to $F$ and $A$ is a subset of $B$ which is a subset of $X$, then $B$ belongs to $F$. Such a family $F$ is called a *filter in X*. To take a specific case, let $X$ designate the set of all positive integers 1, 2, 3, . . . . . Let $F$ designate all the "co-

finite" subsets of $X$—those subsets which omit at most a finite number of elements of $X$. For example, the set $(2, 3, \ldots)$ is in $F$ for it omits 1. The set $(10, 11, \ldots)$ is in $F$, for it omits $1, 2, \ldots, 9$. On the other hand, the set of odd numbers $(1, 3, 5, \ldots)$ is not in $F$ because it omits the infinite number of even numbers. It is not difficult to show that the set of all cofinite subsets of positive integers is a filter. It is often called the Fréchet filter.

Let $X$ be a given set and let $F$ be a filter in $X$. It may be shown (using the principle that every set can be well-ordered) that there exists a maximal filter $V$ that contains $F$. That is to say, $V$ is a filter in $X$ and if a subset of $X$ that is not already there is added to $V$, $V$ ceases to be a filter. Such an object is called an *ultrafilter*.

Thus, as a particularization, there is a Fréchet ultrafilter. How now does one begin to visualize or comprehend or represent or come to grips, with what the Fréchet ultrafilter contains? What is in the ultrafilter and what is not? Yet, this is the starting point for important investigations in contemporary set theory and logic.

At this high altitude of construction and abstraction, one has left behind a good fraction of all mathematicians; the world in which the Fréchet ultrafilter exists is the property of a very limited fraction of mankind.

"How does one know how to set about satisfying oneself on the existence of unicorns?" asked L. Wittgenstein in an article entitled "On Certainty." The unicorn is an animal with the body of a horse and has a long, sharp horn out of the middle of its forehead. There are descriptions of this animal in Ctesias and Aristotle. It is depicted carefully in many works of art including the famous Unicorn Tapestry of the Cloisters in New York and on the coat of arms of the United Kingdom. Many children would recognize a unicorn if they were shown a picture of one; or, given a picture of a random animal, could easily decide whether the picture was or was not that of a unicorn. Many questions can be answered about this animal, for example: how many feet it has, or what, probably, it eats. In poetry, it is a symbol of purity. It has been asserted that its horn, when powdered, is an antidote against poison. So all this and much

*Ludwig Wittgenstein*
*1889–1951*

145

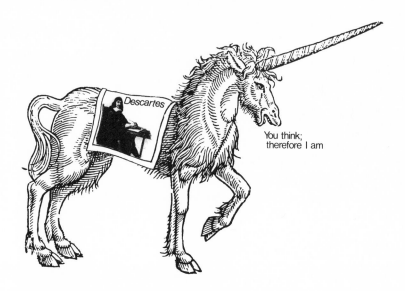

more may be asserted of the unicorn, as well as the fact that there is no such animal.

The unicorn, as a literary legend, exists. As a zoological blueprint it exists. But as a live creature which might potentially be caught and exhibited in a zoo, it does not exist. It is conceivable that it once existed or that it might in the future exist, but it does not now exist.

Like that of the unicorn, there is no single notion of existence of mathematical objects. Existence is intimately related to setting, to demands, to function. $\sqrt{2}$ does not exist as an integer or a fraction, anymore than a tropical fish exists in Arctic waters. But within the milieu of the real numbers, $\sqrt{2}$ is alive and well. The moser does not exist as a completed decimal number; yet it exists as a program or as a set of rules for its construction. The Fréchet ultrafilter exists within a mathematics that accepts the Axiom of Choice. (See Axiom 8 on page 138.)

### Further Readings. See Bibliography

H. Freudenthal [1978]

# Proof

THE ASSERTION has been made that mathematics is uniquely characterized by something known as "proof." The first proof in the history of mathematics is said to have been given by Thales of Miletus (600 B.C.). He proved that the diameter divides a circle into two equal parts. Now this is a statement which is so simple that it appears self evident. The genius of the act was to understand that a proof is possible and necessary. What makes mathematical proof more than mere pedantry is its application in situations where the statements made are far less transparent. In the opinion of some, the name of the mathematics game is proof; no proof, no mathematics. In the opinion of others, this is nonsense; there are many games in mathematics.

*Thales*
*c. 624–548* B.C.

To discuss what proof is, how it operates, and what it is for, we need to have a concrete example of some complexity before us; and there is nothing better than to take a look at what undoubtedly is the most famous theorem in the history of mathematics as it appears in the most famous book in the history of mathematics. We allude to the Pythagorean Theorem, as it occurs in Proposition 47, Book I of Euclid's Elements (300 B.C.). We quote it in the English version given by Sir Thomas Heath. The in-text numbers at the right are references to previously established results or to "common notions."

*Pythagoras*
*c. 580–500* B.C.

***Proposition 47***   In right-angled triangles the square on the side subtending the right angle is equal to the squares on the sides containing the right angle.

Let ABC be a right-angled triangle having the angle BAC right;

I say that the square on BC is equal to the squares on BA, AC.

For let there be described on BC the square BDEC, and on BA, AC the squares GB, HC; [1.46] through A let AL

147

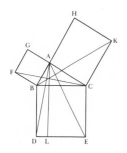

be drawn parallel to either BD or CE, and let AD, FC be joined.

Then, since each of the angles BAC, BAG is right, it follows that with a straight line BA, and at the point A on it, the two straight lines AC, AG not lying on the same side make the adjacent angles equal to two right angles; therefore CA is in a straight line with AG. [1.14]

For the same reason BA is also in a straight line with AH.

And, since the angle DBC is equal to the angle FBA; for each is right: let the angle ABC be added to each; therefore the whole angle DBA is equal to the whole angle FBC. [C.N. 2]

And, since DB is equal to BC, and FB to BA, the two sides AB, BD are equal to the two sides FB, BC respectively, and the angle ABD is equal to the angle FBC: therefore the base AD is equal to the base FC, and the triangle ABD is equal to the triangle FBC. [1.4]

Now the parallelogram BL is double of the triangle ABD, for they have the same base BD and are in the same parallels BD, AL. [1.41]

And the square GB is double of the triangle FBC, for they again have the same base FB and are in the same parallels FB, GC. [1.41]

[But the doubles of equals are equal to one another.]

Therefore the parallelogram BL is also equal to the square GB.

Similarly, if AE, BK be joined, the parallelogram CL can also be proved equal to the square HC; therefore the whole square BDEC is equal to the two squares GB, HC. [C.N. 2]

And the square BDEC is described on BC, and the squares GB, HC on BA, AC.

Therefore the square on the side BC is equal to the squares on the sides BA, AC.

Therefore etc. Q.E.D.

Now, assuming that we have read Euclid up to Proposition 47, and assuming we are able intellectually to get through this material, what are we to make of it all? Perhaps the most beautiful recorded reaction is that ascribed to Thomas Hobbes (1588–1679) by John Aubrey in his "Brief Lives":

He was 40 yeares old before he looked on Geometry; which happened accidentally. Being in a Gentleman's Library, Euclid's Elements lay open, and 'twas the 47 *El. libri* I. He read the Proposition. *By G_,* sayd he (he would now and then sweare an emphaticall Oath by way of emphasis) *this is impossible!* So he reads the Demonstration of it, which referred him back to such a Proposition; which proposition he read. That referred him back to another, which he also read. *Et sic deinceps* [and so on] that at last he was demonstratively convinced of that trueth. This made him in love with Geometry.

What appears initially as unintuitive, dubious, and somewhat mysterious ends up, after a certain kind of mental process, as gloriously true. Euclid, one likes to think, would have been proud of Hobbes and would use him as Exhibit A for the vindication of his long labors in compiling the Elements. Here is the proof process, discovered and promulgated by Greek mathematics, in the service of validation and certification. Now that the statement has been proved, we are to understand that the statement is true beyond the shadow of a doubt.

The backward referral to previous propositions, alluded to by Hobbes, is characteristic of the method of proof, and as we know, this can't go on forever. It stops with the so-called axioms and definitions. Whereas the latter are mere linguistic conventions, the former represent rock bottom self-evident facts upon which the whole structure is to rest, held together by the bolts of logic.

Also characteristic of the method is the considerable degree of abstraction that has occurred in the refinement of such concepts as triangle, right angle, square, etc. The figure itself appears here as a very necessary adjunct to the verbalization. In Euclid's presentation we cannot wholly follow the argumentation without the figure, and unless we are strong enough to imagine the figure in our mind's eye, we would be reduced to supplying our own figure if the author had not done it for us. Notice also that the language of the proof has a formal and severely restricted quality about it. This is not the language of history, nor of drama, nor of day to day life; this is language that has been sharpened and refined so as to serve the precise needs of a precise but limited intellectual goal.

One response to this material was recorded by the poet Edna Millay, in her line, "Euclid alone has looked on Beauty bare." A shudder might even run down our spines if we believe that with a few magic lines of proof we have compelled all the right triangles in the universe to behave in a regular Pythagorean fashion.

Abstraction, formalization, axiomatization, deduction—here are the ingredients of proof. And the proofs in modern mathematics, though they may deal with different raw material or lie at a deeper level, have essentially the same feel to the student or the researcher as the one just quoted.

Further reading of Euclid's masterpiece brings up additional issues. Notice that in the figure certain lines, e.g. BK, AL seem to be extraneous to a minimal figure drawn as an expression of the theorem itself. Such a figure is illustrated here: a right angled triangle with squares drawn upon each of its three sides. The extraneous lines which in high school are often called "construction lines," complicate the figure, but form an essential part of the deductive process. They reorganize the figure into subfigures and the reasoning takes place precisely at this sublevel.

Now, how does one know where to draw these lines so as to reason with them? It would seem that these lines are accidental or fortuitous. In a sense this is true and constitutes the genius or the trick of the thing. Finding the lines is part of finding a proof, and this may be no easy matter. With experience come insight and skill at finding proper construction lines. One person may be more skillful at it than another. There is no guaranteed way to arrive at a proof. This sad truth is equally rankling to schoolchildren and to skillful professionals. Mathematics as a whole may be regarded as a systematization of just those questions which have been pursued successfully.

Mathematics, then, is the subject in which there are proofs. Traditionally, proof was first met in Euclid; and millions of hours have been spent in class after class, in country after country, in generation after generation, proving and reproving the theorems in Euclid. After the introduction of the "new math" in the mid-nineteen fifties, proof spread to other high school mathematics such as algebra, and subjects such as set theory were deliberately introduced so as to be a vehicle for the axiomatic method and proof. In college, a typical lecture in advanced mathematics, especially a lecture given by an instructor with "pure" interests, consists entirely of definition, theorem, proof, definition, theorem, proof, in solemn and unrelieved concatenation. Why is this? If, as claimed, proof is validation and certification, then one might think that once a proof has been accepted by a competent group of scholars, the rest of the scholarly world would be glad to take their word for it and to go on. Why do mathematicians and their students find it worthwhile to prove again and yet again the Pythagorean theorem or the theorems of Lebesgue or Wiener or Kolmogoroff?

Proof serves many purposes simultaneously. In being exposed to the scrutiny and judgment of a new audience, the proof is subject to a constant process of criticism and revalidation. Errors, ambiguities, and misunderstandings are cleared up by constant exposure. Proof is respectability. Proof is the seal of authority.

Proof, in its best instances, increases understanding by revealing the heart of the matter. Proof suggests new mathematics. The novice who studies proofs gets closer to the creation of new mathematics. Proof is mathematical power, the electric voltage of the subject which vitalizes the static assertions of the theorems.

Finally, proof is ritual, and a celebration of the power of pure reason. Such an exercise in reassurance may be very necessary in view of all the messes that clear thinking clearly gets us into.

## Further Readings. See Bibliography

R. Blanché; J. Dieudonné [1971]; G. H. Hardy; T. Heath; R. Wilder, [1944].

# Infinity, or the Miraculous Jar of Mathematics

ATHEMATICS, IN ONE VIEW, is the science of infinity. Whereas the sentences "2 + 3 = 5", "$\frac{1}{2} + \frac{1}{3} = \frac{5}{6}$", "seventy-one is a prime number" are instances of finite mathematics, significant mathematics is thought to emerge when the universe of its discourse is enlarged so as to embrace the infinite. The contemporary stockpile of mathematical objects is full of infinities. The infinite is hard to avoid. Consider a few typical sentences: "there are an infinite number of points on the real line," "$\lim_{n \to \infty} \dfrac{n}{n + 1} = 1$," "$\sum_{n=1}^{\infty} 1/n^2 = \pi^2/6$," "$\displaystyle\int_0^{\infty} \dfrac{\sin x}{x}\, dx = \pi/2$," "$\aleph_0 + 1 = \aleph_0$," "the number of primes is infinite," "is the number of twin primes infinite?," "the tape on a Turing machine is considered to be of infinite extent," "let $N$ be an infinite integer extracted from the set of hyperreals." We have infinities and infinities upon infinities; infinities galore, infinities beyond the dreams of conceptual avarice.

The simplest of all the infinite objects is the system of positive integers $1, 2, 3, 4, 5, \ldots$. The "dot dot dot" here indicates that the list goes on forever. It never stops. This system is commonplace and so useful that it is convenient to give it a name. Some authors call it $N$ (for the numbers) or $Z$ (for *Zahlen*, if they prefer a continental flavor). The set $N$ has the property that if a number is in it, so is its successor. Thus, there can be no greatest number, for we can always add one and get a still greater one. Another property that $N$ has is that you can never exhaust $N$ by excluding its members one at a time. If you delete 6 and 83 from $N$, what remains is an infinite set. The set $N$ is an inexhaust-

152

ible jar, a miraculous jar recalling the miracle of the loaves and the fishes in Matthew 15:34.

This miraculous jar with all its magical properties, properties which appear to go against all the experiences of our finite lives, is an absolutely basic object in mathematics, and thought to be well within the grasp of children in the elementary schools. Mathematics asks us to believe in this miraculous jar and we shan't get far if we don't.

It is fascinating to speculate on how the notion of the infinite breaks into mathematics. What are its origins? The perception of great stretches of time? The perception of great distances such as the vast deserts of Mesopotamia or the straight line to the stars? Or could it be the striving of the soul towards realization and perception, or the striving towards ultimate but unrealizable explanations?

The infinite is that which is without end. It is the eternal, the immortal, the self-renewable, the *apeiron* of the Greeks, the *ein-sof* of the Kabbalah, the cosmic eye of the mystics which observes us and energizes us from the godhead.

Observe the equation

$$\tfrac{1}{2} + \tfrac{1}{4} + \tfrac{1}{8} + \tfrac{1}{16} + \cdots = 1,$$

or, in fancier notation, $\Sigma_{n=1}^{\infty} 2^{-n} = 1$. On the left-hand side we seem to have incompleteness, infinite striving. On the right-hand side we have finitude, completion. There is a tension between the two sides which is a source of power and paradox. There is an overwhelming mathematical desire to bridge the gap between the finite and the infinite. We want to complete the incomplete, to catch it, to cage it, to tame it.*

---

* Kunen describes a game in which two mathematicians with a deep knowledge of the infinite try to outbrag each other in naming a greater cardinal number than their opponent. Of course, one can always add one to get a yet higher number, but the object of the game as played by these experts is to score by breaking through with an altogether new paradigm of cardinal formation. The playoff goes something along these lines:

XVII

1,295,387

$10^{10^{10}}$

$w$

Mathematics thinks it has succeeded in doing this. The unnamable is named, operated on, tamed, exploited, finitized, and ultimately trivialized. Is, then, the mathematical infinite a fraud? Does it represent something that is not really infinite at all? Mathematics is expressible in language that employs a finite number of symbols strung together in sentences of finite length. Some of these sentences seem to express facts about the infinite. Is this a mere trick of language wherein we simply identify certain types of sentences as speaking about "infinite things"? When infinity has been tamed it enjoys a symbolic life.

Cantor introduced the symbol $\aleph_0$ ("aleph nought") for the infinite cardinal number represented by the set of natural numbers, $N$. He showed that this number obeyed laws of arithmetic quite different from those of finite numbers; for instance, $\aleph_0 + 1 = \aleph_0$, $\aleph_0 + \aleph_0 = \aleph_0$, etc.

Now, one could easily manufacture a hand-held computer with a $\aleph_0$ button to obey these Cantorian laws. But if $\aleph_0$ has been encased algorithmically with a finite structure, in what, then, consists its infinity? Are we dealing only with so-called infinities? We think big and we act small. We think infinities and we compute finitudes. This reduction, after the act, is clear enough, but the metaphysics of the act is far from clear.

Mathematics, then, asks us to believe in an infinite set. What does it mean that an infinite set exists? Why should one believe it? In formal presentation this request is institutionalized by axiomatization. Thus, in *Introduction to Set*

---

$w_{(w_w)}$
The first inaccesible cardinal
The first hyper-inaccessible cardinal
The first Mahlo cardinal
The first hyper-Mahlo
The first weakly compact cardinal
The first ineffable cardinal.

Obviously it would not be cricket to name a number that doesn't exist, and a central problem in large cardinal theory is precisely to explicate the sense in which the above mentioned hypers and ineffables exist.

154

*Theory,* by Hrbacek and Jech, we read on page 54:

"Axiom of Infinity. An inductive (i.e. infinite) set exists."

Compare this against the axiom of God as presented by Maimonides (Mishneh Torah, Book 1, Chapter 1):

The basic principle of all basic principles and the pillar of all the sciences is to realize that there is a First Being who brought every existing thing into being.

Mathematical axioms have the reputation of being self-evident, but it might seem that the axioms of infinity and that of God have the same character as far as self-evidence is concerned. Which is mathematics and which is theology? Does this, then, lead us to the idea that an axiom is merely a dialectical position on which to base further argumentation, the opening move of a game without which the game cannot get started?

*Exposure to the ideas of modern mathematics has led artists to attempt to depict graphically the haunting qualities of the infinite.* DE CHIRICO. *Nostalgia of the Infinite.* 1913–14? (dated 1911 on the painting)Courtesy: Museum of Modern Art

Where there is power, there is danger. This is as true in mathematics as it is in kingship. All arguments involving the infinite must be scrutinized with especial care, for the infinite has turned out to be the hiding place of much that is strange and paradoxical. Among the various paradoxes that involve the infinite are Zeno's paradox of Achilles and the tortoise, Galileo's paradox, Berkeley's paradox of the infinitesimals (see Chapter 5, Nonstandard Analysis), a large variety of paradoxes that involve manipulations of infinite sums or infinite integrals, paradoxes of noncompactness, Dirac's paradox of the function that is useful but doesn't exist, etc. From each of these paradoxes we have learned something new about how mathematical objects behave, about how to talk about them. From each we have extracted the venom of contradiction and have reduced paradox to merely standard behavior in a nonstandard environment.

The paradox of Achilles and the tortoise asserts that Achilles cannot catch up with the tortoise, for he must first arrive at the point where the tortoise has just left, and therefore the tortoise is always in the lead.

Galileo's paradox says there are as many square num-

*Galileo Galilei*
*1564–1642*

bers as there are integers and vice versa. This is exhibited

$$\begin{matrix} 1 & 2 & 3 & 4 & 5 & \ldots \\ \updownarrow & \updownarrow & \updownarrow & \updownarrow & \updownarrow & \\ 1 & 4 & 9 & 16 & 25 & \ldots \end{matrix}$$

in the correspondence ⟡ ⟡ ⟡ ⟡ ⟡    . Yet, how can

this be when not every number is a square?

The paradoxes of rearrangements say that the sum of an infinite series may be changed by rearranging its terms.

For example,

$0 = (1 - 1) + (1 - 1) + (1 - 1) + \ldots = 1 + (-1 + 1) + (-1 + 1) + \ldots = 1 + 0 + 0 + \ldots = 1.$

Dirac's function $\delta(x)$ carries the definition

$$\delta(x) = 0 \quad \text{if } x \neq 0, \quad \delta(0) = \infty, \quad \int_{-\infty}^{\infty} \delta(x)dx = 1,$$

which is self-contradictory within the framework of classical analysis.

Achilles is an instance of irrelevant parameterization:

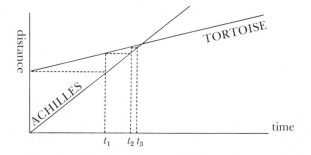

Of course the tortoise is always ahead at the infinite sequence of time instants $t_1, t_2, t_3, \ldots$ where Achilles has just managed to catch up to where the tortoise was at the last time instant. So what? Why limit our discussion to the convergent sequence of times $t_1, t_2, \ldots$? This is a case of the necessity of keeping one's eye on the doughnut and not on the hole.

Galileo's paradox is regularized by observing that the phenomenon it describes is a distinguishing characteristic of an infinite set. An infinite set is, simply, a set which can be put into one-to-one correspondence with a proper subset of itself. Infinite arithmetic is simply not the same as fi-

156

nite arithmetic. If Cantor tells us that $\aleph_0 + 1 = \aleph_0$ is a true equation, one has no authorization to treat $\aleph_0$ as a finite quantity, subtract it from both sides of the equation to arrive at the paradoxical statement $1 = 0$.

Berkeley's paradox of the infinitesimals was first ignored, then bypassed by phrasing all calculus in terms of limiting processes. In the last decade it has been regularized by "nonstandard" analysis in a way which seems to preserve the original flavor of the creators of the calculus.

The paradoxes of rearrangement, aggregation, non-compactness are now dealt with on a day-to-day basis by qualification and restrictions to absolutely convergent series, absolutely convergent integrals, uniform convergence, compact sets. The wary mathematician is hedged in, like the slalom skier, by hundreds of flags within whose limits he must run his course.

The Dirac paradox which postulates the existence of a function with contradictory properties is squared away by the creation of a variety of operational calculi such as that of Temple-Lighthill or Mikusinski or, the most notable, Schwartz's theory of distributions (generalized functions).

By a variety of means, then, the infinite has been harnessed and then housebroken. But the nature of the infinite is that it is open-ended and the necessity for further cosmetic acts will always reappear.

### Further Readings. See Bibliography

B. Bolzano; D. Hilbert; K. Kuhnen; L. Zippin

# The
# Stretched String

I N THE STANDARD FORMALIZATION of mathematics, geometry is reduced, by the use of coordinates, to algebra and analysis. These in turn, by familiar constructions of set theory, are reduced to numbers. This is done because it came to be thought by the end of the nineteenth century that the intuitive notion of the natural numbers was crystal-clear, whereas the notion of a continuous straight-line segment seems more subtle and obscure the more we study it. Nevertheless, it seems clear that historically and psychologically the intuitions of geometry are more primitive than those of arithmetic.

In some primitive cultures there are no number words except one, two, and many. But in every human culture that we will ever discover, it is important to go from one place to another, to fetch water or dig roots. Thus human beings were forced to discover—not once, but over and over again, in each new human life—the concept of the straight line, the shortest path from here to there, the activity of going directly towards something.

In raw nature, untouched by human activity, one sees straight lines in primitive form. The blades of grass or stalks of corn stand erect, the rock falls down straight, objects along a common line of sight are located rectilinearly. But nearly all the straight lines we see around us are human artifacts put there by human labor. The ceiling meets the wall in a straight line, the doors and window-panes and tabletops are all bounded by straight lines. Out the window one sees rooftops whose gables and corners meet in straight lines, whose shingles are layered in rows and rows, all straight.

The world, so it would seem, has compelled us to create the straight line so as to optimize our activity, not only by the problem of getting from here to there as quickly and easily as possible, but by other problems as well. For example, when one goes to build a house of adobe blocks, one finds quickly enough that if they are to fit together nicely, their sides must be straight. Thus the idea of a straight line is intuitively rooted in the kinesthetic and the visual imaginations. We feel in our muscles what it is to go straight toward our goal, we can see with our eyes whether someone else is going straight. The interplay of these two sense intuitions gives the notion of straight line a solidity that enables us to handle it mentally as if it were a real physical object that we handle by hand.

By the time a child has grown up to become a philosopher, the concept of a straight line has become so intrinsic and fundamental a part of his thinking that he may imagine it as an Eternal Form, part of the Heavenly Host of Ideals which he recalls from before birth. Or, if his name be not Plato but Aristotle, he imagines that the straight line is an aspect of Nature, an abstraction of a common quality he has observed in the world of physical objects. He tends to forget the extent to which the straight lines that are observed are there because we have invented and built them. In the course of human physical activity, the straight line enters the human mind, and becomes there a shared concept, the straight line about which we can reason, the mathematical straight line. As mathematicians, we discover that there are theorems to be proved having to do with geodesics—the shortest paths, the solutions of the problem of minimizing the distance from here to there—and having to do with paths of constant curvature—paths which are "unchanged" by sliding them back and forth along themselves.

Just what constitutes the "straightness" of the straight line? There is undoubtedly more in this notion than we know and more than we can state in words or formulas. Here is an instance of this "more." Suppose a, b, c, d are four points on a line. Suppose b is between a and c, and c is between b and d. Then what can we conclude about a, b, and d? It will not take you long to conclude that b must lie between a and d.

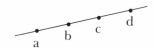

This fact, surprisingly, cannot be proved from Euclid's axioms; it has to be added as an additional axiom in geometry. This omission of Euclid was first noticed 2000 years after Euclid, by M. Pasch in 1882! Moreover, there are important theorems in Euclid whose complete proof requires Pasch's axiom; without it, the proofs are not valid.* This example shows that an intuitive notion may not be completely described by the axioms which are stated in a theory.

The same thing happens in modern mathematics. It was discovered by the Norwegian logician Thorolf Skolem that there are mathematical structures which satisfy the axioms of arithmetic, but which are much larger and more complicated than the system of natural numbers. These "nonstandard arithmetics" may include infinitely large integers. In reasoning about the natural numbers, we rely on our complete mental picture of these numbers. Skolem's example shows that there is more information in that picture than is contained in the usual axioms of arithmetic.

The conclusion that b is between a and d in Pasch's theorem may seem to be a trivial one. For we get this answer simply by making a little picture with pencil and paper. We follow the directions on how to arrange the dots, and we see that in the end b is between a and d. In other words, we use a line on paper to determine the properties of the ideal line, the mathematical line. What could be simpler? But there are two points that are worth bringing up front, out of the background where they usually lurk. First of all, we know that the ideal line is different from the line on paper. Some of the properties of the line on paper are "accidental," not shared by the mathematical line. How do we know which ones?

In the figure for Pasch's axiom, we put a, b, c, d *some-*

---

* H. Guggenheimer has shown that another version of Pasch's axiom can be derived as a theorem using Euclid's Fifth postulate.

*where,* and get our picture. We could have drawn the figure in many other ways, for we have complete freedom as to how far apart any two points should be. Yet we are sure that the answer will always be the same, b is between a and d. We draw only one picture, but we believe that it is representative of all possible pictures. How do we know that?

The answer has to do with our actually sharing a definite intuitive notion, about which we have some reliable knowledge. However, our knowledge of this intuitive notion is by no means complete—neither in the sense of explicit knowledge, nor even in the sense of providing a basis from which to derive complete information, possibly by very long, difficult arguments.

One of the oldest questions about the straight line (as old as Zeno) is this: Can a segment of finite length be divided into infinitely many parts? To the mathematically-trained reader, the answer is an unhesitating "Yes." It may be suspected that few persons, if any, could be found today who would dispute this point of gospel. But 200 years ago, it was a different story.

Listen to the opinion of George Berkeley on this subject:

> . . . to say a finite quantity or extension consists of parts infinite in number is so manifest a contradiction, that everyone at first sight acknowledges it to be so; and it is impossible it should ever gain the assent of any reasonable creature who is not brought to it by gentle and slow degrees, as a converted Gentile to the belief of transubstantiation.

If we accept Berkeley as a competent witness as to the eighteenth-century intuitive notion of the straight line, perhaps we should conclude that the question of infinite divisibility was undecidable. By the twentieth century, the weight of two hundred years of successful practice of mathematical analysis has settled the question. "Between any two distinct points on the line there exists a third point" is intuitively obvious today.

Today the continuum hypothesis (see Chapter 5, Non-Cantorian Set Theory) is an undecidable question about the straight line. This is true in two senses. As a mathematical theorem, the combined work of Gödel and Cohen, it is a statement about the Zermelo-Frankel-Skolem axiom system. It has been proved that neither the continuum hypothesis nor its negation can be proved from these axioms.

The continuum hypothesis is also undecidable in a larger sense—that no one has been able to find intuitively compelling arguments to accept or reject it. Set theorists have been searching for over ten years for some appealing or plausible axiom that would decide the truth of the continuum hypothesis and have not yet found one.

It may be that our intuition of the straight line is permanently incomplete with respect to set-theoretical questions involving infinite sets. In that case, one can add as an axiom either the continuum hypothesis or its negation. We would have many different versions of the straight line, none of which was singled out as intuitively right.

The question of divisibility, it seems, was settled by the historical development of mathematics. Perhaps the continuum hypothesis will be settled in the same way. Mathematical concepts evolve, develop, and are incompletely determined at any particular historical epoch. This does not contradict the fact that they also have well-determined properties, both known and unknown, which entitle them to be regarded as definite objects.

### Further Readings. See Bibliography

H. Guggenheimer.

# The Coin of Tyche

HOW MANY REALLY basic mathematical objects are there? One of them is surely the "miraculous jar" of the positive integers 1, 2, 3, . . . . Another is the concept of a fair coin. Though gambling was rife in the ancient world and although prominent Greeks and Romans sacrificed to Tyche, the goddess of luck, her coin did not arrive on the mathematical scene until the Renaissance. Perhaps one of the things that had delayed this was a metaphysical position which held that God speaks to humans through the action of chance. Thus, in the Book of Samuel, one learns how the Israelites around 1000 B.C. selected a king by casting lots:

> Samuel then made all the tribes of Israel come forward, and the lot fell to the tribe of Benjamin. He then made the tribe of Benjamin come forward clan by clan and the lot fell to the clan of Matri. He then made the clan of Matri come forward man by man and the lot fell to Saul the son of Kish
> 1 Samuel 10:20–21

The implication here is that the casting of lots is not merely a convenient way of arriving at a leader, but it is done with divine approbation and the result is an expression of the divine will. Tossing a coin is still not an outmoded device for decision making in the face of uncertainty.

The modern theory begins with the expulsion of Tyche from the Pantheon. There emerges the vision of the fair coin, the unbiased coin. This coin exists in some mental universe and all modern writers on probability theory have access to it. They toss it regularly and they speculate about what they "observe." What do they report? Well, the fair coin is a two-sided affair in splendid mint condition. For convenience, one side is designated as heads (H) and the other as tails (T). The fair coin is tossed over and over again and one observes the outcome. There is also the notion of a *fair tosser,* i.e., an individual who takes a fair coin and without subterfuge, sleight of hand, or other shady physical manipulation tosses the fair coin. But, for simplicity, let us agree to roll the tossed and the tosser into one abstract concept.

What does one observe? (Remember, this is not a real coin tossed by a real tosser, it is an abstraction.) One observes the axioms for "randomness." The fair coin having been tossed $n$ times, one counts the number of heads and the number of tails that show up. Call them H($n$) and T($n$). Then, of course,

$$H(n) + T(n) = n \qquad (1)$$

since each toss yields either an H or a T. But more than that,

$$\lim_{n\to\infty} \frac{H(n)}{n} = \frac{1}{2} \quad \text{and} \quad \lim_{n\to\infty} \frac{T(n)}{n} = \frac{1}{2}. \qquad (2)$$

This limiting value of $\frac{1}{2}$ is called the probability of tossing H and T respectively:

$$p(\text{H}) = \tfrac{1}{2}, \quad p(\text{T}) = \tfrac{1}{2}. \qquad (3)$$

The fair coin, then, need not show 500 heads and 500 tails when it has been tossed, say, 1,000 times; the probability $\frac{1}{2}$ is established only in the limit.

The fair coin exhibits more than just this. Suppose that we observe only the 1st, 3rd, 5th, . . . tosses and compute the probabilities on this basis. The outcome is the same. Suppose we observe the 1st, 4th, 9th, 16th, . . . tosses. The outcome is the same.

Suppose that, in response to the popular idea that "in the long run things even themselves out," we observe only those tosses which follow immediately after a run of four successive heads. Result: again, half and half. We don't get a predominance of tails. Apparently, things don't even themselves out so soon. At least, not in the sense that if you bet on tails after a long run of heads you have an advantage.

This leads us to the idea that the tosses of a fair coin are insensitive to place selection. That is, if one examines the result of a subsequence of the tosses which have been arrived at by any policy or a rule $R$ of selection which depends on the prior history up to the selected item, we still get a probability of $\frac{1}{2}$.

The insensitivity to place selection can be expressed in an alternative way in what the physicists call a *principle of impotence:* One cannot devise a successful gambling system to be used against a fair coin.

This feeling of impotence, institutionalized as an axiom for randomness, must have been perceived early. Perhaps it explains why Girolamo Cardano, who was an inveterate gambler and who wrote *de Ludo Aleae* (On Dicing), one of the first books on probability theory, gives in this book practical advice on how to cheat. Which raises a metaphysical question: It is possible to cheat against a fair coin?

The above characterizations of the fair coin constitute the foundations of frequency probability. One goes on from these, making further assumptions about probabilities under mixing and combining, and arrives at a full mathematical theory. What is the relationship between this theory and the behavior of real-world coins?

We are led to a paradoxical situation. In an infinite mathematical sequence of H's and T's, the probabilities depend only upon what happens at "the end of the sequence," the "infinite part." No finite amount of information at the beginning of the sequence has any effect in computing a probability! Even if initially we have one million H's followed by H's and T's in proper proportion, the limits will still be half and half. But, of course, in practise if we toss a million consecutive heads, we will conclude that the coin is loaded.

Thus, we are led to the silent assumption, which is not part of the formalized theory, but which is essential for applied probability, that convergence in equation (2) is sufficiently fast that one may judge whether a particular real coin is or is not modelled by the fair coin.

The vision and the intuition of the fair coin are fogged over, and the road to axiomatization is beset with pitfalls. So also are the philosophical and psychological bases of probability. There are a good dozen different definitions

HHHTHTHHTTTTTTHHHTHTHHTTTHTHTTTH
HHTHTHTHTHTTTTHTTTTTTTHHHTTHHTTT
TTTHTTTHTHTTHHTHHHHTTHTTTTHHTTHHT
HTTTHTTHTTTTTTTTTTTTTHHTTTTHTTHTT
THHHHTHTTHTTTHTTHHHHHHTTTHHTHHHHH
HHTTTHHTHHHTHTHHHTHHHHTTTHTTHTHH
HHHTHTHHTTTTTTHHHTHTHHTTTHTHTTTH
HHTHTHTHTHTTTTHTTTTTTTHHHTTHHTTT
TTTHTTTHTHTTHHTHHHHTTHTTTTHHTTHHT
HTTTHTTHTTTTTTTTTTTTTHHTTTTHTTHTT
THHHHTHTTHTTTHTTHHHHHHTTTHHTHHHHH
HHTTTHHTHHHTHTHHHTHHHHTTTHTTHTHH
HHTTHTHHHTHTTTHTTTTHHHTHHHHTHHTH
THHHTHTHHHHHHTHTTTTHTTTHTTTHHHHHHH
TTTTHTHHHTTHHHHHHHHHHTHTTHHTHTTTT
TTTTTTTTHTTTTHTTHTHHHHHHHHTHTHTH
HHHHHTHHTHTHHHTHTHHTTTHTHHTHTTH
THTTHHHTH                         HHTTTTTHH
HHHHTTTTH  **Are These the Tosses**  TTTHTHTHH
HTTHHTTHH      **of a Fair Coin?**    HTHHTTHTH
HTTTTHHHH                         THHTHHHHH
THTHHHTTTHHHHTTTTHHHTHTTTHTHTHTH
THHTHHTTHTTTTHTHTHHHHHHHHHHTTTTH
HHHTTTTHHTHHTTHTTTTTHHTTTHHHHHTT
THTHHHTHTTTHHTTHHTHTHHHHHHHTHTHT
TTTHHTTTHTTTHTHTHTTTHTTTHHTTTHHHHTH
HTTHHHHHTTHHHHTTHHTTTTTHHHTTTTHT
HTHTTHTHTHHHTHTTTTHHHHHHHTTTTHTHT
HHTHHHTTHHHHHHHHTHTTHHHHHHHTTTTHHT
HHTHHHTTTTHHHTHTTHTTTHTTHHHHHHTTHHT
HHHHTTHTHTHHTTTHHTTTTHHTTHHHHHHHHTH
THTHHHTHTTTHHTTHHTHTHHHHHHHTHTHT
TTTHHTTTHTTTHTHTHTTTHTTTHHTTTHHHHTH
HTTHHHHHTTHHHHTTHHTTTTTHHHTTTTHT
HTHTTHTHTHHHTHTTTTHHHHHHHTTTTHTHT
HHTHHHTTHHHHHHHHTHTTHHHHHHHTTTTHHT
HHTHHHTTTTHHTHTTHTTHTTHHHHHHTTHHT
HHHHTTHTHTHHTTTHHTTTTHHTTHHHHHHHHTH

of a random sequence. The one we have been describing above is that of Richard von Mises, 1919. Let us be a little more precise about this definition. We should like a fair coin to have the H's and T's distributed, in the limit, in a 50%–50% fashion. But we should like more. We should like that if we look at the results of two successive tosses, then each of the four possible outcomes HH, HT, TH, TT occurs, in the limit, with probability $\frac{1}{4}$. We should like, shouldn't we, something similar to occur for the eight possible outcomes of three tosses:

*Richard von Mises*
*1883–1953*

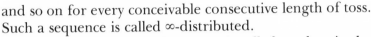

HHH, HHT, HTH, HTT, THH, THT, TTH, TTT,

and so on for every conceivable consecutive length of toss. Such a sequence is called ∞-distributed.

An infinite sequence $x_1, x_2, \ldots$, is called random in the sense of von Mises if every infinite sequence $x_{n_1}, x_{n_2}, x_{n_3}$, $\ldots$ extracted from it and determined by a policy or a rule $R$ is ∞-distributed. Now comes the shocker. It has been established by Joseph Doob that there are no sequences that are random in the sense of von Mises. The requirement is logically self-contradictory.

So one must back down and require less. For practical computing—and "random" sequences are employed quite a bit in programs—one demands quite a bit less. In practise one demands a sequence of integers $x_1, x_2, \ldots$ which is easily programmed and inexpensive to produce. The sequence will be periodic but its period should be sufficiently large with respect to the number of random numbers required. Finally, the sequence should be sufficiently crazy and mixed up so as to pass a number of statistical tests for randomness such as frequency tests, runs test, poker hand tests, spectral tests, etc. The program producing such a sequence is called a random number generator, although the successive integers are given by a completely deterministic program and there is nothing, in principle, that is unpredictable about them.

A popular random number generator is given by the formula

$$x_{n+1} = kx_n \pmod{m}, \; x_0 = 1$$

167

This means that each member $x_{n+1}$ of our random sequence is obtained from its predecessor $x_n$ in the sequence by multiplying with a certain number $k$ and then dividing by a certain other number $m$; $x_{n+1}$ is the remainder left by this division.

Suppose the computer has a word length of $b$ binary digits or "bits." For $k$, select a number of the form $8t \pm 3$, and close to $2^{b/2}$. Choose $m = 2^b$. The multiplication $k \cdot x_0$ produces a product of $2b$ bits. The $b$ higher order bits are discarded, leaving the $b$ lower order bits, which is the residue $x_1$. Then repeat the process,

$$x_2 = kx_1 \pmod{m}$$

and so on. The sequence will not start to repeat itself until the iteration has been carried out $2^{b-2}$ times. If the computing machine has a word length of 35 bits, this gives a "random" sequence of roughly $8.5 \times 10^9$ numbers.

### Further Readings. See Bibliography

H. P. Edmundson; R. von Mises

# The Aesthetic Component

"The mathematical sciences particularly exhibit order, symmetry, and limitation; and these are the greatest forms of the beautiful."

Aristotle, Metaphysics, M 3, 1078 b.

THE AESTHETIC APPEAL of mathematics, both in passive contemplation and in actual research pursuit, has been attested by many authors. Classic and medieval authors, such as Kepler, rhapsodized over the "Divine or Golden Proportion." Poincaré asserted that the aesthetic rather than the logical is the dominant element in mathematical creativity. G. H.

Hardy wrote that "The mathematician's patterns, like the painter's or the poet's, must be beautiful. . . ." The great theoretical physicist P. A. M. Dirac wrote that it is more important to have beauty in one's equations than to have them fit the experiment.

*Henri Poincaré*
*1854–1912*

Blindness to the aesthetic element in mathematics is widespread and can account for a feeling that mathematics is dry as dust, as exciting as a telephone book, as remote as the laws of infangthief of fifteenth century Scotland. Contrariwise, appreciation of this element makes the subject live in a wonderful manner and burn as no other creation of the human mind seems to do.

Beauty in art and in music has been an object of discussion at least since the time of Plato and has been analyzed in terms of such vague concepts as order, proportion, balance, harmony, unity, and clarity. In recent generations, attempts have been made to assign mathematical measures of aesthetic quality to artistic creations. When such measures are built into criteria for construction of musical pieces, say with the computer, it is found that these programs can recapture to a small degree the Mozart-like qualities of Mozart. But the notion of the underlying aesthetic quality remains elusive. Aesthetic judgments tend to be personal, they tend to vary with cultures and with the generations, and philosophical discussions of aesthetics have in recent years tended less toward the dogmatic prescription of what is beautiful than toward discussion of how aesthetic judgments operate and function.

Aesthetic judgment exists in mathematics, is of importance, can be cultivated, can be passed from generation to generation, from teacher to student, from author to reader. But there is very little formal description of what it is and how it operates. Textbooks and monographs are devoid of comments on the aesthetic side of their topics, yet the aesthetic resides in the very manner of doing and the selection of what is done. A work of art, say a piece of colonial Rhode Island cabinetmaking, does not have a verbal description of its unique beauty carved into its mouldings. It is part of an aesthetic tradition and this suffices, except for the scholar.

169

Attempts have been made to analyze mathematical aesthetics into components—alternation of tension and relief, realization of expectations, surprise upon perception of unexpected relationships and unities, sensuous visual pleasure, pleasure at the juxtaposition of the simple and the complex, of freedom and constraint, and, of course, into the elements familiar from the arts, harmony, balance, contrast, etc. Further attempts have been made to locate the source of these feelings at a deeper level, in psychophysiology or in the mystical collective unconscious of Jung. While most practitioners feel strongly about the importance of aesthetics and would augment this list with their own aesthetic categories, they would tend to be skeptical about deeper explanations.

Aesthetic judgements may be transitory and may be located within the traditions of a particular mathematical age and culture. Their validity is similar to that of a school or period of art. It was once maintained that the most beautiful rectangle has its sides in the golden ratio $\Phi = \frac{1}{2}(1 + \sqrt{5})$. Such a statement would not be taken seriously today by a generation brought up on nonclassical art and architecture, despite experiments of Fechner (1876) or of Thorndike (1917) which are said to bear it out.* The aesthetic delight in the golden ratio $\Phi$ appears nowadays to derive from the diverse and unexpected places in which it arises.

There is, first, the geometry of the regular pentagon. If the side AB of a regular pentagon has unit length, then any line such as AC has length $\Phi = \frac{1}{2}(1 + \sqrt{5}) = 2 \cos \pi/5 = 1.61803. \ldots$

It appears, next, in difference equations. Take two numbers at random, say 1 and 4. Add them to get 5. Add 4 and 5 to get 9. Add 5 and 9 to get 14. Keep this process

---

* Surprisingly, the golden ratio seems to have been taken seriously as an aesthetic principle as recently as 1962 in G. E. Duckworth, *Structural Patterns and Proportions in Vergil's Aeneid*, Univ. of Michigan Press, 1962. Duckworth analyzes the Aeneid in terms of the ratio $M/(M + m)$, where $M$ is the line length of the "major passages" and $m$ is that of the "minor passages." Duckworth claims that this ratio is .618 $(=\frac{1}{2}(\sqrt{5} - 1))$ with fair accuracy.

up indefinitely. Then, the ratio of successive numbers approachs $\Phi$ as a limit. Witness:

$$
\begin{array}{ll}
1 + 4 = 5 & 5/4 = 1.250 \\
4 + 5 = 9 & 9/5 = 1.800 \\
5 + 9 = 14 & 14/9 = 1.555 \\
9 + 14 = 23 & 23/14 = 1.643 \\
14 + 23 = 37 & 37/23 = 1.608 \\
23 + 37 = 60 & 60/37 = 1.622 \\
37 + 60 = 97 & 97/60 = 1.617 \\
60 + 97 = 157 & 157/97 = 1.618
\end{array}
$$

Finally, continued fractions yield the following beautiful formula:

$$
\Phi = 1 + \cfrac{1}{1 + \cfrac{1}{1 + \ldots}}.
$$

What on earth, asks the novice, do these diverse situations have to do with one another that they all lead to $\Phi$? And amazement gives way to delight and delight gives way to feelings that the universe is united in a wondrous way.

But feelings of a different sort crop up with study and experience. If one works intensively in the theory of difference equations, the unexpected ceases to be unexpected and changes into solid working intuition, and the corresponding aesthetic delight is possibly diminished and certainly transformed. It might even be claimed that situations of "surprise" pose an uncomfortable mystery, which we try to remove by creating a general theory that contains all the particular systems. Thus, all three examples just given for $\Phi$ can be embraced within a general theory of the eigenvalues of certain matrices. In this way, the attempt to explain (and hence to kill) the surprise is converted into pressure for new research and understanding.

### Further Readings. See Bibliography

H. E. Huntley; L. Pacioli

# Pattern, Order, and Chaos

A SENSE OF STRONG personal aesthetic delight derives from the phenomenon that can be termed order out of chaos. To some extent the whole object of mathematics is to create order where previously chaos seemed to reign, to extract structure and invariance from the midst of disarray and turmoil.*

In early usage, the word chaos referred to the dark, formless, gaping void out of which, according to Genesis 1:2, the universe was fashioned. This is what John Milton meant when he used the word in *Paradise Lost* in paraphrase of the Old Testament:

> In the beginning how the heavens and earth
> Rose out of Chaos.

In the years since Milton, the word chaos has altered its meaning; it has become humanized. Chaos has come to mean a confused state of affairs, disorder, mixed-upness. When things are chaotic, they are helter-skelter, random, irregular. The opposite of chaos is order, arrangement, pattern, regularity, predictability, understanding. Chaos, in the opinion of the skeptic, is the normal state of affairs in life; in the opinion of the thermodynamicist, it is the state to which things tend if left to themselves.

In view of the ubiquity of chaos, it would be very difficult indeed to give a blanket description to tell whether order or arrangement is or is not present in any given situation; yet it is this presence which makes life comprehensible.

---

* Otto Neugebauer told the writer the following legend about Einstein. It seems that when Einstein was a young boy he was a late talker and naturally his parents were worried. Finally, one day at supper, he broke into speech with the words "Die Suppe ist zu heiss." (The soup is too hot.) His parents were greatly relieved, but asked him why he hadn't spoken up to that time. The answer came back: "Bisher war Alles in Ordnung." (Until now everything was in order.)

172

When it is there, we feel it "in our bones," and this is probably a response to the necessities of survival. To create order—particularly intellectual order—is one of the major human talents, and it has been suggested that mathematics is the science of total intellectual order.

Graphic or visual order, pattern, or symmetry has been defined and analyzed in terms of the invariants of transformation groups. Thus, a plane figure possesses axial symmetry around the line $y = 0$ if it is unchanged (invariant) by the transformation.

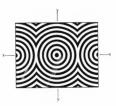

$$\begin{cases} x' = -x \\ y' = y \end{cases}.$$

"A mathematician, like a painter or a poet," wrote G. H. Hardy, "is a master of pattern."

But what can one say about pattern as it occurs within mathematical discourse itself? Are mathematical theories of pattern self-referring?

When scientists propose laws of wide generality they set forth rules of law in place of primeval chaos. When the artist draws his line, or the composer writes his measure, he separates, out of the infinitude of possible shapes and sounds, one which he sets before us as ordered, patterned, and meaningful.

Consider four possibilities: (1) order out of order; (2) chaos out of order; (3) chaos out of chaos; (4) order out of chaos. The first, order out of order, is a reasonable thing. It is like the college band marching in rank and file and then making neat patterns on the football field during halftime. It is pretty, and the passage from one type of order to the other is interesting, but not necessarily exciting. The second possibility, chaos out of order, is, alas, too common. Some sort of intelligence ends up in a mess. This is the bull in the china shop, an exciting and painful happening. The third possibility, chaos out of chaos, is the bull stomping in the town dump. Despite his intelligence, nothing fundamentally happens, no damage is done, and we hardly notice him at all. The fourth, order out of chaos, is our natural striving, and when we achieve it we treasure it.

173

These four types of transformation will be illustrated with some mathematical patterns or theorems.

### *Order Out of Order*

(a) $1^3 = 1^2$
$1^3 + 2^3 = (1 + 2)^2$
$1^3 + 2^3 + 3^3 = (1 + 2 + 3)^2$
$1^3 + 2^3 + 3^3 + 4^3 = (1 + 2 + 3 + 4)^2$
.
.
.

(b) $\frac{4}{9} = .4444 \ldots$
$\frac{5}{37} = .135135135 \ldots$

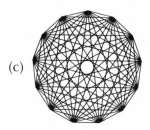

(c)

(d) The equation of the tangent through $(x_1, y_1)$ on the conic $Ax^2 + 2Bxy + Cy^2 + 2Dx + 2Ey + F = 0$ is $Axx_1 + B(xy_1 + x_1 y) + Cyy_1 + D(x + x_1) + E(y + y_1) + F = 0$.

### *Chaos Out of Order*

(a) $12 \times 21 = 252$
$123 \times 321 = 39483$
$1234 \times 4321 = 5332114$
$12345 \times 54321 = 670592745$
$123456 \times 654321 = 80779853376$
$1234567 \times 7654321 = 9449772114007$
$12345678 \times 87654321 = 1082152022374638$
$123456789 \times 987654321 = 121932631112635269$
(b) $\sqrt{2} = 1.4142\ 13562\ 37309\ 50488 \ldots$
(c) $\pi = 3.1415\ 92653\ 58979\ 32384 \ldots$

### *Chaos Out of Chaos*

(a) $53278 \times 2147 = 114387866$

(b)

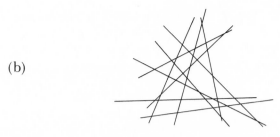

(c) $(1 + 3x^4 - 4x^5)(2 - x + 2x^2) = 2 - x + 2x^2 + 6x^4$
$- 11x^5 + 10x^6 - 8x^7$.

## Order Out of Chaos

(a) Pappus' theorem. On two arbitrary lines $l_1$ and $l_2$ select six points $P_1, \ldots, Q_3$ at random, three on a line. Then the intersection of $P_1Q_2 \ P_2Q_1$; $P_1Q_3 \ P_3Q_1$; $P_2Q_3 \ P_3Q_2$ are always collinear.

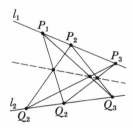

(b) The prime number theorem. The sequence of prime numbers 2, 3, 5, 7, 11, 13, 17, 19, 23, . . . occurs in a seemingly chaotic fashion. Yet, if $\pi(x)$ designates the number of primes that do not exceed $x$, it is known that $\lim_{x \to \infty} \pi(x) \div (x/\log x) = 1$. That is, the size of $\pi(x)$ is approximately $x/\log x$ when $x$ is large. *Example:* if $x = 1,000,000,000$ then it is known that $\pi(x) = 50,847,478$. Now $10^9/\log 10^9 = 48,254,942.43$ . . . . The function

$$\mathrm{li}(x) = \int_0^x \frac{du}{\log u}$$

gives even closer results. (See Chapter 5, The Prime Number Theorem).

(c) A polygon whose vertices are random is transformed

175

by replacing it by the polygon formed from the mid-points of its sides. Upon iteration of this transformation, an ellipse-like convex figure almost always emerges. (From: P.J. Davis, *Circulant Matrices.*)

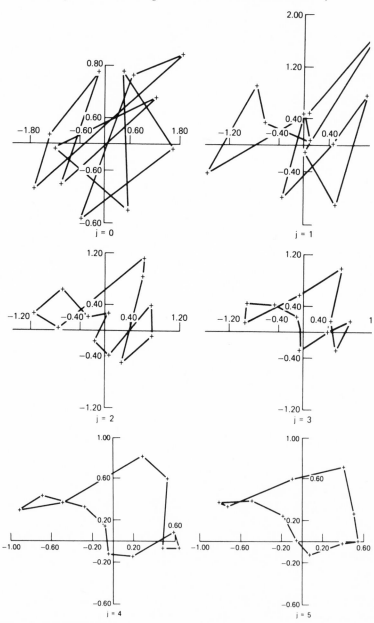

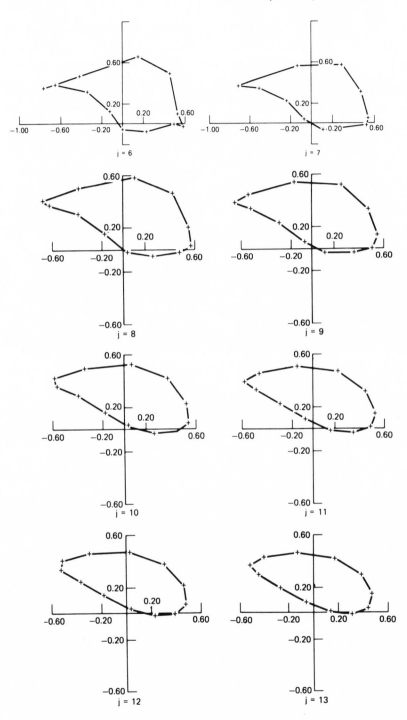

j = 6

j = 7

j = 8

j = 9

j = 10

j = 11

j = 12

j = 13

(d) Order out of chaos is not always arrived at cheaply. According to an unproved (1979) conjecture of Goldbach (1690–1764), every even number is the sum of two prime numbers. For example, $24 = 5 + 19$. This may occur in several different ways: $24 = 7 + 17 = 11 + 13$. The accompanying computer listing is of the decomposition of even numbers into the sum of two primes, where the first entry is as small as possible and the last is as large as possible. The chaos is clear. But what is the underlying order? The proof of Goldbach's conjecture, if and when it is forthcoming, may bring order to this chaos.

ILLUSTRATING GOLDBACH'S CONJECTURE

| | | | | | | |
|---|---|---|---|---|---|---|
| 4 | = 2 | + 2 | | 20882 | = 3 | + 20879 |
| 6 | = 3 | + 3 | | 20884 | = 5 | + 20879 |
| 8 | = 3 | + 5 | | 20886 | = 7 | + 20879 |
| 10 | = 3 | + 7 | | 20888 | = 31 | + 20857 |
| 12 | = 5 | + 7 | | 20890 | = 3 | + 20887 |
| 14 | = 3 | + 11 | | 20892 | = 5 | + 20887 |
| 16 | = 3 | + 13 | | 20894 | = 7 | + 20887 |
| 18 | = 5 | + 13 | | 20896 | = 17 | + 20879 |
| 20 | = 3 | + 17 | | 20898 | = 11 | + 20887 |
| 22 | = 3 | + 19 | | 20900 | = 3 | + 20897 |
| 24 | = 5 | + 19 | | 20902 | = 3 | + 20899 |
| 26 | = 3 | + 23 | | 20904 | = 5 | + 20899 |
| 28 | = 5 | + 23 | | 20906 | = 3 | + 20903 |
| 30 | = 7 | + 23 | | 20908 | = 5 | + 20903 |
| 32 | = 3 | + 29 | | 20910 | = 7 | + 20903 |
| 34 | = 3 | + 31 | | 20912 | = 13 | + 20899 |
| 36 | = 5 | + 31 | | 20914 | = 11 | + 20903 |
| 38 | = 7 | + 31 | | 20916 | = 13 | + 20903 |
| 40 | = 3 | + 37 | | 20918 | = 19 | + 20899 |
| 42 | = 5 | + 37 | | 20920 | = 17 | + 20903 |
| 44 | = 3 | + 41 | | 20922 | = 19 | + 20903 |
| 46 | = 3 | + 43 | | 20924 | = 3 | + 20921 |
| 48 | = 5 | + 43 | | 20926 | = 5 | + 20921 |
| 50 | = 3 | + 47 | | 20928 | = 7 | + 20921 |
| 52 | = 5 | + 47 | | 20930 | = 31 | + 20899 |
| 54 | = 7 | + 47 | | 20932 | = 3 | + 20929 |

| | | | | |
|---|---|---|---|---|
| 56 | = 3 | + 53 | 20934 = 5 | + 20929 |
| 58 | = 5 | + 53 | 20936 = 7 | + 20929 |
| 60 | = 7 | + 53 | 20938 = 17 | + 20921 |
| 62 | = 3 | + 59 | 20940 = 11 | + 20929 |
| 64 | = 3 | + 61 | 20942 = 3 | + 20939 |
| 66 | = 5 | + 61 | 20944 = 5 | + 20939 |
| 68 | = 7 | + 61 | 20946 = 7 | + 20939 |
| 70 | = 3 | + 67 | 20948 = 19 | + 20929 |
| 72 | = 5 | + 67 | 20950 = 3 | + 20947 |
| 74 | = 3 | + 71 | 20952 = 5 | + 20947 |
| 76 | = 3 | + 73 | 20954 = 7 | + 20947 |
| 78 | = 5 | + 73 | 20956 = 17 | + 20939 |
| 80 | = 7 | + 73 | 20958 = 11 | + 20947 |
| 82 | = 3 | + 79 | 20960 = 13 | + 20947 |
| 84 | = 5 | + 79 | 20962 = 3 | + 20959 |
| 86 | = 3 | + 83 | 20964 = 5 | + 20959 |
| 88 | = 5 | + 83 | 20966 = 3 | + 20963 |
| 90 | = 7 | + 83 | 20968 = 5 | + 20963 |
| 92 | = 3 | + 89 | 20970 = 7 | + 20963 |
| 94 | = 5 | + 89 | 20972 = 13 | + 20959 |
| 96 | = 7 | + 89 | 20974 = 11 | + 20963 |
| 98 | = 19 | + 79 | 20976 = 13 | + 20963 |
| 100 | = 3 | + 97 | 20978 = 19 | + 20959 |
| 102 | = 5 | + 97 | 20980 = 17 | + 20963 |
| 104 | = 3 | + 101 | 20982 = 19 | + 20963 |
| 106 | = 3 | + 103 | 20984 = 3 | + 20981 |
| 108 | = 5 | + 103 | 20986 = 3 | + 20983 |
| 110 | = 3 | + 107 | 20988 = 5 | + 20983 |
| 112 | = 3 | + 109 | 20990 = 7 | + 20983 |
| 114 | = 5 | + 109 | 20992 = 11 | + 20981 |
| 116 | = 3 | + 113 | 20994 = 11 | + 20983 |
| | | | 20996 = 13 | + 20983 |
| | | | 20998 = 17 | + 20981 |
| | | | 21000 = 17 | + 20983 |

## Further Readings. See Bibliography

L. A. Steen [1975]

# Algorithmic
# vs. Dialectic
# Mathematics

I N ORDER TO understand the difference in the point
of view between algorithmic and dialectic mathematics
we shall work with an example. Let us suppose that we
have set the problem of finding a solution to the equation $x^2 = 2$. This is a problem for which the Babylonians
around 1700 B.C. found the excellent approximation
$\sqrt{2} = 1$; 24, 51, 10 in their base 60 notation, or
$\sqrt{2} = 1.414212963$ in decimals. This is the identical problem which Pythagoras (550 B.C.) asserted had no fractional
solution and in whose honor he was supposed to have sacrificed a hecatomb of oxen—the problem which caused
the existentialist crisis in ancient Greek mathematics. The
$\sqrt{2}$ exists (as the diagonal of the unit square); yet it does
not exist (as a fraction)! We will now present two solutions
to this problem.

***Solution I.*** Notice that if the number $x$ is the solution to
$x^2 = 2$ then it would follow that $x = 2/x$. Now, if $x$ is slightly
incorrect, say underestimated, then $2/x$ will be overesti-
mated. It might occur to anyone after a moment's thought
that halfway between the underestimate and the overesti-
mate should be a better estimate than either $x$ or $2/x$. Formalizing, let $x_1, x_2, \ldots$ be a sequence of numbers defined
successively by

$$x_{n+1} = \frac{1}{2} \left( x_n + \frac{2}{x_n} \right), n = 1, 2, \ldots$$

If $x_1$ is any positive number, the sequence $x_1, x_2, \ldots$ converges to $\sqrt{2}$ with quadratic rapidity.

For example, start with $x_1 = 1$. Then successively, $x_2 = 1.5$, $x_3 = 1.416666 \ldots$, $x_4 = 1.414215686 \ldots$. The value of $x_4$ is already correct to 5 figures after the decimal point. Quadratic convergence means that the number of correct decimals doubles with each iteration. Here is a recipe or an algorithm for the solution of the problem. The algorithm can be carried out with just addition and division and without a complete theory of the real number system.

***Solution II.*** Consider the graph of the function $y = x^2 - 2$. The graph is actually a parabola, but this is not important. When $x = 1$, $y = -1$. When $x = 2$, $y = 2$. As $x$ moves continuously from 1 to 2, $y$ moves continuously from a negative to a positive value. Hence, somewhere between 1 and 2 there must be a value for $x$ where $y = 0$, or, equivalently, where $x^2 = 2$. The solution is now complete. The details of the argument are supplied by the properties of the real number system and of continuous functions defined on that system.

Solution I is algorithmic mathematics. Solution II is the dialectic solution. In a certain sense, neither solution I nor solution II is a solution at all. Solution I gives us a better and better approximation, but whenever we stop we do not yet have an exact solution in decimals. Solution II tells us that an exact solution "exists." It tells us that it is located between 1 and 2 and, that is all it has to say. The dialectic solution might very well be called an existential solution.

Dialectics brings insight and freedom. Our knowledge of what exists may go far beyond what we are able to calculate or even approximate. Here is a simple instance of this. We are given a triangle with three unequal sides. We ask, is there a vertical line which bisects the area of the triangle? Within algorithmic mathematics one might pose the problem of finding such a line, by ruler and compass, or by more generous means. Within dialectic mathematics, one can answer, yes, such a line exists without doing any work at all. One need only notice that if one moves a knife across the figure from left to right, the fraction of the triangle to the left of the knife varies continuously from 0% to 100%, so there must be an intermediate position where the fraction is precisely 50%.

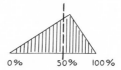

Having arrived at this solution, we may notice with a shock that the specific properties of the triangle weren't used at all; the same argument would work for any kind of an area! And so we assert the existence of a vertical bisector of any given figure, without knowing how to find it, without knowing how to compute the area cut off by the knife, and without even needing to know how to do it. (We return to this idea in Chapter 6, The Classic Classroom Crisis.)

The mathematics of Egypt, of Babylon, and of the ancient Orient was all of the algorithmic type. Dialectical mathematics—strictly logical, deductive mathematics—originated with the Greeks. But it did not displace the algorithmic. In Euclid, the role of dialectic is to justify a construction—i.e., an algorithm.

It is only in modern times that we find mathematics with little or no algorithmic content, which we could call purely dialectical or existential.

One of the first investigations to exhibit a predominantly dialectic spirit was the search for the roots of a polynomial of degree $n$. It was long surmised that a polynomial of degree $n$, $p_n(z) = a_0 z^n + a_1 z^{n-1} + \ldots + a_n$, must have $n$ roots, counting multiplicities. But a closed formula like the quadratic formula or the cubic formula was not found. (It was later shown that it is not possible to find a similar formula for $n > 4$.) The question then became, what other resources can we bring to bear on the problem of finding approximate roots? Ultimately, what assurance do we have for the existence of the roots? The theorems which guarantee this, originally proved by Gauss, are dialectic. The algorithmic aspect is still under discussion.

In most of the twentieth century, mathematics has been existence-oriented rather than algorithm-oriented. Recent years seem to show a shift back to a constructive or algorithmic view point.

Henrici points out that

> *Dialectic mathematics* is a rigorously logical science, where statements are either true or false, and where objects with specified properties either do or do not exist. *Algorithmic mathematics* is a tool for solving problems. Here we are concerned not only with the existence of a mathematical object, but also with the credentials of its existence. *Dialectic* mathematics is an intellectual game played according to rules about which there is a high degree of consensus. The rules of the game of *algorithmic* mathematics may vary according to the urgency of the problem on hand. We never could have put a man on the moon if we had insisted that the trajectories should be computed with dialectic rigor. The rules may also vary according to the computing equipment available. *Dialectic* mathematics invites contemplation. Algorithmic mathematics invites action. *Dialectic* mathematics generates insight. *Algorithmic* mathematics generates results.

There is a distinct paradigm shift that distinguishes the algorithmic from the dialectic, and people who have worked in one mode may very well feel that solutions within the second mode are not "fair" or not "allowed." They experience paradigm shock. P. Gordan who worked algorithmically in invariant theory reputedly felt this shock when confronted with the brilliant work of Hilbert who

183

worked dialectically. "This is not mathematics," said Gordan, "it is theology."

Certainly the algorithmic approach is called for when the problem at hand requires a numerical answer which is of importance for subsequent work either inside or outside mathematics. Numerical analysis is the science and the art of obtaining numerical answers to certain mathematical problems. Some authorities claim that the "art" of numerical analysis merely sweeps under the rug all the inadequacies of the "science." Numerical analysis is simultaneously a branch of applied mathematics and a branch of the computer sciences.

We consider a typical instance. A problem (let us say in physics) has led to a system of ordinary differential equations for the variables $u_1(t)$, $u_2(t)$, . . . , $u_n(t)$, where the independent variable $t$ ranges from $t = 0$ to $t = 1$. This system is to be solved subject to the conditions that the unknowns $u_i(t)$ take on prescribed values at $t = 0$ and $t = 1$. This is the so-called two-point boundary-value problem. A casual inspection of the problem has revealed that there probably is no elementary closed-form expression that solves the problem, and it has therefore been decided to proceed numerically and to compute a table of values, $u_i(t_j)$ $j = 1, 2, . . . , p$, $i = 1, 2, . . . , n$, which will be accepted as the solution. Numerical analysis tells us how to proceed.

The proper procedure may very well depend on the mechanical means for computation at our disposal. If we have pencil and paper, and perhaps a hand-held computer, we should proceed along certain lines. If we have a large computer, there may be different lines. If the computer has certain features of memory and programming or if certain software is available, these may suggest economies of procedure.

Computation on a digital computer involves replacing the continuous variables $u_i(t)$ with discrete variables. But this can be done by a variety of methods. Will the differential equation be replaced by a difference equation? If so, this also can be done in many ways. What, then, is an appropriate way? If we adopt a finite difference policy, we shall be led to systems of algebraic equations that may be linear or nonlinear, depending on the original differential equation. How shall these equations be solved? Can we use direct methods? Or are we compelled to successive approximations via iterative methods? We are confronted with many possible ways to proceed. Numerical analysis will have a word to say about each. Each separate mode of procedure is known as an "algorithm."

Having obtained some kind of answer to the problem by means of an algorithm, numerical analysis attempts to set bounds on how far this answer can differ from the true but unknown answer. Barring blunders* (that is, machine malfunctions, erroneous programming, and other human errors), errors will arise from the fact that continuous variables have been made discrete, that infinite mathematical expressions or processes have been made finite or truncated, and that a computing machine does not do arithmetic with infinite exactitude but with, say, eight figures. Numerical analysis attempts to make an error analysis for each algorithm. There are great difficulties with this type of problem, and the resulting bounds, when obtainable, may be deterministic or statistical. They may be a priori bounds or a posteriori bounds, that is, bounds that can be computed prior to the main computation or bounds requiring that the whole computation be carried out beforehand. They may be only approximate or asymptotic bounds. Finally, they may be bounds that the computer itself has computed.

As part of error analysis, numerical analysis must even consider how to recognize a good answer when it meets one. What is the criterion of a good answer? There may be several. As a very simple example, suppose that we are solving a single equation $f(x) = 0$ and that $x^*$ is the mathematically exact answer. An answer $\bar{x}$ has been produced by a computation. Is $\bar{x}$ a good answer if $|x^* - \bar{x}|$ is small? Or is it a good answer if $f(\bar{x})$ is approximately zero? Considering

---

* Blunders should not be lightly dismissed. They occur with sufficient frequency that the practicing numerical analyst must learn how to recognize them and how to deal with them.

that we probably cannot compute $f$ exactly but can compute only an approximate $\bar{f}$, should we perhaps say that $\bar{x}$ is a good answer if $\bar{f}(\bar{x})$ is approximately zero? Various criteria may lead to widely differing answers.

Having obtained a reasonable answer to a problem, we may be interested in improving the answer. Can we obtain answers of increasing accuracy by a fixed type of algorithm? Numerical analysis has information on this, and the resulting theory, known as convergence theory, is one of the major aspects of the subject.

Numerical analysis therefore comprises the strategy of computation as well as the evaluation of what has been accomplished. A complete numerical analysis of any problem (not often achieved, unfortunately) would consist of

1. The formation of algorithms.
2. Error analysis, including trucation and roundoff error.
3. The study of convergence including the rate of convergence.
4. Comparison of algorithms, to judge the relative utility of different algorithms in different situations.

The pure algorithmic spirit would be content with steps 1 and 4—to invent algorithms, and to try them out in typical problems to see how they work.

When we ask for a careful error analysis, for convergence proofs and estimates of the rate of convergence, we are paying our respects to the dialectical viewpoint.

The algorithmic attitude need not negate the dialectic; but it refuses to be subordinated to it. That is to say, a good algorithm will be used, even if we have no rigorous proof, but only computational experience, to tell us that it is good.

### Further Readings. See Bibliography

P. J. Davis [1969]; P. Henrici [1972], [1974]; J. F. Traub

# The Drive to Generality and Abstraction. The Chinese Remainder Theorem: A Case Study

WE PRESENT HERE highlights in the history of a simple theorem of arithmetic that has been known for at least 2000 years. We shall emphasize the changing nature over the millenia of the statement of the theorem, and brush aside the interesting questions of historical priority, influence, proof, applications.

I first came across the Chinese remainder theorem at the age of ten in *Mathematical Recreations and Essays* by W. W. Rouse Ball. This book, which is easy to read and is a treasure of mathematical delights, has influenced three or four generations of young mathematicians. On page 6 of the 11th edition, we read,

> Ask anyone to select a number less than 60. Request him to perform the following operations: (i) To divide it by 3 and mention the remainder; suppose it to be $a$. (ii) To divide it by 4, and mention the remainder; suppose it to be $b$. (iii) To divide it by 5, and mention the remainder; suppose it to be $c$. Then, the number selected is the remainder obtained by dividing $40a + 45b + 36c$ by 60.

This is followed by a generalization and an algebraic proof which I could not, at the time, understand.

1. The earliest known formulation of the theorem appears to be in the Sun Tzu Suan-ching (i.e., the mathematical classic of Sun Tzu) which has been dated between 280 and 473.

187

We have things of which we do not know the number; if we count them by threes, the remainder is 2; if we count them by fives the remainder is 3; if we count them by sevens the remainder is 2. How many things are there? Answer: 23. Method:

If you count by threes and have the remainder 2, put 140.

If you count by fives and have the remainder 3, put 63.

If you count by sevens and have the remainder 2, put 30.

Add these (numbers) and you get 233.

From this subtract 210 and you have the result.

For each unity as remainder when counting by threes, put 70.

For each unity as remainder when counting by fives, put 21.

For each unity as remainder when counting by sevens, put 15.

If (sum) is 106 or more, subtract 105 from this and you get the result.

2. The highest point of classical Chinese mathematics is the Shu-shu Chiu-chang of Ch'in Chiu-shao. This book, whose title means mathematical treatise in nine sections, appeared in 1247 and contains an extensive treatment of the remainder problem. The relevant text consists of thirty-seven algorithmic instructions (the Ta-Yen Rule) and occupies five pages in Libbrecht. Though of great historical importance, we pass now to the western formulations.

3. Here is the Chinese remainder theorem in the hands of Leonardo Pisano (Fibonacci). In his *Liber Abbaci* (1202) he writes

*Leonardo of Pisa (Fibonacci)*
*c. 1180–1250*

Let a contrived number be divided by 3, also by 5, also by 7; and ask each time what remains from each division. For each unity that remains from the division by 3, retain 70; for each unity that remains from the division by 5, retain 21; and for each unity that remains from the division by 7, retain 15. And as much as the number surpasses 105, subtract from it 105; and what remains to you is the contrived number. Example: suppose from the division by 3 the remainder is 2; for this you retain twice 70, or 140; from which you subtract 105, and 35 remains. From the division

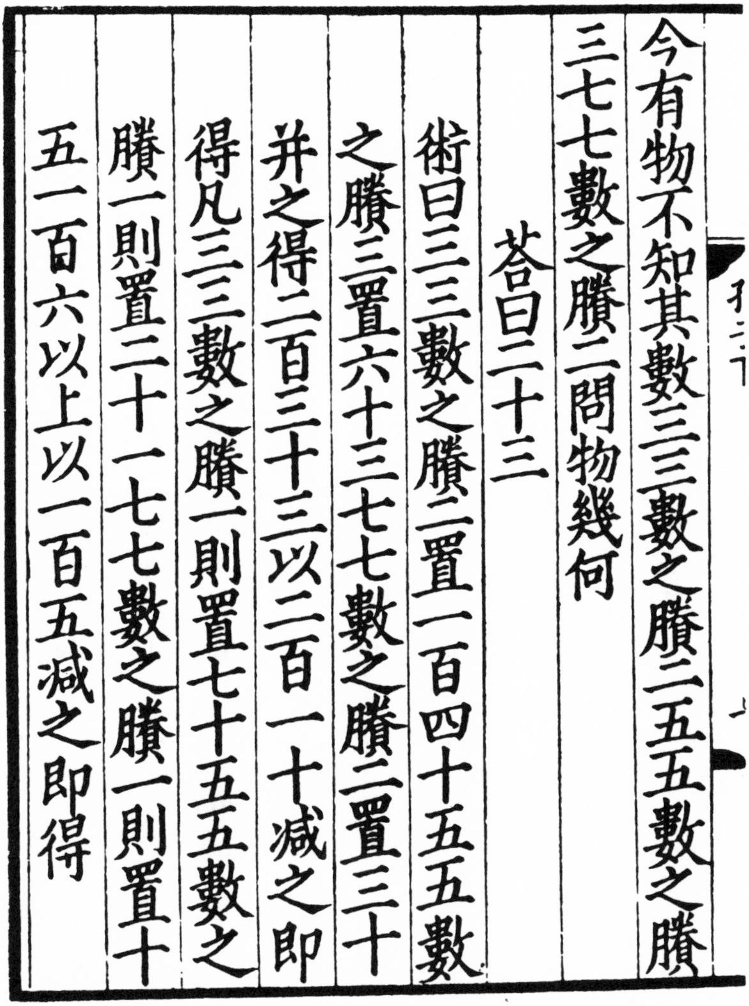

今有物不知其數三三數之賸二五五數之賸
三七七數之賸二問物幾何

荅曰二十三

術曰三三數之賸二置一百四十五五數
之賸三置六十三七七數之賸二置三十
并之得二百三十三以二百一十減之即
得凡三三數之賸一則置七十五五數之
賸一則置二十一七七數之賸一則置十
五一百六以上以一百五減之即得

*The famous* Sun Tzu *problem, the oldest instance of the remainder theorem. From the* Sun Tzu suan-ching

by 5, the remainder is 3; for which you retain three times 21, or 63, which you add to the above 35; you get 98; and from the division by 7, the remainder is 4, for which you retain four times 15, or 60; which you add to the above 98, and you get 158; from which you subtract 105, and the remainder is 53, which is the contrived number.

From this rule comes a pleasant game, namely if someone has learned this rule with you; if somebody else should say some number privately to him, then your companion, not interrogated, should silently divide the number for

189

himself by 3, by 5, and by 7 according to the above-mentioned rule; the remainders from each of these divisions, he says to you in order; and in this way you can know the number said to him in private.

4. Here is the same thing in the hands of Leonhard Euler (*Commentarii Academiae Scientiarum Petropolitanae,** 7 (1734/5)).

> A number is to be found that, when divided by $a$, $b$, $c$, $d$, $e$, which numbers I suppose to be relatively prime, leaves respectively the remainders $p$, $q$, $r$, $s$, $t$. For this problem the following numbers satisfy:
>
> $$Ap + Bq + Cr + Ds + Et + m \times abcde$$
>
> in which $A$ is a number that divided by $bcde$ has no remainder, by $a$, however, has the remainder 1; $B$ is a number that divided by $acde$ has no remainder, by $b$, however, has the remainder 1 . . . which numbers can consequently be found by the rule given for two divisors.

5. The Chinese Remainder Theorem from J. E. Shockley, *Introduction to Number Theory*, 1967.

**Theorem.**  Suppose that $m_1, m_2, \ldots, m_n$ are pairwise relatively prime. Let $M = m_1 \ldots m_n$. We define numbers $b_1$, $b_2, \ldots, b_n$ by choosing $y = b_j$ as a solution of

$$y \frac{M}{m_j} \equiv 1 \bmod(m_j) \qquad (j = 1, 2, \ldots, n.).$$

Then, the general solution of the system

$$x \equiv a_1 (\bmod m_1)$$
$$x \equiv a_2 (\bmod m_2)$$
$$\cdot$$
$$\cdot$$
$$\cdot$$
$$x \equiv a_n (\bmod m_n)$$

is $x \equiv a_1 b_1 \dfrac{M}{m_1} + \ldots + a_n b_n \dfrac{M}{m_n} \pmod{M}$.

---

* Leningrad, in current parlance.

6. The Chinese Remainder Theorem in the hands of a contemporary computer scientist (R. E. Prather, 1976).

"If $n = p_1^{\alpha_1} p_2^{\alpha_2} \ldots p_r^{\alpha_r}$ is the decomposition of the integer $n$ into distinct prime powers: $p_i^{\alpha_i} = q_i$, then the cyclic group $Z_n$ has the product representation

$$Z_n \simeq Z_{q_1} \times Z_{q_2} \times \ldots \times Z_{q_r}."$$

7. Our last example comes from E. Weiss, *Algebraic Number Theory*, 1963.

Axiom IIb. If $S = \{P_1, \ldots, P_r\}$ is any finite subset of $\mathscr{S}$, then for any elements $a_1, \ldots, a_r \in F$ and any integers $m_1$, $m_2, \ldots, m_r$; there exists an element $a \in F$ such that

| | |
|---|---|
| $v_{P_i}(a - a_i) \geq m_i$ | $i = 1, \ldots, r$ |
| $v_{P_i}(a) \geq 0$ | $P \not\subseteq S, P \in \mathscr{S}.$ |

We are further invited to consider the "OAF$\{Q, \mathscr{S}\}$ as described in (4–1–2). Translating to the language of congruences, one observes that our axioms become rather familiar statements. In particular, the content of Axiom IIb is precisely that of the Chinese remainder theorem."

We shall now comment on these presentations.

1. Sun Tzu. The first thing that strikes us today is that the formulation is both specific and algorithmic. The author begins with the particular case of remainders 2, 3, 2, and works out the answer. Up to the words "From this subtract 210 and you have the result," the modern reader is up in the air as to what the general method is or whether we are confronted with an arithmetic rigamarole that is merely happenstance. He is encouraged, of course, by the word "Method" which promises more. The second part beginning "For each unity . . ." displays the general method of which the first part is a special instance and serves to clarify the first part completely.

The complete formulation is arithmetic. Mysteries remain. Where do the magic numbers 70, 21, 15 and 105 come from? What if we don't count them by threes, fives and sevens, but count them by threes, fours, and fives, or indeed by any set of integers. What then?

2. (No comment.)

3. Fibonacci's formulation has not moved far from that of Sun Tzu. The formulation is still arithmetic, and hinged to a specific set of divisors. The recreational aspect of the problem is a pleasant touch and has persisted for years in the recreational literature. Notice that one of the implications of the recreational approach (which is not so clear in the Sun Tzu formulation) is that whatever the remainders are, the magician is able to come up with an answer. The question of the uniqueness of the answer is not dealt with in either the Fibonacci or the Sun Tzu, but the implication seems to be that the possibility of nonuniqueness is to be ignored.

4. Five hundred years later, Euler's presentation is from a different symbolic world. The specific integers are now replaced by the general indeterminate or arbitrary quantities $a, b, \ldots, t$. Modern algebraic notation is firmly established and all the solutions to the problem are exhibited. Though written out for five remainders, the method is, by implication, perfectly general. The last statement refers to a previous result for the solution of the special congruences needed for the determination of the constants $A, \ldots, E$.

5. Shockley's presentation might be called an updated version of what is in Gauss' *Disquisitiones Arithmeticae* (1801). The Gaussian notation for congruences is fully established and lends a degree of elegance previously unknown. The final representation and the auxiliary problems to be solved to arrive at the "Ta-yen" constants are set up in an economical way. This formulation may be considered a high point within the framework of classical algebraized theory of numbers.

6. The difference in tone between this and the previous formulation is intense. Here we have a complete rewriting of the theorem under the influence of the structuralist conception of mathematics.

The finite set of integers $0, 1, 2, \ldots, n - 1$ considered under addition mod $n$ (i.e., ignoring multiples of $n$) constitutes the so-called additive cyclic group, designated by $Z_n$. (See also Chapter 5, Group Theory.) Thus, $Z_4$, for example, is the set of integers 0, 1, 2, 3 with the addition table

| + | 0 | 1 | 2 | 3 |
|---|---|---|---|---|
| 0 | 0 | 1 | 2 | 3 |
| 1 | 1 | 2 | 3 | 0 |
| 2 | 2 | 3 | 0 | 1 |
| 3 | 3 | 0 | 1 | 2 |

The direct product of two such groups $Z_4 \times Z_3$, for example, consists of pairs of integers $(a, b)$ wherein the first integer is a number of $Z_4$ and the second is a number of $Z_3$. Thus, the elements of $Z_4 \times Z_3$ are the twelve pairs

| (0,0) | (1,0) | (2,0) | (3,0) |
|-------|-------|-------|-------|
| (0,1) | (1,1) | (2,1) | (3,1) |
| (0,2) | (1,2) | (2,2) | (3,2) |

Addition of elements of $Z_4 \times Z_3$ is defined as addition of the corresponding integers, the first carried out modulo 4 and the second modulo 3. Thus, for example,

$$(2, 2) + (3, 2) = ((2 + 3)\bmod 4, (2 + 2)\bmod 3) = (1, 1).$$

Now, each pair $(a, b)$ can be identified with the unique number of 0, 1, . . ., 11 which on division by 4 yields $a$ and on division by 3 yields $b$. Under this identification the above table is

| 0 | 9 | 6 | 3 |
|---|---|----|----|
| 4 | 1 | 10 | 7 |
| 8 | 5 | 2 | 11 |

Thus, $(1, 1) = (2, 2) + (3, 2) \rightarrow 1 = (2 + 11) \bmod 12$, which is a particular instance of the isomorphism of the two tables under their individual definitions of $+$.

The present formulation of the remainder theorem asserts that this scheme is true in the case of general $n$, provided we separate $n$ into its prime powers.

Note that this formulation of the remainder theorem simultaneously gives us less and more than the previous formulation. It emphasizes structure at the expense of the algorithm. It provides a complete analysis of modular addition ($Z_n$) in terms of simpler additions ($Z_{q_r}$). It bypasses the question of how the identification of $Z_n$ and $Z_{q_1} \times \ldots \times Z_{q_r}$ can be set up (though this identification occurs in the body of the proof) and it totally ignores the historically motivating question of how, given the remainders, can we expeditiously compute the number which gives rise to the remainders.

In a way, this is very strange in view of Prather's comment at the end of his exposition that the Chinese remainder theorem has proved useful in the design of fast arithmetic units for digital computers. One might think this would call for the knowledge of a concrete algorithm. But then it is true that computer science in its theoretical formulation is dominated by a spirit of abstraction which defers to no other branch of mathematics in its zealotry.

7. Here we are dealing with arithmetic, not over the integers, but over arbitrary fields. At this stage of generalization, the statement just given is such that it is probably not immediately comprehensible to the average professional mathematician. The meaning has receded from what is located in the common experience pool and the statement makes sense only to a very limited and specialized audience.

Even if we were, as the author of the statement suggests, to "translate to the language of congruences," we run into some difficulties. Let us try to translate, using the old dictionary trick of running a definition back to simpler definitions. What we fear is not that we run into a vicious circle of definitions—mathematics avoids this—but that we run into a general statement which is not backed up by any personal operational or intuitional experience.

Let us start with considering the OAF$\{Q, \mathscr{S}\}$. An OAF (ordinary arithmetic field), we are informed, is "a pair {F,

$\mathscr{S}$} where F is a field, and $\mathscr{S}$ is a non-empty collection of discrete prime divisors of F and such that the following axioms are satisfied: I . . . II . . . ." Now, a field is a notion which is part of the common vocabulary of every educated mathematician. A field, briefly, is any system of objects which can be added, subtracted, multiplied, and divided one by the other, according to the ordinary rules of arithmetic. A prime divisor of a field, on the other hand, is not such a notion. Referring back to an earlier part of the book, we learn that the prime divisors of a field F are the sets of equivalence classes of valuations on F, wherein two valuations are considered equivalent if they determine the same topology on F. Referring all the way back to page 1 of Weiss, we learn that a valuation on a field is a function from a field into the nonnegative real numbers which satisfies three axioms and which therefore acts as a generalization of the absolute value. Insofar as there is an extreme paucity of examples cited in the text, it rapidly becomes clear even to the professional, that illumination will occur only after a great deal of mulling and pondering, and after several hours of study, the simple phraseology of the classic Chinese remainder theorem may be as remote as ever.

The paucity of examples also leads the nonspecialist to wonder whether there are any really significant instances of the remainder theorem other than what occurs within the natural integers. If there are, what are they? Or is much of the theory vacuous posturing?

Professor John Wermer tells the story of how, when he was an undergraduate, he took a course in projective geometry from Oscar Zariski, one of the foremost figures in the field of algebraic geometry. Zariski's course was exceedingly general and Wermer, as a young student, was occasionally in need of clarification. "What would you get," he asked his teacher, "if you specialized the field F to the complex numbers?" Zariski answered: "Yes, just take F as the complex numbers."

### Further Readings. See Bibliography

M. J. Crowe; L. E. Dickson; U. Libbrecht; R. E. Prather; W. W. Rouse Ball; J. E. Shockley; E. Weiss; R. L. Wilder [1968]

# Mathematics
# as Enigma

T HE STRAIGHT AND NARROW course of formal computation often leads directly to the stone wall of enigma. Consider the case of Cardano's formula for the solution of the cubic equation. This formula was published in 1545 by Girolamo Cardano in his *Ars Magna* and gave the solution to the cubic equation

$$x^3 + mx = n.$$

Cardano's formula was probably the first great achievement in algebra since the Babylonians, and in its day was considered to be a great breakthrough.

We give Cardano's development in a slightly different arrangement.

Suppose that $t$ and $u$ are two numbers such that

$$\begin{cases} t - u = n \\ tu = (m/3)^3 \end{cases} \qquad (*)$$

*Girolamo Cardano*
*1501–1576*

hold simultaneously. Now define $x$ to be the number

$$x = \sqrt[3]{t} - \sqrt[3]{u} = t^{1/3} - u^{1/3}. \qquad (**)$$

Then, cubing both sides of the equation, we obtain

$$\begin{aligned} x^3 &= (t^{1/3} - u^{1/3})^3 = t - 3t^{2/3}u^{1/3} + 3t^{1/3}u^{2/3} - u \\ &= (t - u) - (3t^{1/3}u^{1/3})(t^{1/3} - u^{1/3}) \\ &= n - mx. \end{aligned}$$

Hence (*) and (**) imply that $x$ satisfies the cubic

$$x^3 + mx = n. \qquad (***)$$

Now we know how to solve (*) for $t$ and $u$ in terms of $n$ and $m$. For, $u = t - n$. Hence $t(t - n) = (m/3)^3$. This leads to the quadratic equation $t^2 - nt - (m/3)^3 = 0$. By the quadratic formula, this equation has the solution

$$t = \frac{n + \sqrt{n^2 + 4(m/3)^3}}{2} = \frac{n}{2} + \sqrt{(n/2)^2 + (m/3)^3}$$

Hence,

$$u = \frac{-n}{2} + \sqrt{(n/2)^2 + (m/3)^3}.$$

Substituting this information in (**), one obtains

$$x = \left(\frac{n}{2} + \sqrt{(n/2)^2 + (m/3)^3}\right)^{1/3} \qquad (* * * *)$$
$$- \left(-\frac{n}{2} + \sqrt{(n/2)^2 + (m/3)^3}\right)^{1/3}.$$

This is the famous cubic formula which Cardano is reputed to have weaseled out of his fellow mathematician Tartaglia under an oath of secrecy.

Let's try it out. Take the equation $x^3 + x = 2$ which obviously has $x = 1$ as a solution. In this equation $m = 1$ and $n = 2$, so that Cardano's formula yields

$$x = (1 + \sqrt{1 + \tfrac{1}{27}})^{1/3} - (-1 + \sqrt{1 + \tfrac{1}{27}})^{1/3}.$$

A hand-held computer in no time figures this out as

$$x = 1.263762616 - .2637626158$$

which is $x = 1$ to within $2 \times 10^{-10}$. Pretty good.

Made bold by this success, let's try it again. Take the equation $x^3 - 15x = 4$ which obviously has the solution $x = 4$. Cardano's formula yields

$$x = (2 + \sqrt{-121})^{1/3} - (-2 + \sqrt{-121})^{1/3}.$$

Hmmm! What's going on here?

We have to give a "meaning" to $\sqrt{-121}$. In particular, we have to explain how to add $\sqrt{-121}$ to a real number (2 or $-2$ in this case) and then how to take the cube root of the resulting sum. We can't go to the hand-held computer so readily. Now put yourself in Cardano's shoes. The year is 1545. Square roots of negative numbers have no legitimacy; the theory of complex numbers is nonexistent. How to interpret these meaningless symbols?

Here is incompleteness and enigma. The internal necessities of mathematics have set up pressures for explication. We are curious. We need to understand. Our methodology has led us to a new problem. It would be almost three centuries before an adequate theory would be available to interpret properly and to legitimize this work.

The solution to the mystery was ultimately given around 1800 by interpreting complex numbers as points in a coordinate plane, whose horizontal axis is the real axis and whose vertical axis is the "imaginary" axis or i-axis.

Once we conceive of the real line as embedded in a plane of complex numbers, we have entered a whole new domain of mathematics. All our old knowledge of real algebra and analysis becomes enlarged and enriched when reinterpreted in the complex domain. In addition, we immediately see countless new problems and questions which could not even have been raised in the context of real numbers alone.

In Cardano's formula, the unwitting algebraist comes to a window through which he sees a tantalizing and bewildering glimpse of a still undiscovered country.

### Further Readings. See Bibliography

E. Borel; R. L. Wilder [1974]

# Unity within Diversity

UNIFICATION, the establishment of a relationship between seemingly diverse objects, is at once one of the great motivating forces and one of the great sources of aesthetic satisfaction in mathematics. It is beautifully illustrated by the formula of Euler which unifies the trigonometric functions with the "power" or "exponential" functions. The trigonometric ratios and the sequences of exponential growth are both of ancient origin. Recall the legend of the magician who was to be paid in grains of wheat—one on the first square of the chessboard, with the number of grains doubling with each square. No doubt the sequence 1, 2, 4, 8, 16, . . . is the oldest exponential sequence. Now what on earth do these ideas have to do with one another?

It would be a nice piece of mathematical history to trace the growth of these notions until they fuse. We would see the extension of the sine and cosine functions to periodic functions, the switch-over to $e^x$ as the basic exponential, where $e$ is the mysterious transcendental number 2.718281828459 . . . , the development of the theory of power series, the bold but entirely natural extension of the range of applicability of power series to admit complex variables, the derivation of the three expansions

$$\sin x = x - \frac{x^3}{3!} + \frac{x^5}{5!} + \ldots ,$$

$$\cos x = 1 - \frac{x^2}{2!} + \frac{x^4}{4!} - \ldots ,$$

$$e^x = 1 + x + \frac{x^2}{2!} + \frac{x^3}{3!} + \ldots ,$$

leading to the final unification, Euler's formula,

$$e^{ix} = \cos x + i \sin x, \text{ where } i = \sqrt{-1}.$$

Thus, the exponential emerges as trigonometry in disguise. Reciprocally, by solving backwards one has

$$\cos x = \tfrac{1}{2}(e^{ix} + e^{-ix}),$$

$$\sin x = \frac{1}{2i}(e^{ix} - e^{-ix}),$$

so that trigonometry is equally exponential algebra in disguise. The special case where $x = \pi = 3.14159 \ldots$ leads to

$$e^{\pi i} = \cos \pi + i \sin \pi = -1, \quad \text{or} \quad e^{\pi i} + 1 = 0.$$

There is an aura of mystery in this last equation, which links the five most important constants in the whole of analysis: 0, 1, $e$, $\pi$, and $i$.

199

Within the same story one moves forward from this mystic landing to Fourier analysis, periodogram analysis, Fourier analysis over groups, differential equations, and one comes rapidly to great theories, great technological applications, and always a sense of the actual and potential unities that lurk in the corners of the universe. (See Chapter 5, Fourier Analysis.)

# 5

# SELECTED TOPICS IN MATHEMATICS

# Selected Topics
# in Mathematics

THE HEART of the mathematical experience is, of course, mathematics itself. This is the material in the technical journals, monographs, and, if deemed sufficiently interesting and important, the material that is taught. While it is not the purpose of this book to teach any portion of mathematics in a systematic way, it would be a serious omission if we did not expound a number of individual topics. We have selected six.

The theory of simple finite groups is one of the most active and successful areas of current mathematical research. It is also an area that is worth discussing from a methodological point of view because of the unprecedented length and detail of its proofs.

The problem of twin primes exhibits the importance of experience and calculation in arriving at theoretical judgements.

Non-Euclidean geometry represents one of the major breakthroughs in mathematics and is a turning point in the history of ideas.

Non-Cantorian set theory has had a major bearing on the whole question of mathematical existence and reality, on the choice between Platonism and formalism. In recent years, its ideas have both been fruitful and influential.

Nonstandard analysis exhibits a striking application of modern mathematical logic to problems in analysis. By its rehabilitation of discarded ideas, it shows how completely inadequate it is to limit the history of mathematics to the history of what has been formalized and made rigorous. The unrigorous and the contradictory play important parts in this history.

Fourier analysis is a subject that is absolutely central to a good deal of modern pure and applied mathematics. Our discussion shows the genesis of its basic ideas and how it shaped the concepts surrounding the notions of functions, integrals, and infinite dimensional spaces.

Each of these sections starts out very gently, and becomes somewhat more technical as it goes on. The reader who finds the going heavy is encouraged to skip forward.

# Group Theory and the Classification of Finite Simple Groups

I T IS INTERESTING that the most famous problem of the century in the theory of groups can be stated from scratch in just a few lines, and that this statement, in principle, contains all the material one needs to work on the problem. Here are the lines.

*Laws or Axioms for Mathematical Groups*

1. A group is a set $G$ of elements which can be combined with one another to get other elements of $G$. The combination of two elements $a$ and $b$ in that order is indicated by $a \cdot b$. For every $a$ and $b$ in $G$, $a \cdot b$ is defined and is in $G$.

2. For all elements $a$, $b$, $c$ in $G$,

$$a \cdot (b \cdot c) = (a \cdot b) \cdot c$$

203

*Decoration based upon cyclic groups.*

3. There is an identity element $e$ in $G$ such that $e \cdot a = a \cdot e = a$ for all $a$ in $G$.

4. For each element $a$ in $G$, there is an inverse element $a^{-1}$ in $G$ such that $a \cdot a^{-1} = a^{-1} \cdot a = e$.

Order: the number of elements in $G$ is called the *order* of $G$. If the number of elements of $G$ is finite then $G$ is called a *finite* group.

Subgroup: A subset of $G$ which is itself a group under the rule of combination in $G$ is called a *subgroup* of $G$.

Normal subgroup: If $H$ is a subgroup of $G$, $H$ is called *normal* if for every $g$ in $G$, $ghg^{-1}$ is in $H$ for every $h$ in $H$.

Simple Group: A group $G$ is called *simple* if it has no normal subgroups other than the identity or $G$ itself.

Cyclic Group: A finite group is called *cyclic* if its elements can be arranged in such a way that in the group multiplication table, each row is the previous row moved to the left one space and wrapped around.

## The Most Famous Problem of the Century in Group Theory

Prove that every finite simple group is either cyclic or of even order.

Now what would a little man from Galaxy X-9 (or even a mathematically unsophisticated earthling) really understand of what is going on here even though we have asserted that this is a complete statement? How can these words be translated back to something palpable? Would he understand why great value was set on this problem? Or would he have first to be indoctrinated into the whole mathematical culture with its history, motivations, methodology, theorems, and value systems before this understanding could be achieved? As abstraction is piled upon abstraction, meaning recedes and becomes remote. Let us see the extent to which we can flesh out the bare bones of our minimal statement in a few additional paragraphs.

*Groups.* A "group" is an abstract mathematical structure, one of the simplest and the most pervasive in the whole of mathematics. The notion finds applications, for example, to the theory of equations, to number theory, to differential geometry, to crystallography, to atomic and particle physics. The latter applications are particularly interesting in view of the fact that in 1910 a board of experts including Oswald Veblen and Sir James Jeans, upon reviewing the mathematics curriculum at Princeton, concluded that group theory ought to be thrown out as useless. So much for the crystal ball of experts.

Groups started out life as the theory of permutations or "substitutions." Consider, for example, three different objects numbered 1, 2, 3. These three objects may be permuted in six different ways:

$$e: 1\ 2\ 3 \to 1\ 2\ 3$$
$$a: 1\ 2\ 3 \to 1\ 3\ 2$$
$$b: 1\ 2\ 3 \to 2\ 1\ 3$$
$$c: 1\ 2\ 3 \to 3\ 2\ 1$$
$$p: 1\ 2\ 3 \to 2\ 3\ 1$$
$$q: 1\ 2\ 3 \to 3\ 1\ 2$$

Suppose that the operations of transforming the symbols 1, 2, 3 into those occurring on the right-hand side are designated by the letters placed at the left. Thus $e$, $a$, $b$, $c$, $p$, $q$ represent the six permutations of three objects. Suppose, next, that these permutations are combined or "multiplied" in an obvious way. We designate this operation by $\cdot$ We mean by this that $a \cdot b$ is the permutation which results from carrying out $a$, then $b$. To identify $a \cdot b$ as a permutation, we may compute

$$
\begin{array}{ccc}
a & b & \\
1 \to & 1 \to & 2 \\
2 \to & 3 \to & 3 \\
3 \to & 2 \to & 1
\end{array}
$$

Thus, $a \cdot b$ sends 1, 2, 3 into 2, 3, 1 which is precisely the permutation designated by $p$. We may therefore write $a \cdot b = p$.

All of this combinatorial information may be summed up neatly in the adjoining multiplication table.

|   | $e$ | $p$ | $q$ | $a$ | $b$ | $c$ |
|---|---|---|---|---|---|---|
| $e$ | $e$ | $p$ | $q$ | $a$ | $b$ | $c$ |
| $p$ | $p$ | $q$ | $e$ | $c$ | $a$ | $b$ |
| $q$ | $q$ | $e$ | $p$ | $b$ | $c$ | $a$ |
| $a$ | $a$ | $b$ | $c$ | $e$ | $p$ | $q$ |
| $b$ | $b$ | $c$ | $a$ | $q$ | $e$ | $p$ |
| $c$ | $c$ | $a$ | $b$ | $p$ | $q$ | $e$ |

The mathematical structure consisting of the elements $e$, $a$, $b$, $c$, $p$, $q$, combined by the operation $\cdot$ as indicated in the above table, may now be seen (by a systematic verification, if need be) to satisfy the four axioms and therefore constitute a group. The axioms, of course, were extracted and refined over several generations from many examples as the ones which were appropriate for the unification of the examples into a general theory.

Within a group, a certain amount of elementary algebra may be performed. Thus, in any group, the equation $x \cdot a = b$ has a unique solution $x = b \cdot a^{-1}$. The equation $ca = da$ implies $c = d$, etc.

*Order.* In the example just given, the number of elements is six. Finite group theory is confined to the study of those groups with a finite number of elements (finite order). There are groups with an infinite number of elements and the mathematical issues there are rather different.

It should be clear that the symbols for the group elements can be quite arbitrary and that the group multiplication table may be rewritten in permuted order. Two groups are *isomorphic* (or essentially identical) if they have the same multiplication table. Below is a table which gives the numbers of different groups of different orders.

| Order | Number of Groups |
|-------|------------------|
| 1 | 1 |
| 2 | 1 |
| 3 | 1 |
| 4 | 2 |
| 5 | 1 |
| 6 | 2 |
| 7 | 1 |
| 8 | 5 |
| 9 | 2 |
| 10 | 2 |
| 11 | 1 |
| 12 | 5 |

There is at the time of writing no known systematic way of generating all the groups of a given order.

Each particular group is a particular mathematical object with its own peculiarities, and people who work in group theory get quite chummy with them and give them special names and symbols. Thus, e.g., one has the dihedral group of order 4, the alternating group of order 12, etc.

*Subgroup.* The set of elements $e$, $p$, $q$, extracted from the multiplication table given above, has its own simple multiplication table:

*Joseph Louis Lagrange*
*1736–1813*

|   | $e$ | $p$ | $q$ |
|---|-----|-----|-----|
| $e$ | $e$ | $p$ | $q$ |
| $p$ | $p$ | $q$ | $e$ |
| $q$ | $q$ | $e$ | $p$ |

This system is easily found to verify the four axioms and therefore constitutes a subgroup of the original group. By a famous theorem of Lagrange one knows that the order of a subgroup always divides the order of the group itself. A consequence of this is that a group of prime order has no subgroups (other than ($e$) itself).

*Cyclic Group.* Start with the identity element $e$ and other elements in arbitrary order: $e$, $a$, $b$, . . . , $c$. Make a multi-

207

plication table by moving the elements successively to the left with wraparound.

|   | *e* | *a* | *b* | *c* |
|---|-----|-----|-----|-----|
| *e* | *e* | *a* | *b* | *c* |
| *a* | *a* | *b* | *c* | *e* |
| *b* | *b* | *c* | *e* | *a* |
| *c* | *c* | *e* | *a* | *b* |

This is a group, and is known as a cyclic group. The above table shows the cyclic group of order 4. A similar pattern goes through for any order. A consequence of this is that there is always at least one group of a given order, for there is always the cyclic group.

A cyclic group of order $n$ can be conceptualized geometrically as the rotation of a plane through multiples of the angle $\frac{1}{n}$ (360°). It may be conceptualized in terms of classical algebra as the set of the $n$ complex numbers $w^k = \left(\cos\frac{2\pi k}{n} + i\sin\frac{2\pi k}{n}\right)$, $k = 0, 1, \ldots, n - 1$, $i = \sqrt{-1}$, under ordinary (complex) multiplication.

## Normal Subgroups; Simple Groups

The theory of finite groups is analogous to number theory in the following sense: just as every positive integer has a unique factorization into a product of primes, so every finite group can be "factored" in a certain sense; it can be represented as a "product" of a normal subgroup and a "quotient group." In this way, an arbitrary finite group can be built up out of "simple groups"—groups that have no normal subgroups except the trivial ones (the whole group itself, or the single identity element). These "simple" groups are analogous to prime numbers which have no factors except the trivial ones—the number 1 and the original number itself.

So the study of simple groups plays the same central role in finite group theory that the study of primes plays in

208

number theory. The main goal of finite group theory is to give a complete classification of all the "simple groups."

A major breakthrough occurred in 1963 when Walter Feit and John Thompson proved that every simple group is either cyclic or has an even number of elements. This had been conjectured by Burnside many years earlier. Following the inspiration of the Feit-Thompson success, a tremendous surge of new activity erupted in finite group theory. Today specialists in this area believe they are within a stone's throw of a complete classification of the simple groups.

**Further Readings. See Bibliography**

D. Gorenstein.

# The Prime Number Theorem

THE THEORY of numbers is simultaneously one of the most elementary branches of mathematics in that it deals, essentially, with the arithmetic properties of the integers 1, 2, 3, . . . and one of the most difficult branches insofar as it is laden with difficult problems and difficult techniques.

Among the advanced topics in theory of numbers, three may be selected as particularly noteworthy: the theory of partitions, Fermat's "Last Theorem," and the prime number theorem. The theory of partitions concerns itself with the number of ways in which a number may be broken up into smaller numbers. Thus, including the "null" partition, two may be broken up as 2 or 1 + 1. Three may be broken up as 3, 2 + 1, 1 + 1 + 1, four may be broken up as 4, 3 + 1, 2 + 2, 2 + 1 + 1, 1 + 1 + 1 + 1. The number of ways that a given number may be broken up is far from a simple matter, and has been the object of study since the mid-seventeen hundreds. The reader might like to experi-

*Pierre de Fermat*
*1601–1665*

ment and see whether he can systematize the process and verify that the number 10 can be broken up in 42 different ways.

Fermat's "Last Theorem" asserts that if $n > 2$, the equation $x^n + y^n = z^n$ cannot be solved in integers $x$, $y$, $z$, with $xyz \neq 0$. This theorem has been proved (1979) for all $n < 30,000$, but the general theorem is remarkably elusive. Due to the peculiar history of this problem, it has attracted more than its share of mathematical crackpottery and most mathematicians ardently wish that the problem would be settled.

The prime number theorem, which is the subject of this section, has great attractions and mystery and is related to some of the central objects of mathematical analysis. It is also related to what is probably the most famous of the unsolved mathematical problems—the so-called Riemann Hypothesis. It is one of the finest examples of the extraction of order from chaos in the whole of mathematics.

Soon after a child learns to multiply and divide, he notices that some numbers are special. When a number is factored, it is decomposed into its basic constituents—its prime factors. Thus, $6 = 2 \times 3$, $28 = 2 \times 2 \times 7$, $270 = 2 \times 3 \times 3 \times 3 \times 5$ and these decompositions cannot be carried further. The numbers $2, 3, 5, 7, \ldots$ are the prime numbers, numbers that cannot themselves be split into further multiplications. Among the integers, the prime numbers play a role that is analogous to the elements of chemistry.

Let us make a list of the first few prime numbers:

| 2 | 3 | 5 | 7 | 11 | 13 | 17 | 19 | 23 | 29 |
|---|---|---|---|----|----|----|----|----|----|
| 31 | 37 | 41 | 43 | 47 | 53 | 59 | 61 | 67 | 71 |
| 73 | 79 | 83 | 89 | 97 | 101 | 103 | 107 | 109 | 113 . . . |

This list never ends. Euclid already had proved that there are an infinite number of primes. Euclid's proof is easy and elegant and we will give it.

Suppose we have a complete list of all the prime numbers up to a certain prime $p_m$. Consider the integer $N = (2 \cdot 3 \cdot 5 \cdot \cdot \cdot p_m) + 1$, formed by adding 1 to the product

210

Table of the First 2500 Prime Numbers

| n | 0 | 1 | 2 | 3 | 4 | 5 | 6 | 7 | 8 | 9 | 10 | 11 | 12 | 13 | 14 | 15 | 16 | 17 | 18 | 19 | 20 | 21 | 22 | 23 | 24 |
|---|---|---|---|---|---|---|---|---|---|---|---|---|---|---|---|---|---|---|---|---|---|---|---|---|---|
| 1 | 2 | 547 | 1229 | 1993 | 2749 | 3581 | 4421 | 5281 | 6143 | 7001 | 7927 | 8837 | 9739 | 10663 | 11677 | 12569 | 13513 | 14533 | 15413 | 16411 | 17393 | 18329 | 19427 | 20359 | 21391 |
| 2 | 3 | 557 | 1231 | 1997 | 2753 | 3583 | 4423 | 5297 | 6151 | 7013 | 7933 | 8839 | 9743 | 10667 | 11681 | 12577 | 13523 | 14537 | 15427 | 16417 | 17401 | 18341 | 19429 | 20369 | 21397 |
| 3 | 5 | 563 | 1237 | 1999 | 2767 | 3593 | 4441 | 5303 | 6163 | 7019 | 7937 | 8849 | 9749 | 10687 | 11689 | 12583 | 13537 | 14543 | 15439 | 16421 | 17417 | 18353 | 19433 | 20389 | 21401 |
| 4 | 7 | 569 | 1249 | 2003 | 2777 | 3607 | 4447 | 5309 | 6173 | 7027 | 7949 | 8861 | 9767 | 10691 | 11699 | 12589 | 13553 | 14549 | 15443 | 16427 | 17419 | 18367 | 19441 | 20393 | 21407 |
| 5 | 11 | 571 | 1259 | 2011 | 2789 | 3613 | 4451 | 5323 | 6197 | 7039 | 7951 | 8863 | 9769 | 10709 | 11701 | 12601 | 13567 | 14551 | 15451 | 16433 | 17431 | 18371 | 19447 | 20399 | 21419 |
| 6 | 13 | 577 | 1277 | 2017 | 2791 | 3617 | 4457 | 5333 | 6199 | 7043 | 7963 | 8867 | 9781 | 10711 | 11717 | 12611 | 13577 | 14557 | 15461 | 16447 | 17443 | 18379 | 19457 | 20407 | 21433 |
| 7 | 17 | 587 | 1279 | 2027 | 2797 | 3623 | 4463 | 5347 | 6203 | 7057 | 7993 | 8887 | 9787 | 10723 | 11719 | 12613 | 13591 | 14561 | 15467 | 16451 | 17449 | 18397 | 19463 | 20411 | 21467 |
| 8 | 19 | 593 | 1283 | 2029 | 2801 | 3631 | 4481 | 5351 | 6211 | 7069 | 8009 | 8893 | 9791 | 10729 | 11731 | 12619 | 13597 | 14563 | 15473 | 16453 | 17467 | 18401 | 19469 | 20431 | 21481 |
| 9 | 23 | 599 | 1289 | 2039 | 2803 | 3637 | 4483 | 5381 | 6217 | 7079 | 8011 | 8923 | 9803 | 10733 | 11743 | 12637 | 13613 | 14591 | 15493 | 16477 | 17471 | 18413 | 19471 | 20441 | 21487 |
| 10 | 29 | 601 | 1291 | 2053 | 2819 | 3643 | 4493 | 5387 | 6221 | 7103 | 8017 | 8929 | 9811 | 10739 | 11777 | 12641 | 13619 | 14593 | 15497 | 16481 | 17477 | 18427 | 19477 | 20443 | 21491 |
| 11 | 31 | 607 | 1297 | 2063 | 2833 | 3659 | 4507 | 5393 | 6229 | 7109 | 8039 | 8933 | 9817 | 10753 | 11779 | 12647 | 13627 | 14621 | 15511 | 16487 | 17483 | 18433 | 19483 | 20477 | 21493 |
| 12 | 37 | 613 | 1301 | 2069 | 2837 | 3671 | 4513 | 5399 | 6247 | 7121 | 8053 | 8941 | 9829 | 10771 | 11783 | 12653 | 13633 | 14627 | 15527 | 16493 | 17489 | 18439 | 19489 | 20479 | 21499 |
| 13 | 41 | 617 | 1303 | 2081 | 2843 | 3673 | 4517 | 5407 | 6257 | 7127 | 8059 | 8951 | 9833 | 10781 | 11789 | 12659 | 13649 | 14629 | 15541 | 16519 | 17491 | 18443 | 19501 | 20483 | 21517 |
| 14 | 43 | 619 | 1307 | 2083 | 2851 | 3677 | 4519 | 5413 | 6263 | 7129 | 8069 | 8963 | 9839 | 10789 | 11801 | 12671 | 13669 | 14633 | 15551 | 16529 | 17497 | 18451 | 19507 | 20507 | 21521 |
| 15 | 47 | 631 | 1319 | 2087 | 2857 | 3691 | 4523 | 5417 | 6269 | 7151 | 8081 | 8969 | 9851 | 10799 | 11807 | 12689 | 13679 | 14639 | 15559 | 16547 | 17509 | 18457 | 19531 | 20509 | 21523 |
| 16 | 53 | 641 | 1321 | 2089 | 2861 | 3697 | 4547 | 5419 | 6271 | 7159 | 8087 | 8971 | 9857 | 10831 | 11813 | 12697 | 13681 | 14653 | 15569 | 16553 | 17519 | 18461 | 19541 | 20521 | 21529 |
| 17 | 59 | 643 | 1327 | 2099 | 2879 | 3701 | 4549 | 5431 | 6277 | 7177 | 8089 | 8999 | 9859 | 10837 | 11821 | 12703 | 13687 | 14657 | 15581 | 16561 | 17539 | 18481 | 19543 | 20533 | 21557 |
| 18 | 61 | 647 | 1361 | 2111 | 2887 | 3709 | 4561 | 5437 | 6287 | 7187 | 8093 | 9001 | 9871 | 10847 | 11827 | 12713 | 13691 | 14669 | 15583 | 16567 | 17551 | 18493 | 19553 | 20543 | 21559 |
| 19 | 67 | 653 | 1367 | 2113 | 2897 | 3719 | 4567 | 5441 | 6299 | 7193 | 8101 | 9007 | 9883 | 10853 | 11831 | 12721 | 13693 | 14683 | 15601 | 16573 | 17569 | 18503 | 19559 | 20549 | 21563 |
| 20 | 71 | 659 | 1373 | 2129 | 2903 | 3727 | 4583 | 5443 | 6301 | 7207 | 8111 | 9011 | 9887 | 10859 | 11833 | 12739 | 13697 | 14699 | 15607 | 16603 | 17573 | 18517 | 19571 | 20551 | 21569 |
| 21 | 73 | 661 | 1381 | 2131 | 2909 | 3733 | 4591 | 5449 | 6311 | 7211 | 8117 | 9013 | 9901 | 10861 | 11839 | 12743 | 13709 | 14713 | 15619 | 16607 | 17579 | 18521 | 19577 | 20563 | 21569 |
| 22 | 79 | 673 | 1399 | 2137 | 2917 | 3739 | 4597 | 5471 | 6317 | 7213 | 8123 | 9029 | 9907 | 10867 | 11863 | 12757 | 13711 | 14717 | 15629 | 16619 | 17581 | 18523 | 19583 | 20593 | 21577 |
| 23 | 83 | 677 | 1409 | 2141 | 2927 | 3761 | 4603 | 5477 | 6323 | 7219 | 8147 | 9041 | 9923 | 10883 | 11867 | 12763 | 13721 | 14723 | 15641 | 16631 | 17597 | 18539 | 19597 | 20599 | 21587 |
| 24 | 89 | 683 | 1423 | 2143 | 2939 | 3767 | 4621 | 5479 | 6329 | 7229 | 8161 | 9043 | 9929 | 10889 | 11887 | 12781 | 13723 | 14731 | 15643 | 16633 | 17599 | 18541 | 19603 | 20611 | 21589 |
| 25 | 97 | 691 | 1427 | 2153 | 2953 | 3769 | 4637 | 5483 | 6337 | 7237 | 8167 | 9049 | 9931 | 10891 | 11897 | 12791 | 13729 | 14737 | 15647 | 16649 | 17609 | 18553 | 19609 | 20627 | 21599 |
| 26 | 101 | 701 | 1429 | 2161 | 2957 | 3779 | 4639 | 5501 | 6343 | 7243 | 8171 | 9059 | 9941 | 10903 | 11903 | 12799 | 13751 | 14741 | 15649 | 16651 | 17623 | 18583 | 19661 | 20639 | 21601 |
| 27 | 103 | 709 | 1433 | 2179 | 2963 | 3793 | 4643 | 5503 | 6353 | 7247 | 8179 | 9067 | 9949 | 10909 | 11909 | 12809 | 13757 | 14747 | 15661 | 16657 | 17627 | 18587 | 19681 | 20641 | 21611 |
| 28 | 107 | 719 | 1439 | 2203 | 2969 | 3797 | 4649 | 5507 | 6359 | 7253 | 8191 | 9091 | 9967 | 10937 | 11923 | 12821 | 13759 | 14753 | 15667 | 16661 | 17657 | 18593 | 19687 | 20663 | 21613 |
| 29 | 109 | 727 | 1447 | 2207 | 2971 | 3803 | 4651 | 5519 | 6361 | 7283 | 8209 | 9103 | 9973 | 10939 | 11927 | 12823 | 13763 | 14759 | 15671 | 16673 | 17659 | 18617 | 19697 | 20681 | 21617 |
| 30 | 113 | 733 | 1451 | 2213 | 2999 | 3821 | 4657 | 5521 | 6367 | 7297 | 8219 | 9109 | 10007 | 10949 | 11933 | 12829 | 13781 | 14767 | 15679 | 16691 | 17669 | 18637 | 19699 | 20693 | 21647 |
| 31 | 127 | 739 | 1453 | 2221 | 3001 | 3823 | 4663 | 5527 | 6373 | 7307 | 8221 | 9127 | 10009 | 10957 | 11939 | 12841 | 13789 | 14771 | 15683 | 16699 | 17681 | 18661 | 19709 | 20707 | 21649 |
| 32 | 131 | 743 | 1459 | 2237 | 3011 | 3833 | 4673 | 5531 | 6379 | 7309 | 8231 | 9133 | 10037 | 10973 | 11941 | 12853 | 13799 | 14779 | 15727 | 16703 | 17683 | 18671 | 19717 | 20717 | 21661 |
| 33 | 137 | 751 | 1471 | 2239 | 3019 | 3847 | 4679 | 5557 | 6389 | 7321 | 8233 | 9137 | 10039 | 10979 | 11953 | 12889 | 13807 | 14783 | 15731 | 16707 | 17707 | 18679 | 19727 | 20719 | 21673 |
| 34 | 139 | 757 | 1481 | 2243 | 3023 | 3851 | 4691 | 5563 | 6397 | 7331 | 8237 | 9151 | 10061 | 10987 | 11959 | 12893 | 13829 | 14797 | 15733 | 16729 | 17713 | 18691 | 19739 | 20731 | 21683 |
| 35 | 149 | 761 | 1483 | 2251 | 3037 | 3853 | 4703 | 5569 | 6421 | 7333 | 8243 | 9157 | 10067 | 10993 | 11969 | 12899 | 13831 | 14813 | 15737 | 16741 | 17729 | 18701 | 19751 | 20743 | 21701 |
| 36 | 151 | 769 | 1487 | 2267 | 3041 | 3863 | 4721 | 5573 | 6427 | 7349 | 8263 | 9161 | 10069 | 11003 | 11971 | 12907 | 13841 | 14821 | 15739 | 16747 | 17737 | 18713 | 19753 | 20747 | 21713 |
| 37 | 157 | 773 | 1489 | 2269 | 3049 | 3877 | 4723 | 5581 | 6449 | 7351 | 8269 | 9173 | 10079 | 11027 | 11981 | 12911 | 13859 | 14827 | 15749 | 16759 | 17747 | 18719 | 19759 | 20749 | 21727 |
| 38 | 163 | 787 | 1493 | 2273 | 3061 | 3881 | 4729 | 5591 | 6451 | 7369 | 8273 | 9181 | 10091 | 11047 | 11987 | 12917 | 13873 | 14831 | 15761 | 16763 | 17749 | 18731 | 19763 | 20753 | 21737 |
| 39 | 167 | 797 | 1499 | 2281 | 3067 | 3889 | 4733 | 5623 | 6469 | 7393 | 8287 | 9187 | 10093 | 11057 | 12007 | 12919 | 13877 | 14843 | 15767 | 16787 | 17761 | 18743 | 19777 | 20759 | 21739 |
| 40 | 173 | 809 | 1511 | 2287 | 3079 | 3907 | 4751 | 5639 | 6473 | 7411 | 8291 | 9199 | 10099 | 11059 | 12011 | 12923 | 13879 | 14851 | 15773 | 16811 | 17783 | 18749 | 19793 | 20771 | 21751 |
| 41 | 179 | 811 | 1523 | 2293 | 3083 | 3911 | 4759 | 5641 | 6481 | 7417 | 8293 | 9203 | 10103 | 11069 | 12037 | 12941 | 13883 | 14867 | 15787 | 16823 | 17789 | 18757 | 19801 | 20773 | 21757 |
| 42 | 181 | 821 | 1531 | 2297 | 3089 | 3917 | 4783 | 5647 | 6491 | 7433 | 8297 | 9209 | 10111 | 11071 | 12041 | 12953 | 13901 | 14869 | 15791 | 16829 | 17791 | 18773 | 19813 | 20789 | 21767 |
| 43 | 191 | 823 | 1543 | 2309 | 3109 | 3919 | 4787 | 5651 | 6521 | 7451 | 8311 | 9221 | 10133 | 11083 | 12043 | 12959 | 13903 | 14879 | 15797 | 16831 | 17807 | 18787 | 19819 | 20807 | 21773 |
| 44 | 193 | 827 | 1549 | 2311 | 3119 | 3923 | 4789 | 5653 | 6529 | 7457 | 8317 | 9227 | 10139 | 11087 | 12049 | 12967 | 13907 | 14887 | 15803 | 16843 | 17827 | 18793 | 19841 | 20809 | 21787 |
| 45 | 197 | 829 | 1553 | 2333 | 3121 | 3929 | 4793 | 5657 | 6547 | 7459 | 8329 | 9239 | 10141 | 11093 | 12071 | 12973 | 13913 | 14891 | 15809 | 16871 | 17837 | 18797 | 19843 | 20849 | 21799 |
| 46 | 199 | 839 | 1559 | 2339 | 3137 | 3931 | 4799 | 5659 | 6551 | 7477 | 8353 | 9241 | 10151 | 11113 | 12073 | 12979 | 13921 | 14897 | 15817 | 16879 | 17839 | 18803 | 19853 | 20857 | 21803 |
| 47 | 211 | 853 | 1567 | 2341 | 3163 | 3943 | 4801 | 5669 | 6553 | 7481 | 8363 | 9257 | 10159 | 11117 | 12097 | 12983 | 13931 | 14923 | 15823 | 16883 | 17851 | 18839 | 19861 | 20873 | 21817 |
| 48 | 223 | 857 | 1571 | 2347 | 3167 | 3947 | 4813 | 5683 | 6563 | 7487 | 8369 | 9277 | 10163 | 11119 | 12101 | 13001 | 13933 | 14929 | 15859 | 16889 | 17863 | 18859 | 19867 | 20879 | 21821 |
| 49 | 227 | 859 | 1579 | 2351 | 3169 | 3967 | 4817 | 5689 | 6569 | 7489 | 8377 | 9281 | 10169 | 11131 | 12107 | 13003 | 13963 | 14939 | 15877 | 16901 | 17881 | 18869 | 19889 | 20887 | 21839 |
| 50 | 229 | 863 | 1583 | 2357 | 3181 | 3989 | 4831 | 5693 | 6571 | 7499 | 8387 | 9283 | 10177 | 11149 | 12109 | 13007 | 13967 | 14947 | 15881 | 16903 | 17891 | 18899 | 19891 | 20897 | 21841 |
| 51 | 233 | 877 | 1597 | 2371 | 3187 | 4001 | 4861 | 5701 | 6577 | 7507 | 8389 | 9293 | 10181 | 11159 | 12113 | 13009 | 13997 | 14951 | 15887 | 16927 | 17903 | 18911 | 19913 | 20899 | 21851 |
| 52 | 239 | 881 | 1601 | 2377 | 3191 | 4003 | 4871 | 5711 | 6581 | 7517 | 8419 | 9311 | 10193 | 11161 | 12119 | 13033 | 13999 | 14957 | 15889 | 16931 | 17909 | 18913 | 19919 | 20903 | 21859 |
| 53 | 241 | 883 | 1607 | 2381 | 3203 | 4007 | 4877 | 5717 | 6599 | 7523 | 8423 | 9319 | 10211 | 11171 | 12143 | 13037 | 14009 | 14969 | 15901 | 16937 | 17911 | 18917 | 19927 | 20921 | 21863 |
| 54 | 251 | 887 | 1609 | 2383 | 3209 | 4013 | 4889 | 5737 | 6607 | 7529 | 8429 | 9323 | 10223 | 11173 | 12149 | 13043 | 14011 | 14983 | 15907 | 16943 | 17921 | 18919 | 19937 | 20929 | 21871 |
| 55 | 257 | 907 | 1613 | 2389 | 3217 | 4019 | 4903 | 5741 | 6619 | 7537 | 8431 | 9337 | 10243 | 11177 | 12157 | 13049 | 14029 | 15013 | 15913 | 16963 | 17923 | 18947 | 19949 | 20939 | 21881 |
| 56 | 263 | 911 | 1619 | 2393 | 3221 | 4021 | 4909 | 5743 | 6637 | 7541 | 8443 | 9341 | 10247 | 11197 | 12161 | 13063 | 14033 | 15017 | 15919 | 16963 | 17929 | 18959 | 19961 | 20947 | 21893 |
| 57 | 269 | 919 | 1621 | 2399 | 3229 | 4027 | 4919 | 5749 | 6653 | 7547 | 8447 | 9343 | 10253 | 11213 | 12163 | 13093 | 14051 | 15031 | 15923 | 16979 | 17939 | 18973 | 19963 | 20959 | 21911 |
| 58 | 271 | 929 | 1627 | 2411 | 3251 | 4049 | 4931 | 5779 | 6659 | 7549 | 8461 | 9349 | 10259 | 11239 | 12197 | 13099 | 14057 | 15053 | 15937 | 16981 | 17957 | 18979 | 19973 | 20963 | 21929 |
| 59 | 277 | 937 | 1637 | 2417 | 3253 | 4051 | 4933 | 5783 | 6661 | 7559 | 8467 | 9371 | 10267 | 11243 | 12203 | 13103 | 14071 | 15061 | 15959 | 16987 | 17959 | 19001 | 19979 | 20981 | 21937 |
| 60 | 281 | 941 | 1657 | 2423 | 3257 | 4057 | 4937 | 5791 | 6673 | 7561 | 8501 | 9377 | 10271 | 11251 | 12211 | 13109 | 14081 | 15073 | 15971 | 16993 | 17971 | 19009 | 19991 | 20983 | 21943 |
| 61 | 283 | 947 | 1663 | 2437 | 3259 | 4073 | 4943 | 5801 | 6679 | 7573 | 8513 | 9391 | 10273 | 11257 | 12227 | 13121 | 14083 | 15077 | 15973 | 17011 | 17977 | 19013 | 19993 | 21001 | 21961 |
| 62 | 293 | 953 | 1667 | 2441 | 3271 | 4079 | 4951 | 5807 | 6689 | 7577 | 8521 | 9397 | 10289 | 11261 | 12239 | 13127 | 14087 | 15083 | 15991 | 17021 | 17981 | 19031 | 19997 | 21011 | 21977 |
| 63 | 307 | 967 | 1669 | 2447 | 3299 | 4091 | 4957 | 5813 | 6691 | 7583 | 8527 | 9403 | 10301 | 11273 | 12241 | 13147 | 14107 | 15091 | 16001 | 17027 | 17987 | 19037 | 20011 | 21013 | 21991 |
| 64 | 311 | 971 | 1693 | 2459 | 3301 | 4093 | 4967 | 5821 | 6701 | 7589 | 8537 | 9413 | 10303 | 11279 | 12251 | 13159 | 14143 | 15101 | 16007 | 17029 | 17989 | 19051 | 20021 | 21017 | 21997 |
| 65 | 313 | 977 | 1697 | 2467 | 3307 | 4099 | 4969 | 5827 | 6703 | 7591 | 8539 | 9419 | 10313 | 11287 | 12253 | 13163 | 14149 | 15107 | 16033 | 17033 | 18013 | 19069 | 20023 | 21019 | 22003 |
| 66 | 317 | 983 | 1699 | 2473 | 3313 | 4111 | 4973 | 5839 | 6709 | 7603 | 8543 | 9421 | 10321 | 11299 | 12263 | 13163 | 14153 | 15121 | 16057 | 17041 | 18041 | 19073 | 20029 | 21023 | 22013 |
| 67 | 331 | 991 | 1709 | 2477 | 3319 | 4127 | 4987 | 5843 | 6719 | 7607 | 8563 | 9431 | 10331 | 11311 | 12269 | 13171 | 14159 | 15131 | 16061 | 17047 | 18043 | 19079 | 20047 | 21031 | 22027 |
| 68 | 337 | 997 | 1721 | 2503 | 3323 | 4129 | 4993 | 5849 | 6733 | 7621 | 8573 | 9433 | 10333 | 11317 | 12277 | 13177 | 14173 | 15137 | 16063 | 17053 | 18047 | 19081 | 20051 | 21059 | 22031 |
| 69 | 347 | 1009 | 1723 | 2521 | 3329 | 4133 | 4999 | 5851 | 6737 | 7639 | 8581 | 9437 | 10337 | 11321 | 12281 | 13183 | 14177 | 15139 | 16067 | 17077 | 18049 | 19087 | 20063 | 21061 | 22037 |
| 70 | 349 | 1013 | 1733 | 2531 | 3331 | 4139 | 5003 | 5857 | 6761 | 7643 | 8597 | 9439 | 10343 | 11329 | 12289 | 13187 | 14197 | 15149 | 16069 | 17093 | 18059 | 19121 | 20071 | 21067 | 22039 |
| 71 | 353 | 1019 | 1741 | 2539 | 3343 | 4153 | 5009 | 5861 | 6763 | 7649 | 8599 | 9461 | 10357 | 11351 | 12301 | 13217 | 14207 | 15161 | 16073 | 17099 | 18061 | 19139 | 20089 | 21089 | 22051 |
| 72 | 359 | 1021 | 1747 | 2543 | 3347 | 4157 | 5011 | 5867 | 6779 | 7669 | 8609 | 9463 | 10369 | 11353 | 12323 | 13219 | 14221 | 15173 | 16087 | 17107 | 18077 | 19141 | 20101 | 21101 | 22063 |
| 73 | 367 | 1031 | 1753 | 2549 | 3359 | 4159 | 5021 | 5869 | 6781 | 7673 | 8623 | 9467 | 10391 | 11369 | 12329 | 13229 | 14243 | 15187 | 16091 | 17117 | 18089 | 19157 | 20107 | 21107 | 22067 |
| 74 | 373 | 1033 | 1759 | 2551 | 3361 | 4177 | 5023 | 5879 | 6791 | 7681 | 8627 | 9473 | 10399 | 11383 | 12343 | 13241 | 14249 | 15193 | 16097 | 17123 | 18097 | 19163 | 20113 | 21121 | 22073 |
| 75 | 379 | 1039 | 1777 | 2557 | 3371 | 4201 | 5039 | 5881 | 6793 | 7687 | 8629 | 9479 | 10427 | 11393 | 12347 | 13249 | 14251 | 15199 | 16103 | 17137 | 18119 | 19181 | 20117 | 21139 | 22079 |
| 76 | 383 | 1049 | 1783 | 2579 | 3373 | 4211 | 5051 | 5897 | 6803 | 7691 | 8641 | 9491 | 10429 | 11399 | 12373 | 13259 | 14281 | 15217 | 16111 | 17159 | 18121 | 19183 | 20123 | 21143 | 22091 |
| 77 | 389 | 1051 | 1787 | 2591 | 3389 | 4217 | 5059 | 5903 | 6823 | 7699 | 8647 | 9497 | 10433 | 11411 | 12377 | 13267 | 14293 | 15227 | 16127 | 17167 | 18127 | 19207 | 20129 | 21149 | 22093 |
| 78 | 397 | 1061 | 1789 | 2593 | 3391 | 4219 | 5077 | 5923 | 6827 | 7703 | 8663 | 9511 | 10453 | 11423 | 12379 | 13291 | 14303 | 15233 | 16139 | 17183 | 18131 | 19211 | 20143 | 21157 | 22109 |
| 79 | 401 | 1063 | 1801 | 2609 | 3407 | 4229 | 5081 | 5927 | 6829 | 7717 | 8669 | 9521 | 10457 | 11437 | 12391 | 13297 | 14321 | 15241 | 16141 | 17189 | 18133 | 19213 | 20147 | 21163 | 22111 |
| 80 | 409 | 1069 | 1811 | 2617 | 3413 | 4231 | 5087 | 5939 | 6833 | 7723 | 8677 | 9533 | 10459 | 11443 | 12401 | 13309 | 14323 | 15259 | 16183 | 17191 | 18143 | 19219 | 20149 | 21169 | 22123 |
| 81 | 419 | 1087 | 1823 | 2621 | 3433 | 4241 | 5099 | 5953 | 6841 | 7727 | 8681 | 9539 | 10463 | 11447 | 12409 | 13313 | 14327 | 15263 | 16187 | 17203 | 18149 | 19231 | 20161 | 21179 | 22129 |
| 82 | 421 | 1091 | 1831 | 2633 | 3449 | 4243 | 5101 | 5981 | 6857 | 7741 | 8689 | 9547 | 10477 | 11467 | 12413 | 13327 | 14341 | 15269 | 16189 | 17207 | 18169 | 19237 | 20173 | 21187 | 22133 |
| 83 | 431 | 1093 | 1847 | 2647 | 3457 | 4253 | 5107 | 5987 | 6863 | 7753 | 8693 | 9551 | 10487 | 11471 | 12421 | 13331 | 14347 | 15271 | 16193 | 17209 | 18181 | 19249 | 20177 | 21191 | 22147 |
| 84 | 433 | 1097 | 1861 | 2657 | 3461 | 4259 | 5113 | 6007 | 6869 | 7757 | 8699 | 9587 | 10499 | 11483 | 12433 | 13337 | 14369 | 15277 | 16217 | 17231 | 18191 | 19259 | 20183 | 21193 | 22153 |
| 85 | 439 | 1103 | 1867 | 2659 | 3463 | 4261 | 5119 | 6011 | 6871 | 7759 | 8707 | 9601 | 10501 | 11489 | 12437 | 13339 | 14387 | 15287 | 16223 | 17239 | 18199 | 19267 | 20201 | 21211 | 22157 |
| 86 | 443 | 1109 | 1871 | 2663 | 3467 | 4271 | 5147 | 6029 | 6883 | 7789 | 8713 | 9613 | 10513 | 11491 | 12451 | 13367 | 14389 | 15289 | 16229 | 17257 | 18211 | 19273 | 20219 | 21221 | 22159 |
| 87 | 449 | 1117 | 1873 | 2671 | 3469 | 4273 | 5153 | 6037 | 6899 | 7793 | 8719 | 9619 | 10529 | 11497 | 12457 | 13381 | 14401 | 15299 | 16231 | 17291 | 18217 | 19289 | 20231 | 21227 | 22171 |
| 88 | 457 | 1123 | 1877 | 2677 | 3491 | 4283 | 5167 | 6043 | 6907 | 7817 | 8731 | 9623 | 10531 | 11503 | 12473 | 13397 | 14407 | 15307 | 16249 | 17293 | 18223 | 19301 | 20233 | 21247 | 22189 |
| 89 | 461 | 1129 | 1879 | 2683 | 3499 | 4289 | 5171 | 6047 | 6911 | 7823 | 8737 | 9629 | 10559 | 11519 | 12479 | 13399 | 14411 | 15313 | 16253 | 17299 | 18229 | 19309 | 20249 | 21269 | 22193 |
| 90 | 463 | 1151 | 1889 | 2687 | 3511 | 4297 | 5179 | 6053 | 6917 | 7829 | 8741 | 9631 | 10567 | 11527 | 12487 | 13411 | 14419 | 15319 | 16267 | 17317 | 18233 | 19319 | 20261 | 21277 | 22229 |
| 91 | 467 | 1153 | 1901 | 2689 | 3517 | 4327 | 5189 | 6067 | 6947 | 7841 | 8747 | 9643 | 10589 | 11549 | 12491 | 13417 | 14423 | 15329 | 16273 | 17321 | 18251 | 19333 | 20269 | 21283 | 22247 |
| 92 | 479 | 1163 | 1907 | 2693 | 3527 | 4337 | 5197 | 6073 | 6949 | 7853 | 8753 | 9649 | 10597 | 11551 | 12497 | 13421 | 14431 | 15331 | 16301 | 17327 | 18257 | 19373 | 20287 | 21313 | 22259 |
| 93 | 487 | 1171 | 1913 | 2699 | 3529 | 4339 | 5209 | 6079 | 6959 | 7867 | 8761 | 9661 | 10601 | 11579 | 12503 | 13441 | 14437 | 15349 | 16319 | 17333 | 18269 | 19379 | 20297 | 21317 | 22271 |
| 94 | 491 | 1181 | 1931 | 2707 | 3533 | 4349 | 5227 | 6089 | 6961 | 7873 | 8779 | 9677 | 10607 | 11587 | 12511 | 13451 | 14447 | 15359 | 16333 | 17341 | 18287 | 19381 | 20323 | 21319 | 22273 |
| 95 | 499 | 1187 | 1933 | 2711 | 3539 | 4357 | 5231 | 6091 | 6967 | 7877 | 8783 | 9679 | 10613 | 11593 | 12517 | 13457 | 14449 | 15361 | 16339 | 17351 | 18289 | 19387 | 20327 | 21323 | 22277 |
| 96 | 503 | 1193 | 1949 | 2713 | 3541 | 4363 | 5233 | 6101 | 6971 | 7879 | 8803 | 9689 | 10627 | 11597 | 12527 | 13463 | 14461 | 15373 | 16349 | 17359 | 18301 | 19391 | 20333 | 21341 | 22279 |
| 97 | 509 | 1201 | 1951 | 2719 | 3547 | 4373 | 5237 | 6113 | 6977 | 7883 | 8807 | 9697 | 10631 | 11617 | 12539 | 13469 | 14479 | 15377 | 16361 | 17377 | 18307 | 19403 | 20341 | 21347 | 22283 |
| 98 | 521 | 1213 | 1973 | 2729 | 3557 | 4391 | 5261 | 6121 | 6983 | 7901 | 8819 | 9719 | 10639 | 11621 | 12541 | 13477 | 14489 | 15383 | 16363 | 17383 | 18311 | 19417 | 20347 | 21377 | 22291 |
| 99 | 523 | 1217 | 1979 | 2731 | 3559 | 4397 | 5273 | 6131 | 6991 | 7907 | 8821 | 9721 | 10651 | 11633 | 12547 | 13487 | 14503 | 15391 | 16369 | 17387 | 18313 | 19421 | 20353 | 21379 | 22303 |
| 100 | 541 | 1223 | 1987 | 2741 | 3571 | 4409 | 5279 | 6133 | 6997 | 7919 | 8831 | 9733 | 10657 | 11657 | 12553 | 13499 | 14519 | 15401 | 16381 | 17389 | 18313 | 19423 | 20357 | 21383 | 22307 |

*Table of the First 2500 Prime Numbers*

From D. N. Lehmer, *List of prime numbers from 1 to 10,006,721*, Carnegie Institution of Washington. Publication No. 165, Washington, D.C., 1914.

of all the primes up to $p_m$. Now $N$ is larger than $p_m$ (for it is certainly more than twice its size). When $N$ is divided by 2 it goes $3 \cdot 5 \cdot \cdots \cdot p_m$ times and leaves a remainder 1. When it is divided by 3, it goes $2 \cdot 5 \cdot \cdots \cdot p_m$ times and leaves a remainder 1. Similarly, it leaves a remainder of 1 when divided by any of the primes $2, 3, 5, \ldots, p_m$.

Now $N$ is either a prime number or it isn't. If it is a prime number, it is a prime number greater than $p_m$. If it isn't a prime number, it may be factored into prime numbers. But none of its prime factors can be $2, 3, 5, \ldots, p_m$ as we just saw. Therefore there is a prime number greater than $p_m$.

The logical argument (actually, the dilemma, which forces one to the same conclusion whichever path one is compelled to take) tells us that the list of primes never ends.

The second feature of the list of primes that strikes one is the absence of any noticeable pattern or regularity. Of course all the prime numbers except 2 are odd, so the gap between any two successive primes has to be an even number. But there seems to be no rhyme or reason as to which even number it happens to be.

There are nine prime numbers between 9,999,900 and 10,000,000:

| | | | |
|---|---|---|---|
| 9,999,901 | 9,999,907 | 9,999,929 | 9,999,931 |
| 9,999,937 | 9,999,943 | 9,999,971 | 9,999,973 |
| 9,999,991. | | | |

But among the next hundred integers, from 10,000,000 to 10,000,100, there are only two:

$$10,000,019 \quad \text{and} \quad 10,000,079.$$

"Upon looking at these numbers, one has the feeling of being in the presence of one of the inexplicable secrets of creation," writes Don Zagier in an outburst of modern number mysticism.

What is known about primes and what is not known or conjectural would fill a large book. Here are some samples. The largest known prime in 1979 was $2^{21,701} - 1$. There is a prime between $n$ and $2n$ for every integer $n > 1$. Is there a prime between $n^2$ and $(n + 1)^2$ for every $n > 0$? No one

212

knows. Are there an infinity of primes of the form $n^2 + 1$ where $n$ is an integer? No one knows. There are runs of integers of arbitrary length which are free of primes. No polynomial with integer coefficients can take on only prime values at the integers. There is an irrational number $A$ such that $[A^{3^n}]$ takes on only prime values as $n = 0, 1, 2, \ldots$ . (Here the notation $[x]$ means the greatest integer $\leq x$.) Is every even number the sum of two odd primes? No one knows; this is the notorious Goldbach conjecture. Are there an infinite number of prime pairs, such as 11;13 or 17;19 or 10,006,427;10,006,429 which differ by 2? This is the problem of the twin primes, and no one knows the answer though most mathematicians are convinced that the statement is very likely to be true.

Some order begins to emerge from this chaos when the primes are considered not in their individuality but in the aggregate; one considers the social statistics of the primes and not the eccentricities of the individuals. One first makes a large tabulation of primes. This is difficult and tedious with pencil and paper, but with a modern computer it is easy. Then one counts them to see how many there are up to a given point. The function $\pi(n)$ is defined as the number of primes less than or equal to the number $n$. The function $\pi(n)$ measures the distribution of the prime numbers. Having obtained it, it is only natural to compute the ratio $n/\pi(n)$ which tells us what fraction of the numbers up to a given point are primes. (Actually, it is the reciprocal of this fraction.) Here is the result of a recent computation.

| n | $\pi(n)$ | $n/\pi(n)$ |
|---|---|---|
| 10 | 4 | 2.5 |
| 100 | 25 | 4.0 |
| 1000 | 168 | 6.0 |
| 10,000 | 1,229 | 8.1 |
| 100,000 | 9,592 | 10.4 |
| 1,000,000 | 78,498 | 12.7 |
| 10,000,000 | 664,579 | 15.0 |
| 100,000,000 | 5,761,455 | 17.4 |
| 1,000,000,000 | 50,847,534 | 19.7 |
| 10,000,000,000 | 455,052,512 | 22.0 |

Notice that as one moves from one power of 10 to the next, the ratio $n/\pi(n)$ increases by roughly 2.3. (For example,

*Carl Friedrich Gauss*
*1777–1855*

*Jacques Hadamard*
*1865–1963*

22.0 − 19.7 = 2.3.) At this point, any mathematician worth his salt thinks of $\log_e 10$ (= 2.30258 . . .) and on the basis of this evidence, it is easy to formulate the conjecture that $\pi(n)$ is approximately equal to $\dfrac{n}{\log n}$. The more formal statement that

$$\lim_{n \to \infty} \pi(n)/(n/\log n) = 1$$

is the famous prime number theorem. The discovery of the theorem can be traced as far back as Gauss, at age fifteen (around 1792), but the rigorous mathematical proof dates from 1896 and the independent work of C. de la Vallée Poussin and Jacques Hadamard. Here is order extracted from confusion, providing a moral lesson on how individual eccentricities can exist side by side with law and order.

While the expression $n/\log n$ is a fairly simple approximation for $\pi(n)$, it is not terribly close, and mathematicians have been interested in improving it. Of course, one does this at the price of complicating the approximant. One of the most satisfactory approximants to $\pi(n)$ is the function

$$R(n) = 1 + \sum_{k=1}^{\infty} \frac{1}{k\zeta(k+1)} \frac{(\log n)^k}{k!}$$

where $\zeta(z)$ designates the celebrated Riemann zeta function: $\zeta(z) = 1 + \dfrac{1}{2^z} + \dfrac{1}{3^z} + \dfrac{1}{4^z} + \dots$ . The accompanying table shows what a remarkably good approximation $R(n)$ is to $\pi(n)$:

|  | $\pi(n)$ | $R(n)$ |
|---|---|---|
| 100,000,000 | 5,761,455 | 5,761,552 |
| 200,000,000 | 11,078,937 | 11,079,090 |
| 300,000,000 | 16,252,325 | 16,252,355 |
| 400,000,000 | 21,336,326 | 21,336,185 |
| 500,000,000 | 26,355,867 | 26,355,517 |
| 600,000,000 | 31,324,703 | 31,324,622 |
| 700,000,000 | 36,252,931 | 36,252,719 |
| 800,000,000 | 41,146,179 | 41,146,248 |
| 900,000,000 | 46,009,215 | 46,009,949 |
| 1,000,000,000 | 50,847,534 | 50,847,455 |

Let us turn, finally, to the question of twin prime pairs. It is thought that there are an infinite number of such pairs, though this is still an open question.

Why do we believe it is true, even though there is no proof? First of all, there is numerical evidence; we find more prime pairs whenever we look for them; there does not seem to be a region of the natural number system so remote that it lies beyond the largest prime pair. But more than that, we have an idea *how many* prime pairs there are. We can get this idea by noticing that the occurrence of prime pairs in a table of prime numbers seem to be unpredictable or *random*. This suggests the conjecture that the chance of two numbers $n$ and $n + 2$, both being prime, acts like the chance of getting a head on two successive tosses of a coin. If two successive random experiments are independent, the chance of success on both is the product of the chances of success on either; for example, if one coin has probability $\frac{1}{2}$ of coming up heads, two coins have probability $\frac{1}{2} \times \frac{1}{2} = \frac{1}{4}$ of coming up a pair of heads.

Now the prime number theorem, which *has* been proved, says that if $n$ is a large number, and we choose a number $x$ at random between 0 and $n$, the chance that $x$ is prime will be "about" $\dfrac{1}{\log n}$. The bigger $n$ is, the better is the approximation given by $\dfrac{1}{\log n}$ to the proportion of primes in the numbers up to $n$.

If we trust our feeling that the occurrence of twin primes is like two coins coming up heads, then the chance that both $x$ and $x + 2$ are prime would be about $\dfrac{1}{(\log n)^2}$.

In other words, there would be about $\dfrac{n}{(\log n)^2}$ prime pairs to be found between 0 and $n$. This fraction approaches infinity as $n$ goes to infinity, so this would provide a quantitative version of the prime pair conjecture.

For reasons involving the dependence of $x + 2$ being prime on the supposition that $x$ is already prime, one should modify the estimate $\dfrac{n}{(\log n)^2}$ to $\dfrac{(1.32032..)n}{(\log n)^2}$.

Appended is a comparison between what has been found and what is predicted by this simple formula. The agreement is remarkably good, but the final Q.E.D. is yet to be written.

| Interval | Prime twins expected | Prime twins found |
|---|---|---|
| 100,000,000–<br>100,150,000 | 584 | 601 |
| 1,000,000,000–<br>1,000,150,000 | 461 | 466 |
| 10,000,000,000–<br>10,000,150,000 | 374 | 389 |
| 100,000,000,000–<br>100,000,150,000 | 309 | 276 |
| 1,000,000,000,000–<br>1,000,000,150,000 | 259 | 276 |
| 10,000,000,000,000–<br>10,000,000,150,000 | 221 | 208 |
| 100,000,000,000,000–<br>100,000,000,150,000 | 191 | 186 |
| 1,000,000,000,000,000–<br>1,000,000,000,150,000 | 166 | 161 |

## Further Readings. See Bibliography

E. Grosswald; D. N. Lehmer; D. Zagier.

# Non-Euclidean Geometry

*Euclid*
*c. 300* B.C.

T HE APPEARANCE on the mathematical scene a century and a half ago of non-Euclidean geometries was accompanied by considerable disbelief and shock. The existence of such geometries is now easily explained in a few sentences and will easily be understood. Any mathematical theory such as arithmetic, geometry, algebra, topology, etc., can be presented as an axiomatic scheme wherein consequences are deduced systematically and logically from the axioms. Such a logico-deductive scheme may be compared to a game and the axioms of the scheme to the rules of the game. Anyone who plays games knows that one can invent variations on given games and the consequences will be different. A non-Euclidean geometry is a geometry that is played with axioms that are different from those of Euclid.

Of course, this simple explanation violates the historical order. It borrows from a philosophy of mathematics which came about precisely as a result of the discovery of such geometries. For a fuller understanding of the matter, it is necessary to see what happened chronologically.

Since the Greeks, geometry has had a dual aspect. It is claimed to be an accurate description of the space in which we live and it is also an intellectual discipline, a deductive structure. These two aspects are now viewed as separate, but this was not always the case. The geometry of Euclid was based on a number of axioms and postulates of which we quote the first five postulates.

(The distinction between the words axiom and postulate is fuzzy. Modern mathematics uses the words almost interchangeably.)

1. A straight line may be drawn between any two points.
2. Any terminated straight line may be extended indefinitely.

217

3. A circle may be drawn with any given point as center and any given radius.

4. All right angles are equal.

5. If two straight lines lying in a plane are met by another line, and if the sum of the internal angles on one side is less than two right angles, then the straight lines will meet if extended sufficiently on the side on which the sum of the angles is less than two right angles.

In other words, referring to the figure,

if $\angle A + \angle B < 180°$ the lines $L_1$ and $L_2$ will intersect at some point on the right hand side of $L_3$.

The word axiom or postulate, in an earlier view, meant a self-evident or a universally recognized truth, a truth accepted without proof. Within deductive geometry, the axiom functions as a cornerstone on which further conclusions are based. Within descriptive geometry, the axiom functions as a true and accurate statement of the world of spatial experiences. The former view persists; the latter has had to give way.

If one takes a look at Postulates 1, 2, 3, 4, they appear easy of statement and, indeed, self-evident. Postulate 5 is different. It is complicated to state and rather less self-evident. It seems to transcend direct physical experience. Postulate 5 is known as Euclid's Parallel Postulate or, more familiarly, in a friendly allusion to the Amendments of the United States Constitution, Euclid's Fifth. From the earliest times it attracted special attention.

The historical development of non-Euclidean geometry was a result of attempts to deal with this axiom. Notice, further, that although Euclid's Fifth is known as the parallel

218

axiom, the word "parallel" does not occur in it. The word parallel is expanded in Euclid under Definition 23.

"Parallel straight lines are lines which being in the same plane and being produced indefinitely in both directions do not meet one another in either direction."

The reason for our calling Euclid's Fifth the parallel axiom is that it is totally equivalent to any of the following statements involving the word parallel:

1. If a straight line intersects one of two parallels, it will intersect the other.

2. Straight lines parallel to the same straight line are parallel to each other.

3. Two straight lines which intersect one another cannot be parallel to the same line.

4. Given, in a plane, a line *L* and a point *P* not on *L*. Then through *P* there exists one and only one line parallel to *L*.

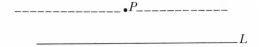

Equivalence means that any of these statements plus the other axioms implies Euclid's Fifth, and vice versa.

Over the years, for reasons which are partly technical and partly aesthetic, the fourth formulation has come to be the standard formulation of Euclid's assertion of parallelism. It is known as the Playfair Axiom, after the Britisher John Playfair, 1748–1819.

The early investigations with Euclid's Fifth tried to assuage the doubts of its validity by attempting to derive it logically from the other axioms, which seemed to be self-evident. It would then become a theorem and its status would be assured. These attempts failed, and for good reason—we now know that it cannot be so derived. This was established by 1868.

With the failure of direct methods, it was inevitable that mathematicians would turn to indirect methods. In such

*Bernhard Riemann*
*1826–1866*

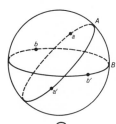

*On surface of a sphere*
*"straight line" is inter-*
*preted to mean "great*
*circle" (A and B at*
*top). Through any pair*
*of diametrically opposite*
*points (aa' and bb')*
*there pass many great*
*circles. If we interpret*
*"point" to mean "point*
*pair," then Euclid's first*

an approach, one denies the Fifth and then tries to derive a contradiction.

Two notable investigators employing *reductio ad absurdam* were Girolamo Saccheri (1667–1733) and Johann Lambert (1728–1777). Both of these denied the Fifth.

Saccheri works with a quadrilateral ABCD which has right angles at A and B and in which AD = BC. This is now known in axiomatic geometry as a Saccheri quadrilateral. It should be noted that within Euclidean geometry AD will be parallel to BC and this makes the angles at D and C both right angles. But Saccheri, not taking the Fifth, concludes that he really has three options:

1. The angles at C and D are both right angles.

2. They are both obtuse angles.

3. They are both acute angles.

Some of the conclusions from (2) and (3) are sufficiently startling to go against "intuition." At this point Saccheri throws in the towel and cries "contradiction."

Lambert, the bolder and more skillful of the two, lasts a few more rounds, and does not give up until he has discovered that within the new and hypothetical system he has devised one could prove the existence of an absolute unit of length. Since, he argues, there can be no absolute unit of length, the whole enterprise must have been fallacious.

It is not our intent here to trace the stream of discovery and its many tributary branches. Many mathematicians played a role—Gauss, Lobachevsky, Bolyai, and Riemann, to name the most important. The discovery was attended by many misunderstandings, doubts, misgivings. It seemed to be at the edge of madness. The birth pains were severe. Thus, e.g., Bolyai's father wrote to him "For God's sake,

220

please give it up. Fear it no less than the sensual passions because it, too, may take up all your time and deprive you of your health, peace of mind and happiness in life."

It was found that there are not one but two non-Euclidean geometries. They currently go by the names of Lobachevskian (or hyperbolic) geometry and Riemannian (or elliptic) geometry. With respect to the Playfair Axiom, these two non-Euclidean geometries correspond to the axioms:

*Lobachevsky:* Given in a plane a line $L$ and a point $P$ not on $L$. Then there are at least two lines through $P$ parallel to $L$.

*Riemann:* Given in a plane a line $L$ and a point $P$ not on $L$. Then there are *no* lines through $P$ parallel to $L$.

The geometries, Euclidean, Lobachevskian, and Riemannian, carry with them three distinct sets of conclusions (theorems), which are worked out in detail in textbooks on the subject. There are different mensuration formulas, different projective aspects. A comparison between the three is most interesting. Some of the elementary differences are summed up in the table adjoined.

We give two more comparisons. The famous theorem of Pythagoras now has three forms.

Euclidean Geometry: $c^2 = a^2 + b^2$

Lobachevskian Geometry: $2(e^{c/k} + e^{-c/k}) = (e^{a/k} + e^{-a/k})(e^{b/k} + e^{-b/k})$ where $k$ is a certain fixed constant and $e = 2.718\ldots$

Riemannian Geometry: Differential form: $ds^2 = \alpha dx^2 + 2\beta dxdy + \gamma dy^2$ where $\left(\begin{smallmatrix} \alpha & \beta \\ \beta & \gamma \end{smallmatrix}\right)$ is positive definite.

The circumference $C$ of a circle whose radius is $r$ has the formulas

*Euclidean Geometry: $C = 2\pi r$*

*Lobatchevskian Geometry: $C = \pi k(e^{r/k} - e^{-r/k})$.*

The Riemannian formula for $C$ is not expressible in simple terms.

It remains for us to discuss the logical consistency of non-Euclidean geometry. In order to do this we merely replace the word "line" everywhere by the phrase "great cir-

*postulate is true. The second postulate is true if one allows the extended "straight line" to have a finite total length, or to retrace itself many times as it goes around the sphere. The third postulate is also true if one understands distance to be measured along great circles that can be retraced several times; here a "circle" means merely the set of points on the sphere at a given great-circle distance from a given point. The fourth postulate is likewise true. Playfair's postulate is false, because any two great circles intersect. Thus the sphere is a model of non-Euclidean geometry. So is the pseudo-sphere (bottom), if straight lines are interpreted as being the shortest curves connecting any two points on the surface. On the surface of the pseudosphere there are many "straight lines" that pass through a given point and do not cross a given straight line.*

*Janos Bolyai*
*1802–1860*

|  | Euclidean | Lobachevskian | Riemannian | |
|---|---|---|---|---|
| Two distinct lines intersect in | at most one | at most one | one (single elliptic) | point |
| | | | two (double elliptic) | points |
| Given line L and point P not on L, there exist | one and only one line | at least two lines | no lines | through P parallel to L |
| A line | is | is | is not | separated into two parts by a point |
| Parallel lines | are equidistant | are never equidistant | do not exist | |
| If a line intersects one of two parallel lines, it | must | may or may not | — | intersect the other |
| The valid Saccheri hypothesis is the | right angle | acute angle | obtuse angle | hypothesis |
| Two distinct lines perpendicular to the same line | are parallel | are parallel | intersect | |
| The angle sum of a triangle is | equal to | less than | greater than | 180 degrees |
| The area of a triangle is | independent | proportional to the defect | proportional to the excess | of its angle sum |
| Two triangles with equal corresponding angles are | similar | congruent | congruent | |

*Table Comparing Euclidean and Non-Euclidean Plane Geometry\**

*\* From Prenowitz and Jordan, Basic Concepts of Geometry.*

cle," a circle formed on the surface of a sphere by a plane passing through the center of the sphere. We now regard the axioms as statements about points and great circles on a given sphere. Moreover, we agree to identify each pair of diametrically opposite points on the sphere as a single point. If the reader prefers, he can imagine the axioms of non-Euclidean geometry rewritten, with the word "line" everywhere replaced by "great circle," the word "point" everywhere replaced by "point pair." Then it is evident that all the axioms are true, at least insofar as our ordinary notions about the surface of a sphere are true. In fact, from the axioms of Euclidean solid geometry one can easily prove as theorems that the surface of a sphere is a non-Euclidean surface in the sense we have just described. In other words, we now see that if the axioms of non-Euclidean geometry led to a contradiction, then so would the ordinary Euclidean geometry of spheres lead to a contradiction. Thus we have a *relative* proof of consistency; if Euclidean three-dimensional geometry is consistent, then so is non-Euclidean two dimensional geometry. We say that the surface of the Euclidean sphere is a model for the axioms of non-Euclidean geometry. (In the particular model we have used the parallel postulate fails because there are no parallel lines. It is also possible to construct a surface, the "pseudosphere," for which the parallel postulate is false because there is more than one line through a point parallel to a given line.)

### Further Readings. See Bibliography

A. D. Alexandroff, Chapter 17; H. Eves and C. V. Newsom; E. B. Golos; M. J. Greenberg; M. Kline [1972], Chapter 36; W. Prenowitz and M. Jordan; C. E. Sjöstedt

# Non-Cantorian Set Theory

THE ABSTRACT theory of sets has recently undergone a revolution that in several ways is analogous to the nineteenth-century revolution in geometry. We propose to use the tale of non-Euclidean geometry to illuminate the story of nonstandard set theory.

A set, of course, is one of the simplest and most primitive

*Georg Cantor*
*1845–1918*

ideas in mathematics, so simple that today it is part of the kindergarten curriculum. No doubt for this very reason its role as the most fundamental concept of mathematics was not made explicit until the 1880s. Only then did Georg Cantor make the first nontrivial discovery in the theory of sets.

Cantor pointed out that for infinite sets it makes sense to talk about the number of elements in the set, or at least to state that two different sets have the same number of elements. Just as with finite sets, we can say that two sets have the same number of elements—the same "cardinality"—if we can match up the elements in the two sets one for one. If this can be done, we call the two sets equivalent.

The set of all natural numbers can be matched up with the set of all even numbers, and also with the set of all fractions. These two examples illustrate a paradoxical property of infinite sets: an infinite set can be equivalent to one of its subsets. In fact, it is easily proved that a set is infinite if, and only if, it is equivalent to some proper subset of itself.

All of this is engaging, but it was not new with Cantor. The notion of the cardinality of infinite sets would be interesting only if it could be shown that not all infinite sets have the same cardinality. It was this that was Cantor's first great discovery in set theory. By his famous diagonal proof he showed that the set of natural numbers is *not* equivalent to the set of points on a line segment. (See Appendix A.)

Thus there are at least two different kinds of infinity. The first, the infinity of the natural numbers (and of any equivalent infinite sets), is called aleph nought ($\aleph_0$). Sets with cardinality $\aleph_0$ are called countable. The second kind of infinity is the one represented by a line segment. Its cardinality is designated by a lower-case German $c$ ($\mathfrak{c}$), for "continuum." *Any* line segment, of arbitrary length, has cardinality $\mathfrak{c}$. So does any rectangle in the plane, any cube in space, or for that matter all of unbounded $n$-dimensional space, whether $n$ is 1, 2, 3 or 1,000.

Once a single step up the chain of infinities has been taken, the next follows naturally. We encounter the notion of the set of all subsets of a given set. If the original set is called $A$, this new set is called the power set of $A$ and is written $2^A$. And just as we obtain the power set $2^A$ from $A$, we can next obtain $2^{(2^A)}$ from $2^A$, and so on as long as we please.

Cantor proved that whether $A$ is finite or infinite, $2^A$ is never equivalent to $A$. Therefore the procedure of forming the set of all subsets generates an endless chain of increasing, nonequivalent infinite sets. In particular, if $A$ is the set of natural numbers, then it is easy to prove that $2^A$ (the set of all sets of natural numbers) is equivalent to the continuum (the set of all points on a line segment). In brief,

$$2^{\aleph_0} = c.$$

At this point a question may occur to the reader. Is there an infinite set with cardinality *between* $\aleph_0$ and $c$? That is, is there on a line segment an infinite set of points that is not equivalent to the whole segment, and also not equivalent to the set of natural numbers?

This question occurred to Cantor, but he was unable to find any such set. He concluded—or rather conjectured—that no such thing exists. This guess of Cantor's acquired the name "the continuum hypothesis." Its proof or disproof was first on the celebrated list of unsolved mathematical problems drawn up by David Hilbert in 1900. Only in 1963 was it finally settled. It was settled, however, in a sense utterly different from what Hilbert had in mind.

To tackle this problem one could no longer rely on Cantor's definition of a set as "any collection into a whole of definite and separate objects of our intuition or our thought." In fact, this definition, seemingly so transparent, turned out to conceal some treacherous pitfalls. The free use of Cantor's intuitive notion of a set can lead to contradictions. Set theory can serve as a secure foundation for mathematics only if a more sophisticated approach is employed to steer clear of antinomies, as contradictions of the type proposed by Russell later came to be known.

It has happened before that unwelcome paradoxes have intruded into a seemingly clear mathematical theory. There are the paradoxes of Zeno, which revealed to the Greeks unsuspected complexities in intuitive concepts of lines and points. We can draw an analogy: As Russell found a contradiction in the unrestricted use of the intuitive concept of set, so Zeno had found contradictions in the unrestricted use of the intuitive concepts of "line" and "point."

In its beginning with Thales in the sixth century B.C. Greek geometry had relied on an unspecified intuitive concept of "line" and "point." Some 300 years later, however, Euclid had given these concepts an axiomatic treatment. For Euclid, geometric objects were still intuitively known real entities, but insofar as they were the subject of geometrical reasoning they were specified by certain unproved assertions ("axioms" and "postulates"), on the basis of which all their other properties were supposed to be proved as "theorems." We do not know if, and to what extent, this development was a response to paradoxes such as Zeno's. There is no doubt, however, that to the Greeks geometry was made much more secure by virtue of depending (at least so they believed and intended) only on logical inference from a small number of clearly stated assumptions.

The analogous development for set theory took not 300 years but only 35. If Cantor played the role of Thales—the founder of the subject, who was able to rely on intuitive reasoning alone—then the role of Euclid was played by Ernst Zermelo, who in 1908 founded axiomatic set theory. Of course, Euclid was really only one of a long succession of Greek geometers who created "Euclidean geometry"; so also Zermelo was only the first of half a dozen great names in the creation of axiomatic set theory.

Just as Euclid had listed certain properties of points and lines and had regarded as proved only those theorems in geometry that could be obtained from these axioms (and not from any possibly intuitive arguments), so in axiomatic set theory a set is regarded simply as an undefined object satisfying a given list of axioms. Of course, we still want to study sets (or lines, as the case may be), and so the axioms are chosen not arbitrarily but in accord with our intuitive notion of a set or a line. Intuition is nonetheless barred

226

from any further formal role; only those propositions are accepted that follow from the axioms. The fact that objects described by these axioms actually may exist in the real world is irrelevant to the process of formal deduction (although it is essential to discovery).

We agree to act as if the symbols for "line," "point" and "angle" in geometry, or the symbols for "set," "is a subset of" and so on in set theory, are mere marks on paper, which may be rearranged only according to a given list of rules (axioms and rules of inference). Accepted as theorems are only those statements that are obtained according to such manipulations of symbols. (In actual practice only those statements are accepted that clearly *could* be obtained in this manner if one took enough time and trouble.)

Now, in the history of geometry, one postulate played a special role. This was the parallel postulate, which says that through a given point there can be drawn precisely one line parallel to a given line. (See the discussion of non-Euclidean geometry earlier in this chapter.) The difficulty with this statement as an axiom is that it does not have the self-evident character one prefers in the foundation stones of a mathematical theory. In fact, parallel lines are defined as lines that never meet, even if they are extended indefinitely ("to infinity"). Since any lines we draw on paper or on a blackboard have finite length, this is an axiom that by its nature cannot be verified by direct observation of the senses. Nonetheless, it plays an indispensable role in Euclidean geometry. For many centuries a leading problem in geometry was to *prove* the parallel postulate, to show that it could be obtained as a theorem from the more self-evident Euclidean axioms.

In abstract set theory, it so happens, there also was a particular axiom that some mathematicians found hard to swallow. This was the axiom of choice, which says the following: If $\alpha$ is any collection of sets $\{A, B, \ldots\}$, and none of the sets in $\alpha$ is empty, then there exists a set $Z$ consisting of precisely one element each from $A$, from $B$ and so on through all the sets in $\alpha$. For instance, if $\alpha$ consists of two sets, the set of all triangles and the set of all squares, then $\alpha$

227

clearly satisfies the axiom of choice. We merely choose some particular triangle and some particular square and then let these two elements constitute $Z$.

Most people find the axiom of choice, like the parallel postulate, intuitively very plausible. The difficulty with it is in the latitude we allow $\alpha$: "any" collection of sets. As we have seen, there are endless chains of ever bigger infinite sets. For such an inconceivably huge collection of sets there is no way of actually choosing one by one from all its member sets. If we accept the axiom of choice, our acceptance is simply an act of faith that such a choice is possible, just as our acceptance of the parallel postulate is an act of faith about how lines would act if they were extended to infinity. It turns out that from the innocent-seeming axiom of choice some unexpected and extremely powerful conclusions follow. For example, we are able to use inductive reasoning to prove statements about the elements in *any* set, in much the same way that mathematical induction can be used to prove theorems about the natural numbers 1, 2, 3, and so on.

*Kurt Gödel*
*1906–1978*

The axiom of choice played a special role in set theory. Many mathematicians thought its use should be avoided whenever possible. Such a form of axiomatic set theory, in which the axiom of choice is *not* assumed to be either true or false, would be one on which almost all mathematicians would be prepared to rely. In what follows we use the term "restricted set theory" for such an axiom system. We use the term "standard set theory" for the theory based on the full set of axioms put forward by Zermelo and Abraham Fraenkel: restricted set theory *plus* the axiom of choice.

In 1938 this subject was profoundly illuminated by Kurt Gödel. Gödel is best known for his great "incompleteness" theorems of 1930–1931. Here we refer to later work by Gödel that is not well known to nonmathematicians. In 1938 Gödel proved the following fundamental result: If restricted set theory is consistent, then so is standard set theory. In other words, the axiom of choice is no more dangerous than the other axioms; if a contradiction can be found in standard set theory, then there must already be a contradiction hidden within restricted set theory.

228

But that was not all Gödel proved. We remind the reader of Cantor's "continuum hypothesis," namely that no infinite cardinal exists that is greater than $\aleph_0$ and smaller than $c$. Gödel also showed that we can safely take the continuum hypothesis as an additional axiom in set theory; that is, if the continuum hypothesis plus restricted set theory implies a contradiction, then again there must already be a contradiction hidden within restricted set theory. This was a half-solution of Cantor's problem; it was not a *proof* of the continuum hypothesis but only a proof that it cannot be disproved.

To understand how Gödel achieved his results we need to understand what is meant by a model for an axiom system. Let us return for a moment to the axioms of plane geometry. If we take these axioms, including the parallel postulate, we have the axioms of Euclidean geometry; if instead we keep all the other axioms as before but replace the parallel postulate by its negation, we have the axioms of a non-Euclidean geometry. For both axiom systems—Euclidean and non-Euclidean—we ask: Can these axioms lead to a contradiction?

To ask the question of the Euclidean system may seem unreasonable. How could there be anything wrong with our familiar, 2,000-year-old high school geometry? On the other hand, to the nonmathematician there certainly is something suspicious about the second axiom system, with its denial of the intuitively plausible parallel postulate. Nonetheless, from the viewpoint of twentieth-century mathematics the two kinds of geometry stand more or less on an equal footing. Both are sometimes applicable to the physical world and both are consistent, in a relative sense.

The invention of non-Euclidean geometry, and the recognition that its consistency is implied by the consistency of Euclidean geometry, was the work of many great nineteenth-century mathematicians; we mention the name of Bernhard Riemann in particular. Only in the twentieth century was the question raised of whether or not Euclidean geometry itself is consistent.

This question was asked and answered by David Hilbert. Hilbert's solution was a simple application of the idea of a coordinate system. To each point in the plane we can associate a pair of numbers: its $x$ and $y$ coordinates. Then with each line or circle we can associate an equation: a relation between the $x$ and $y$ coordinates that is true only for the points on that line or circle. In this way we set up a correspondence between geometry and elementary algebra. For every statement in one subject there is a corresponding statement in the other. It follows that the axioms of Euclidean geometry can lead to a contradiction only if the rules of elementary algebra—the properties of the ordinary real numbers—can lead to a contradiction. Here again we have a relative proof of consistency. Non-Euclidean geometry was consistent if Euclidean geometry was consistent; now Euclidean geometry is consistent if elementary algebra is consistent. The Euclidean sphere was a model for the non-Euclidean plane; the set of pairs of coordinates is in turn a model for the Euclidean plane.

With these examples before us we can say that Gödel's proof of the relative consistency of the axiom of choice and of the continuum hypothesis is analogous to Hilbert's proof of the relative consistency of Euclidean geometry. In both instances the standard theory was justified in terms of a more elementary one. Of course, no one ever seriously doubted the reliability of Euclidean geometry, whereas such outstanding mathematicians as L. E. J. Brouwer, Hermann Weyl and Henri Poincaré had grave doubts about the axiom of choice. In this sense Gödel's result had a much greater impact and significance.

The analogous development with respect to non-Euclidean geometry—what we might call non-Cantorian set theory—has taken place only since 1963, in the work of Paul J. Cohen. What is meant by "non-Cantorian set theory"? Just as Euclidean and non-Euclidean geometry use the same axioms, with the one exception of the parallel postulate, so standard ("Cantorian") and nonstandard ("non-Cantorian") set theory differ only in one axiom. Non-Cantorian set theory takes the axioms of restricted set theory and adds not the axiom of choice but rather one or another form of the negation of the axiom of choice. In particular we can take as an axiom the negation of the continuum hypothesis. Thus, as we shall explain, there now exists a complete solution of the continuum problem. To Gödel's discovery that the continuum hypothesis is not disprovable is added the fact that it is also not provable.

Both Gödel's result and the new discoveries require the construction of a model, just as the consistency proofs for geometry that we have described required a model. In both cases we want to prove that if restricted set theory is consistent, then so is standard set theory (or nonstandard theory).

Gödel's idea was to construct a model for restricted set theory, and to prove that in this model the axiom of choice and the continuum hypothesis were theorems. He proceeded in the following way. Using only the axioms of restricted set theory [*see page* 138] we are guaranteed first the existence of at least one set (the empty set) by Axiom 2; then by Axiom 3 and Axiom 4 we are guaranteed the existence of an infinite sequence of ever larger finite sets; then by Axiom 5, the existence of an infinite set; then by Axiom 7, of an endless sequence of ever larger (nonequivalent) infinite sets, and so on. In essentially this way Gödel specified a class of sets by the manner in which they could actually be constructed in successive steps from simpler sets. These sets he called the "constructible sets"; their existence was guaranteed by the axioms of restricted set theory. Then he showed that within the realm of the constructible sets the axiom of choice and the continuum hypothesis can both be proved. That is to say, first, from any constructible collection $\alpha$ of constructible sets $(A, B, \ldots)$ one can choose a constructible set $Z$ consisting of at least one element each from $A$, $B$ and so on. This is the axiom of choice, which here might more properly be called the theorem of choice. Second, if $A$ is any infinite constructible set, then there is no constructible set "between" $A$ and $2^A$ (bigger than $A$, smaller than the power set of $A$ and equivalent to neither). If $A$ is taken as the first infinite cardinal, this last statement is the continuum hypothesis.

Hence a "generalized continuum hypothesis" was proved in the case of *constructible* set theory. Gödel's work would therefore dispose of these two questions completely if we were prepared to adopt the axiom that only constructible sets exist. Why not do so? Because one feels it is unreasonable to insist that a set must be constructed according to *any* prescribed formula in order to be recognized as a genuine set. Thus in ordinary (not necessarily constructible) set theory neither the axiom of choice nor the continuum hypothesis had been proved. At least this much was certain: either of them could be assumed without causing any contradiction unless the "safe" axioms of restricted set theory already are self-contradictory. Any contradiction they cause must already be present in constructible set theory, which is a model for ordinary set theory. In other words, it was known that neither could be disproved from the other axioms but not whether they could be proved.

Here the analogy with the parallel postulate in Euclidean geometry becomes particularly apt. That Euclid's axioms are consistent was taken for granted until quite recently. The question that interested geometers was whether or not they are independent, that is, whether the parallel postulate could be proved on the basis of the others. A whole series of geometers tried to prove the parallel postulate by showing that its negation led to absurdities. But they were led not to absurdities but to the discovery of "fantastic" geometries that had as much logical consistency as the Euclidean geometry of "the real world." Only after this had happened was it recognized that two-dimensional non-Euclidean geometry was just the ordinary Euclidean geometry of certain curved surfaces (spheres and pseudospheres).

The analogous step in set theory would be to deny the axiom of choice or the continuum hypothesis. By this we mean, of course, that the step would be to prove that such a negation is consistent with restricted set theory, in the same sense in which Gödel had proved that the affirmation was consistent. It is this proof that has been accomplished in the past few years, giving rise to a surge of activity in mathematical logic whose final outcome cannot be guessed.

Since it is a question of proving the relative consistency of an axiom system, we naturally think of constructing a model. As we have seen, the relative consistency of non-Euclidean geometry was established when surfaces in Euclidean three-space were shown to be models of two-dimensional non-Euclidean geometry. In a comparable way, in order to prove the legitimacy of a non-Cantorian set theory in which the axiom of choice or the continuum hypothesis is false we must use the axioms of restricted set theory to construct a model in which the negation of the axiom of choice or the negation of the continuum hypothesis can be proved as theorems.

It must be confessed that construction of this model is a complex and delicate affair. This is perhaps to be expected. In Gödel's constructible sets, his model of Cantorian set theory, the task was to create something essentially the same as our intuitive notion of sets but more tractable. In our present task we have to create a model of something unintuitive and strange, using the familiar building stones of restricted set theory.

Rather than throw up our hands and say it is impossible to describe this model in a nontechnical way, we shall attempt at least to give a descriptive account of one or two of the leading ideas that are involved. Our starting point is ordinary set theory (without the axiom of choice). We hope only to prove the consistency of non-Cantorian set theory in a relative sense. Just as the models of non-Euclidean geometry prove that non-Euclidean geometry is consistent if Euclidean geometry is consistent, so we shall prove that if restricted set theory is consistent, it remains so if we add the statement "The axiom of choice is false" or the statement "The continuum hypothesis is false." We may now assume that we have available as a starting point a model for restricted set theory. Call this model $M$; it can be regarded as Gödel's class of constructible sets.

We know from Gödel's work that in order for the axiom of choice or the continuum hypothesis to fail we must add to $M$ at least one nonconstructible set. How to do this? We introduce the letter $a$ to stand for an object to be added to $M$; it remains to determine what kind of thing $a$ should be. Once we add $a$ we must also add everything that can be formed from $a$ by the permitted operations of restricted

233

set theory: uniting two or more sets to form a new set, forming the power set and so on. The new collection of sets generated in this way by $M + a$ will be called $N$. The problem is how to choose $a$ in such a way that (1) $N$ is a model for restricted set theory, as $M$ was by assumption, and (2) $a$ is not constructible in $N$. Only if this is possible is there any hope of denying the axiom of choice or the continuum hypothesis.

We can get a vague feeling of what has to be done by asking how a geometer of 1850 who was trying to discover the pseudosphere might have proceeded. In a very rough sense, it is as if he had started with a curve $M$ in the Euclidean plane, thought of a point $a$ not in that plane, and then connected that point $a$ to all the points in $M$. Since $a$ is chosen not to lie in the plane of $M$, the resulting surface $N$ will surely not be the same as the Euclidean plane. Thus it is reasonable to think that with enough ingenuity and technical skill one could show that it is really a model for a non-Euclidean geometry.

The analogous thing in non-Cantorian set theory is to choose the new set $a$ as a nonconstructible set, then to generate a new model $N$ consisting of all sets obtained by the operations of restricted set theory applied to $a$ and to the sets in $M$. If this can be done, it will have been proved that one is safely able to negate the axiom of constructibility. Since Gödel showed that constructibility implies the axiom of choice and the continuum hypothesis, this is the necessary first step in negating either of these two statements.

In order to carry out this first step two things must be shown: that $a$ can be chosen so that it remains nonconstructible, not only in $M$ but also in $N$, and that $N$, like $M$, is a model for restricted set theory. To specify $a$ we take a round-about procedure. We imagine that we are going to make a list of all possible statements about $a$, as a set in $N$. Then $a$ will be specified if we give a rule by which we can determine whether or not any such statement is true.

The crucial idea turns out to be to choose $a$ to be a "generic" element, that is, to choose $a$ so that only those statements are true for $a$ that are true for almost all sets in $M$. This is a paradoxical notion. Every set in $M$ has both partic-

234

ular special properties that identify it, and also general typical properties that it shares with almost all the other sets in *M*. It turns out to be possible in a precise way to make this distinction between special and generic properties perfectly explicit and formal. Then when we choose *a* to be a generic set (one with, so to speak, no special properties that distinguish it from any set in *M*), it follows that *N* is still a model for restricted set theory. The new element *a* we have introduced has no troublesome properties that can spoil the *M* we started with. At the same time *a* is nonconstructible. Any constructible set has a special character—the steps by which it can be constructed—and our *a* precisely lacks any such individuality.

To construct a model in which the continuum hypothesis is false we must add to *M* not just one new element *a* but a great many new elements. In fact, we must add an infinite number of them. We can actually do this in such a way that the elements we add have cardinality

$$\aleph_2 = 2^{(2^{\aleph_0})}$$

from the viewpoint of the model *M*. Again a rough geometric analogy may be helpful: To a two-dimensional creature living embedded in a non-Euclidean surface it would be impossible to recognize that his world is part of a three-dimensional Euclidean space. In the present instance we, standing outside *M*, can see that we have thrown in only a countable infinity of new elements. They are such, however, that the counting cannot be done by any apparatus available in *M* itself. Thus we obtain a new model *N'*, in which the continuum hypothesis is false. The new elements, which in *N'* play the role of real numbers (that is, points on a line segment), have cardinality greater than $2^{\aleph_0}$, and so there is now an infinite cardinal—namely $2^{\aleph_0}$—that is greater than $\aleph_0$ and yet smaller than $c$, since in our model *N'*, $c$ is equal to

$$2^{(2^{\aleph_0})}.$$

Since we can construct a model of set theory in which the continuum hypothesis is false, it follows that we can add to our ordinary restricted set theory the assumption of the falsity of the continuum hypothesis; no contradiction can result that was not already present. In the same spirit we can construct models for set theory in which the axiom of choice fails. We can even be quite specific about which infinite sets it is possible to "choose from" and which are "too big to choose from."

Whereas Gödel produced his results with a single model (the constructible sets), we have in non-Cantorian set theory not one but many models, each constructed with a particular purpose in mind. Perhaps more important than any of the models is the technique that enables one to construct them all: the notion of "generic" and the related notion of "forcing." Very roughly speaking, generic sets have only those properties they are "forced" to have in order to be setlike. In order to decide whether $a$ is "forced" to have a certain property we must look at all of $N$. Yet $N$ is not really defined until we have specified $a!$ The recognition of how to make this seemingly circular argument noncircular is another key element in the new theory.

In the *Handbook of Mathematical Logic* (edited by J. Barwise) there is an account by J. P. Burgess of the later ramifications of Cohen's method. He wrote,

> Cohen's method has since been applied to prove consistency for hypotheses in transfinite arithmetic, infinitary combinatorics, measure theory, topology of the real line, universal algebra, and model theory.

The truth of the continuum hypothesis remains undecided. Cohen and Gödel proved that the axioms of Zermelo-Frankel set theory are not sufficient to decide it. If we believe sets are real, then we may be convinced that the continuum hypothesis must be either true or false. What we must do then is discover a new axiom which is intuitively plausible and which is strong enough to settle the question. No one has found such a new axiom, and it therefore remains our free choice to adopt the continuum hypothesis or to reject it.

### Further Readings. See Bibliography

J. Burgess; P. J. Cohen [1966]; K. Gödel.

236

# Appendix A

## Cantor's Diagonal Process

Here is a simple version of it. Consider all the functions $f$ which are defined on the integers $1, 2, 3, \ldots$ *Theorem:* It is not possible to arrange all these functions in a list. *Proof:* Assume that it is possible. Then there would be a first function in the list. Call it $f_1$. There would be a second function $f_2$, etc. Now, for each number $n$, where $n$ takes on the values $1, 2, 3, \ldots$, consider the numbers $f_n(n) + 1$. This sequence of numbers itself constitutes a function defined on the integers and so, by our assumption, it must occur in the list. Call it $f_k$. By definition, $f_k(n) = f_n(n) + 1$, and this is valid for $n = 1, 2, 3, \ldots$. In particular, it is valid for $n = k$, and this yields $f_k(k) = f_k(k) + 1$. Thus, $0 = 1$, a contradiction.

The diagonal process plays a particularly dominant role in the theory of recursive functions.

# Nonstandard Analysis

NONSTANDARD ANALYSIS, a new branch of mathematics invented by the logician Abraham Robinson, marks a new stage of development in several famous and ancient paradoxes. Robinson revived the notion of the "infinitesimal"—a number that is infinitely small yet greater than zero. This concept has roots stretching back into antiquity. To traditional, or "standard," analysis it seemed blatantly self-contradictory. Yet it has been an important tool in mechanics and geometry from at least the time of Archimedes.

In the nineteenth century infinitesimals were driven out

*Abraham Robinson*
*1918–1974*

237

of mathematics once and for all, or so it seemed. To meet the demands of logic the infinitesimal calculus of Isaac Newton and Gottfried Wilhelm von Leibniz was reformulated by Karl Weierstrass without infinitesimals. Yet today it is mathematical logic, in its contemporary sophistication and power, that has revived the infinitesimal and made it acceptable again. Robinson has in a sense vindicated the reckless abandon of eighteenth-century mathematics against the strait-laced rigor of the nineteenth century, adding a new chapter in the never ending war between the finite and the infinite, the continuous and the discontinuous.

In the controversies over the infinitesimal that accompanied the development of the calculus, Euclid's geometry was the standard against which the moderns were measured. In Euclid both the infinite and the infinitesimal are deliberately excluded. We read in Euclid that a point is that which has position but no magnitude. This definition has been called meaningless, but perhaps it is just a pledge not to use infinitesimal arguments. This was a rejection of earlier concepts in Greek thought. The atomism of Democritus had been meant to refer not only to matter but also to time and space. But then the arguments of Zeno had made untenable the notion of time as a row of successive instants, or the line as a row of successive "indivisibles." Aristotle, the founder of systematic logic, banished the infinitely large or small from geometry.

Here is a typical example of the use of infinitesimal arguments in geometry:

> We wish to find the relation between the area of a circle and its circumference. For simplicity we suppose that the radius of the circle is 1. Now, the circle can be thought of as composed of infinitely many straight-line segments, all equal to each other and infinitely short. The circle is then the sum of infinitesimal triangles, all of which have altitude 1. For a triangle the area is half the base times the altitude. Therefore the sum of the areas of the triangles is half the sum of the bases. But the sum of the areas of the triangles is the area of the circle, and the sum of the bases of the

triangles is its circumference. Therefore the area of the circle of radius 1 is equal to one half its circumference.

This argument, which Euclid would have rejected, was published in the fifteenth century by Nicholas of Cusa. The conclusion is of course true, but objections to the argument are not hard to find. The notion of a triangle with an infinitely small base is elusive, to say the least. Surely the base of a triangle must have length either zero or greater than zero. If it is zero, then the area is zero, and no matter how many terms we add we can get nothing but zero. On the other hand, if it is greater than zero, no matter how small, we will get an infinitely great sum if we add infinitely many terms. In neither case can we get a circle of finite circumference as a sum of infinitely many identical pieces.

*Archimedes*
*c. 287* B.C.–*212* B.C.

The essence of this rebuttal is the assertion that even a very small nonzero number becomes arbitrarily large if it is added to itself enough times. Because the assertion was first made explicit by Archimedes, it is called the Archimedean property of the real numbers. An infinitesimal, if it existed, would be precisely a non-Archimedean number: a number greater than zero, which nevertheless remained less than 1, say, no matter how (finitely) many times it was added to itself. Archimedes, working in the tradition of Aristotle and Euclid, asserted that every number is Archimedean; there are no infinitesimals. Archimedes, however, was also a natural philosopher, an engineer and a physicist. He used infinitesimals and his physical intuition to solve problems in the geometry of parabolas. Then, since infinitesimals "do not exist," he gave a "rigorous" proof of his results, using the "method of exhaustion," which relies on an indirect argument and purely finite constructions. The rigorous proof is given in his treatise *On the Quadrature of the Parabola,* which has been known since antiquity. The use of infinitesimals, which actually served to discover the answer, is in a paper called "On the Method," which was unknown until its sensational discovery in 1906.

Archimedes' method of exhaustion, which avoids infinitesimals, is in spirit close to the "epsilon-delta" method with

239

which Weierstrass and his followers in the nineteenth century drove infinitesimal methods out of analysis. It is easy to explain if we refer to our example of the circle as an infinite-sided polygon. We wish to get a logically acceptable proof of the formula "The area of a circle with a radius of one unit equals half the circumference," which we discovered by a logically unacceptable argument.

We reason as follows. The formula asserts the equality of two quantities associated with a circle with a radius of 1: its area and half its circumference. Thus if the formula is false, one of these quantities is larger than the other. Let $A$ be the positive number obtained by subtracting the smaller from the larger. Now, we can circumscribe about the circle a regular polygon with as many sides as we wish. Since the polygon is composed of a finite number of finite triangles with altitude 1, we know that its area is half its perimeter. By making the number of sides sufficiently large we can arrange for the polygon's area to differ from the area of the circle by less than half of $A$ (whatever its value is taken to be); at the same time the perimeter of the polygon will differ from the perimeter of the circle by less than half of $A$. But then the area and the semiperimeter of the circle must differ by less than $A$, which contradicts the supposition from which we started. Hence the supposition is impossible and $A$ must be zero, as we wished to prove.

This argument is logically impeccable. Compared with the directness of the first analysis, however, there is something fussy, even pedantic, about it. After all, if the use of infinitesimals gives the right answer, must not the argument be correct in some sense? Even if we cannot justify the concepts it employs, how can it really be wrong if it works?

Such a defense of infinitesimals was not made by Archimedes. Indeed, in "On the Method" he is careful to explain that "the fact here stated is not actually demonstrated by the argument used" and that a rigorous proof had been published separately. On the other hand, Nicholas of Cusa, who was a cardinal of the church, preferred the reasoning by infinite quantities because of his belief that the infinite was "the source and means, and at the same time the unattainable goal, of all knowledge." Nicholas was followed in his mysticism by Johannes Kepler, one of the founders of modern science. In a work less well known nowadays than his discoveries in astronomy, Kepler in 1612 used infinitesimals to find the best proportions for a wine cask. He was not troubled by the self-contradictions in his method; he relied on divine inspiration, and he wrote that "nature teaches geometry by instinct alone, even without ratiocination." Moreover, his formulas for the volumes of wine casks are correct.

*Jakob Bernoulli*
*1654–1705*

The most famous mathematical mystic was no doubt Blaise Pascal. In answering those of his contemporaries who objected to reasoning with infinitely small quantities, Pascal was fond of saying that the heart intervenes to make the work clear. Pascal looked on the infinitely large and the infinitely small as mysteries, something that nature has proposed to man not for him to understand but for him to admire.

*Johann Bernoulli*
*1667–1748*

The full flower of infinitesimal reasoning came with the generations after Pascal: Newton, Leibniz, the Bernoulli brothers (Jakob and Johann) and Leonhard Euler. The fundamental theorems of the calculus were found by Newton and Leibniz in the 1660s and 1670s. The first textbook on the calculus was written in 1696 by the Marquis de L'Hospital, a pupil of Leibniz and Johann Bernoulli. Here it is stated at the outset as an axiom that two quantities differing by an infinitesimal can be considered to be equal. In other words, the quantities are at the same time considered to be equal to each other and not equal to each other! A second axiom states that a curve is "the totality of an infinity of straight segments, each infinitely small." This is an open embracing of methods that Aristotle had outlawed 2,000 years earlier.

*Blaise Pascal*
*1623–1662*

Indeed, wrote L'Hospital, "ordinary analysis deals only with finite quantities; this one penetrates as far as infinity itself. It compares the infinitely small differences of finite quantities; it discovers the relations between these differences, and in this way makes known the relations between

*G. F. A. de L'Hospital*
*1661–1704*

241

*Sir Isaac Newton*
*1642–1727*

*Gottfried Wilhelm Leibniz*
*1646–1716*

finite quantities that are, as it were, infinite compared with the infinitely small quantities. One may even say that this analysis extends beyond infinity, for it does not confine itself to the infinitely small differences but discovers the relations between the differences of these differences."

Newton and Leibniz did not share L'Hospital's enthusiasm. Leibniz did not claim that infinitesimals really existed, only that one could reason without error as if they did exist. Although Leibniz could not substantiate this claim, Robinson's work shows that in some sense he was right after all. Newton tried to avoid the infinitesimal. In his *Principia Mathematica,* as in Archimedes' *On the Quadrature of the Parabola,* results that were originally found by infinitesimal methods are presented in a purely finite Euclidean fashion.

Dynamics had become as important as geometry in providing questions for mathematical analysis. The leading problem was the connection between "fluents" and "fluxions," what would today be called the instantaneous position and the instantaneous velocity of a moving body.

Consider a falling stone. Its motion is described by giving its position as a function of time. As it falls its velocity increases, so that the velocity at each instant is also a variable function of time. Newton called the position function the "fluent" and the velocity function the "fluxion." If either of the two is given, the other can be determined; this connection is the heart of the infinitesimal calculus fashioned by Newton and Leibniz.

In the case of the falling stone the fluent is given by the formula $s = 16t^2$, where $s$ is the number of feet traveled and $t$ is the number of seconds elapsed since the stone was released. As the stone falls its velocity increases steadily. How can we compute the velocity of the falling stone at some instant of time, say at $t = 1$?

We could find the *average* velocity for a finite time by the elementary formula: velocity equals distance divided by time. Can we use this formula to find the instantaneous velocity? In an infinitesimal increment of time the increment of distance would also be infinitesimal; their ratio, the average speed during the instant, should be the finite instantaneous velocity we seek.

We let $dt$ stand for the infinitesimal increment of time and $ds$ for the corresponding increment of distance. (Of course $ds$ and $dt$ must be thought of as single symbols and not as $d$ times $t$ or $d$ times $s$.) We want to find the ratio $ds/dt$, which is to be finite. To find the increment of distance from $t = 1$ to $t = 1 + dt$ we compute the position of the stone when $t = 1$, which is $16 \times 1^2 = 16$, and its position when $t = 1 + dt$, which is $16 \times (1 + dt)^2$. Using a little elementary algebra, we find that $ds$, the increment of distance, which is the difference of these two distances, is $32dt + 16dt^2$. Thus the ratio $ds/dt$, which is the quantity we are trying to find, is equal to $32 + 16dt$.

Have we solved our problem? Since the answer should be a finite quantity, we should like to drop the infinitesimal term, $16dt$, and get the answer, 32 feet per second, for the instantaneous velocity. That is precisely what Bishop Berkeley will not let us do.

*The Analyst*, Berkeley's brilliant and devastating critique of the infinitesimal method, appeared in 1734. The book was addressed to "an infidel mathematician," who is generally supposed to have been Newton's friend the astronomer Edmund Halley. Halley financed the publication of the *Principia* and helped to prepare it for the press. It is said that he also persuaded a friend of Berkeley's of the "inconceivability of the doctrines of Christianity"; the Bishop responded that Newton's fluxions were as "obscure, repugnant and precarious" as any point in divinity.

"I shall claim the privilege of a Free-thinker," wrote the Bishop, "and take the liberty to inquire into the object, principles, and method of demonstration admitted by the mathematicians of the present date, with the same freedom that you presume to treat the principles and mysteries of Religion." Berkeley declared that the Leibniz procedure, simply "considering" $32 + 16dt$ to be "the same" as 32, was unintelligible. "Nor will it avail," he wrote, "to say that [the term neglected] is a quantity exceedingly small; since we are told that *in rebus mathematicis errores quam min-*

*imi non sunt contemnendi.*" If something is neglected, no matter how small, we can no longer claim to have the exact velocity but only in approximation.

Newton, unlike Leibniz, tried in his later writings to soften the "harshness" of the doctrine of infinitesimals by using physically suggestive language. "By the ultimate velocity is meant that with which the body is moved, neither before it arrives at its last place, when the motion ceases, nor after; but at the very instant when it arrives. . . . And, in like manner, by the ultimate ratio of evanescent quantities is to be understood the ratio of the quantities, not before they vanish, nor after, but that with which they vanish." When he proceeded to compute, however, he still had to justify dropping unwanted "negligible" terms from his computed answer. Newton's argument was to find first, as we have done, $ds/dt = 32 + 16dt$, and then to set the increment $dt$ equal to zero, leaving 32 as the exact answer.

But, wrote Berkeley, "it should seem that this reasoning is not fair or conclusive." After all, $dt$ is either equal to zero or not equal to zero. If $dt$ is not zero, then $32 + 16dt$ is not the same as 32. If $dt$ is zero, then the increment in distance $ds$ is also zero, and the fraction $ds/dt$ is not $32 + 16dt$ but a meaningless expression, $0/0$. "For when it is said, let the increments vanish, i.e., let the increments be nothing, or let there be no increments, the former supposition that the increments were something, or that there were increments, is destroyed, and yet a consequence of that supposition, i.e., an expression got by virtue thereof, is retained. Which is a false way of reasoning." Berkeley charitably concluded: "What are these fluxions? The velocities of evanescent increments. And what are these same evanescent increments? They are neither finite quantities, nor quantities infinitely small, nor yet nothing. May we not call them the ghosts of departed quantities?"

Berkeley's logic could not be answered; nevertheless, mathematicians went on using infinitesimals for another century, and with great success. Indeed, physicists and engineers have never stopped using them. In pure mathematics, on the other hand, a return to Euclidean rigor was achieved in the nineteenth century, culminating under the leadership of Weierstrass in 1872. It is interesting to note that the eighteenth century, the great age of the infinitesimal, was the time when no barrier between mathematics and physics was recognized. The leading physicists and the leading mathematicians were the same people. When pure mathematics reappeared as a separate discipline, mathematicians again made sure that the foundations of their work contained no obvious contradictions. Modern analysis secured its foundations by doing what the Greeks had done: outlawing infinitesimals.

To find an instantaneous velocity according to the Weierstrass method we abandon any attempt to compute the speed as a ratio. Instead we define the speed as a limit, which is approximated by ratios of finite increments. Let $\Delta t$ be a variable finite time increment and $\Delta s$ be the corresponding variable space increment. Then $\Delta s/\Delta t$ is the variable quantity $32 + 16\Delta t$. By choosing $\Delta t$ sufficiently small we can make $\Delta s/\Delta t$ take on values as close as we like to the value 32, and so, by definition, the speed at $t = 1$ is exactly 32.

This approach succeeds in removing any reference to numbers that are not finite. It also avoids any attempt directly to set $\Delta t$ equal to zero in the fraction $\Delta s/\Delta t$. Thus we avoid both of the logical pitfalls exposed by Bishop Berkeley. We do, however, pay a price. The intuitively clear and physically measurable quantity, the instantaneous velocity, becomes subject to the surprisingly subtle notion of "limit." If we spell out in detail what that means, we have the following tongue-twister:

> The velocity is $v$ if, for any positive number $\epsilon$, $\Delta s/\Delta t - v$ is less than $\epsilon$ in absolute value for all values of $\Delta t$ less in absolute value than some other positive number $\delta$ (which will depend on $\epsilon$ and $t$).

We have defined $v$ by means of a subtle relation between two new quantities, $\epsilon$ and $\delta$, which in some sense are irrelevant to $v$ itself. At least ignorance of $\epsilon$ and $\delta$ never prevented Bernoulli or Euler from finding a velocity. The truth is that in a real sense we already knew what instantaneous velocity was before we learned this definition; for

245

the sake of logical consistency we accept a definition that is much harder to understand than the concept being defined. Of course, to a trained mathematician the epsilon-delta definition is intuitive; this shows what can be accomplished by proper training.

The reconstruction of the calculus on the basis of the limit concept and its epsilon-delta definition amounted to a reduction of the calculus to the arithmetic of real numbers. The momentum gathered by these foundational clarifications led naturally to an assault on the logical foundations of the real-number system itself. This was a return after two and a half millenniums to the problem of irrational numbers, which the Greeks had abandoned as hopeless after Pythagoras. One of the tools in these efforts was the newly developing field of mathematical, or symbolic, logic.

More recently it has been found that mathematical logic provides a conceptual foundation for the theory of computing machines and computer programs. Hence this prototype of purity in mathematics now has to be regarded as belonging to the applicable part of mathematics.

The link between logic and computing is to a great extent the notion of a formal language, which is the kind of language machines understand. And it is the notion of a formal language that enabled Robinson to make precise Leibniz' claim that one could without error reason as if infinitesimals existed.

Leibniz had thought of infinitesimals as being infinitely small positive or negative numbers that still had "the same properties" as the ordinary numbers of mathematics. On its face the idea seems self-contradictory. If infinitesimals have the same "properties" as ordinary numbers, how can they have the "property" of being positive yet smaller than any ordinary positive number? It was by using a formal language that Robinson was able to resolve the paradox. Robinson showed how to construct a system containing infinitesimals that was identical with the system of "real" numbers with respect to all those properties expressible in a certain formal language. Naturally the "property" of being positive yet smaller than any ordinary positive number will turn out *not* to be expressible in the language, thereby escaping the paradox.

The situation is familiar to anyone who has ever communicated with a computing machine. A computer accepts as inputs only symbols from a certain list that is given in advance to the user, and the symbols must be used in accordance with certain given rules. Ordinary language, as used in human communication, is subject to rules that linguists are still far from understanding. Computers are "stupid," if you have to communicate with them, precisely because unlike humans they work in a formal language with a given vocabulary and a given set of rules. Humans work in a natural language, with rules that have never been made fully explicit.

Mathematics, of course, is a human activity, like philosophy or the design of computers; like these other activities, it is carried on by humans using natural languages. At the same time mathematics has, as a special feature, the ability to be well described by a formal language, which in some sense mirrors its content precisely. It might be said that the possibility of putting a mathematical discovery into a formal language is the test of whether it is fully understood.

In nonstandard analysis one takes as the starting point the finite real numbers and the rest of the calculus as known to standard mathematicians. Call this the "standard universe," designated by the letter $M$. The formal language in which we talk about $M$ can be designated $L$. Any sentence in $L$ is a proposition about $M$, and of course it must be either true or false. That is, any sentence in $L$ is either true or its negation is true. We call the set of all true sentences $K$, and we say $M$ is a "model" for $K$. By this we mean that $M$ is a mathematical structure such that every sentence in $K$, when interpreted as referring to $M$, is true. Of course, we do not "know" $K$ in any effective sense; if we did, we would have the answer to every possible question in analysis. Nevertheless, we regard $K$ as being a well-defined object, about which we can reason and draw conclusions.

The essential fact, the main point, is that in addition to $M$, the standard universe, there are also nonstandard models for $K$. That is, there are mathematical structures $M^*$, essentially different from $M$ (in a sense we shall explain) and that nevertheless are models for $K$ in the natural sense of the term: there are objects in $M^*$ and relations between objects in $M^*$ such that if the symbols in $L$ are reinterpreted to apply to these pseudo-objects and pseudo-relations in the appropriate way, then every sentence in $K$ is still true, although with a different meaning.

A crude analogy may help the intuition. Let $M$ be the set of graduating seniors at Central High School. Suppose, for argument's sake, that all these students had their picture taken for the yearbook, where the students all appear in two-inch squares. Then $M^*$ can be the set of all two-inch squares on any page of the yearbook. Clearly, with an obvious interpretation, any true statement about a student at Central High corresponds to a true statement about a certain two-inch square in the yearbook. Still, there are many two-inch squares in the yearbook that do not correspond to any student. $M^*$ is much bigger than $M$; in addition to members corresponding to the members of $M$, it also contains many other members.

Hence the statement "Harry Smith is thinner than George Klein," when interpreted in $M^*$, is a statement about certain two-inch squares. It is not true if the relation "thinner than" is interpreted in the standard way. Thus "thinner than" has to be reinterpreted, as a pseudorelation, between pseudostudents (pictures of students). We could define the pseudorelation "thinner than" (in quotation marks) by saying that the two-inch square labeled "Harry Smith" is "thinner than" the two-inch square labeled "George Klein" only if Harry Smith is actually thinner than George Klein. In this way true statements about students are reinterpreted as true statements about two-inch squares.

Of course, in this example the entire argument is a bit contrived. If $M$ is the standard universe for the calculus, however, then $M^*$, the nonstandard universe, is a remarkable and interesting place.

The existence of interesting nonstandard models was first discovered by the Norwegian logician Thoralf A. Skolem, who found that the axioms of counting—the axioms that describe the "natural numbers" 1, 2, 3, and so on—have nonstandard models containing "strange" objects not comtemplated in ordinary arithmetic. Robinson's great insight was to see how this exotic offshoot of modern formal logic could be the basis for resurrecting infinitesimal methods in differential and integral calculus. In this resurrection he relied on a theorem first proved by the Russian logician Anatoli Malcev and then generalized by Leon A. Henkin of the University of California at Berkeley. This is the "compactness" theorem. It is related to the famous "completeness" theorem of Kurt Gödel, which states that a set of sentences is logically consistent (no contradiction can be deduced from the sentences) if and only if the sentences have a model, that is, if and only if there is a "universe" in which they are all true.

The compactness theorem states the following: Suppose we have a collection of sentences in the language $L$. Suppose in the standard universe every finite subset of this collection is true. Then there exists a nonstandard universe where the entire collection is true at once.

The compactness theorem follows easily from the completeness theorem: if every finite subset of a collection of sentences of $L$ is true in the standard universe, then every finite subset is logically consistent. So the entire collection of sentences is logically consistent (since any deduction can make use of only a finite number of premises). By the completeness theorem there is a (nonstandard) universe in which the entire collection is true.

A direct consequence of the compactness theorem is the "existence" of infinitesimals. To see how this amazing result follows from the compactness theorem consider the sentences:

"$C$ is a number bigger than zero and less than $\frac{1}{2}$."

"$C$ is a number bigger than zero and less than $\frac{1}{3}$."

"$C$ is a number bigger than zero and less than $\frac{1}{4}$." And so on.

This is an infinite collection of sentences each of which can be written in the formal language $L$. With reference to the standard universe $R$ of real numbers, every finite subset is true, because if you have finitely many sentences of the form "$C$ is a number bigger than zero and less than $1/n$," then one of the sentences will contain the smallest fraction $1/n$, and $1/2n$ will indeed be bigger than zero and smaller than all the fractions in your finite list of sentences. And yet if you consider the entire infinite set of these sentences, it is false with reference to the standard real numbers, because no matter how small a positive real number $c$ you choose, $1/n$ will be smaller than $c$ if $n$ is big enough.

The compactness theorem of Malcev and Henkin states that there is a nonstandard universe containing pseudoreals $R^*$ including a positive pseudoreal number $c$ smaller than any number of the form $1/n$. That is, $c$ is infinitesimal. Moreover, $c$ has all the properties of standard real numbers in a perfectly precise sense: any true statement about the standard reals that you can state in the formal language $L$ is true also about the nonstandard reals, including the infinitesimal $c$—under the appropriate interpretation. (The two-inch square labeled "Harry Smith" is not really thinner than the two-inch square labeled "George Klein," but the statement "Harry Smith" is "thinner than" "George Klein" is true, under our nonstandard interpretation of "thinner than.") On the other hand, properties shared by all the standard real numbers may not apply to the nonstandard pseudonumbers, if these properties cannot be expressed in the formal language $L$.

The Archimedean property (nonexistence of infinitesimals) of $R$ can be expressed by using an infinite set of sentences of $L$ as follows (we use the symbol ">" as usual to mean "is greater than"). For each positive element $c$ of $R$ all but a finite number of the sentences below are true:

$$c > 1$$
$$c + c > 1$$
$$c + c + c > 1, \text{ and so on.}$$

This is not true, however, for the pseudoreals $R^*$: if $c$ is infinitesimal (hence pseudoreal), all these sentences are false. In other words, no finite sum of $c$'s can exceed 1, no matter how many terms we take. The very fact that the Archime-

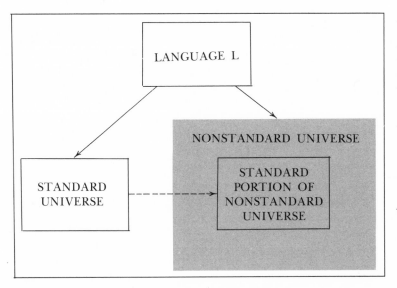

Role of formal language in mediating between standard and nonstandard universes is portrayed. Formal language L describes the standard universe, which includes the real numbers of classical mathematics. Sentences of L that are true in standard universe are also true in nonstandard one, which contains additional mathematical objects such as infinitesimals. Nonstandard analysis thus makes the infinitesimal method precise for the first time.

dean property is true in the standard world but false in the nonstandard one proves that the property cannot be expressed by a sentence of $L$; the statement we have used involves infinitely many sentences. It is precisely this distinction that makes the pseudoobjects useful. They behave "formally" like standard objects and yet they differ with respect to important properties that are not formalized by $L$.

Although the nonstandard universe is conceptually distinct from the standard one, it is desirable to think of it as an enlargement of the standard universe. Since $R^*$ is a model for $L$, every true sentence about $R$ has an interpretation in $R^*$. In particular the names of numbers in $R$ have an interpretation as names of objects in $R^*$. We can simply identify the object in $R^*$ called "2" with the familiar number 2 in $R$. Then $R^*$ contains the standard real numbers in $R$, along with the vast collection of infinitesimal and infinite quantities, in which $R$ is embedded.

An object in $R^*$ (a pseudoreal number) is called infinite if it is pseudogreater than every standard real number; otherwise it is called finite. A positive pseudoreal number is called infinitesimal if it is pseudosmaller than every positive standard real number. If the pseudodifference of two pseudoreals is finite, we say they belong to the same "galaxy"; the pseudoreal axis contains an uncountable infinity of galaxies. If the pseudodifference of two pseudoreals is infinitesimal, we say they belong to the same "monad" (a term Robinson borrowed from Leibniz' philosophical writings). If a pseudoreal $r^*$ is infinitely close to a standard real number $r$, we say $r$ is the standard part of $r^*$. All the standard reals are of course in the same galaxy, which is called the principal galaxy. In the principal galaxy every monad contains one and only one standard real number. This monad is the "infinitesimal neighborhood" of $r$: the set of nonstandard reals infinitely close to $r$. The notion of a monad turns out to be applicable not only to real numbers but also to general metric and topological spaces. Nonstandard analysis therefore is relevant not just to elementary calculus but to the entire range of modern abstract analysis.

When we say infinitesimals or monads exist, it should be clear that we do not mean this at all in the sense it would have been understood by Euclid or by Berkeley. Until 100 years ago it was tacitly assumed by all philosophers and mathematicians that the subject matter of mathematics was objectively real in a sense close to the sense in which the subject matter of physics is real. Whether infinitesimals did or did not exist was a question of fact, not too different from the question of whether material atoms do or do not exist. Today many, perhaps most, mathematicians have no such conviction of the objective existence of the objects they study. Model theory entails no commitment one way or the other on such ontological questions. What mathematicians want from infinitesimals is not material existence but rather the right to use them in proofs. For this all one needs is the assurance that a proof using infinitesimals is no worse than one free of infinitesimals.

The employment of nonstandard analysis in research goes something like this. One wishes to prove a theorem involving only standard objects. If one embeds the standard objects in the nonstandard enlargement, one may be able to find a much shorter and more "insightful" proof by using nonstandard objects. The theorem has then been proved actually with reference to the nonstandard interpretation of its words and symbols. Those nonstandard objects that correspond to standard objects have the feature that sentences about them are true (in the nonstandard interpretation) only if the same sentence is true with reference to the standard object (in the standard interpretation). Thus we prove theorems about standard objects by reasoning about nonstandard objects.

For example, recall Nicholas of Cusa's "proof" that the area of a circle with a radius of 1 equals half its circumference. In Robinson's theory we see in what sense Nicholas' argument is correct. Once infinitesimal and infinite numbers are available (in the nonstandard universe) it can be proved that the area of the circle is the standard part of the sum (in the nonstandard universe) of infinitely many infinitesimals.

Here is how the falling-stone problem would look according to Robinson. We define the instantaneous velocity not as the ratio of infinitesimal increments, as L'Hôpital did, but rather as the standard part of that ratio; then $ds$, $dt$ and their ratio $ds/dt$ are nonstandard real numbers. We have as before $ds/dt = 32 + 16dt$, but now we immediately conclude, rigorously and without any limiting argument, that $v$, the standard part of $ds/dt$, equals 32. A slight modification in the Leibniz method of infinitesimals, distinguishing carefully between the nonstandard number $ds/dt$ and its standard part $v$, avoids the contradiction, which L'Hôpital simply ignored.

Of course, a proof is required that the Robinson definition gives the same answer in general as the Weierstrass definition. The proof is not difficult, but we shall not attempt to give it here.

What is achieved is that the infinitesimal method is for the first time made precise. In the past mathematicians had to make a choice. If they used infinitesimals, they had to rely on experience and intuition to reason correctly. "Just go on," Jean Le Rond d'Alembert is supposed to have assured a hesitating mathematical friend, "and faith will soon return." For rigorous certainty one had to resort to the cumbersome Archimedean method of exhaustion or its modern version, the Weierstrass epsilon-delta method.

Now the method of infinitesimals, or more generally the method of monads, is elevated from the heuristic to the rigorous level. The approach of formal logic succeeds by totally evading the question that excited Berkeley and all the other controversialists of former times, that is, whether or not infinitesimal quantities really exist in some objective sense.

From the viewpoint of the working mathematician the important thing is that he regains certain methods of proof, certain lines of reasoning, that have been fruitful since before Archimedes. The notion of an infinitesimal neighborhood is no longer a self-contradictory figure of speech but a precisely defined concept, as legitimate as any other in analysis.

The applications we have discussed are elementary, in fact trivial. Nontrivial applications have been and are being made. Work has appeared on nonstandard dynamics and nonstandard probability. Robinson and his pupil Allen Bernstein used nonstandard analysis to solve a previously unsolved problem on compact linear operators. It must nonetheless be said that many analysts remain skeptical about the ultimate importance of Robinson's method. It is quite true that whatever can be done with infinitesimals can in principle be done without them. Perhaps, as with other radical innovations, the full use of the new ideas will be made by a new generation of mathematicians who are not too deeply embedded in standard methods to enjoy the freedom and power of nonstandard analysis.

### Further Readings. See Bibliography

A. Robinson [1966]. M. Davis [1977] K. D. Stroyan and W. A. J. Luxemburg

# Fourier Analysis

TO GIVE A PERFORMANCE of Verdi's opera *Aida*, one could do without brass and woodwinds, strings and percussion, baritones and sopranos; all that is needed is a complete collection of tuning forks, and an accurate method for controlling their loudness.

*Hermann von Helmholtz*
*1821–1894*

This is an application to acoustics of "Fourier's theorem," one of the most useful facts in many branches of physics and engineering. A physical "proof" of the theorem was given by Hermann von Helmholtz when he demonstrated the production of complex musical sounds by suitable combinations of electrically driven tuning forks. (Nowadays, devices of this kind are called electronic music synthesizers.)

In mathematical terms, each tuning fork gives off a vibration whose graph as a function of time is a sine wave:

The distance from one peak to the next, the wave-length or the period, would be $\frac{1}{264}$ of a second if the note were middle C. The height of each peak is the amplitude, and roughly measures the loudness. The physical basis of any musical sound is a periodic variation in air pressure whose graph might be a curve like this:

Fourier's theorem says, in graphical terms, that a curve

255

like the one just shown can be obtained by adding up graphs like the first one:

*A combination of three pure tones whose frequency ratios are small whole numbers.*

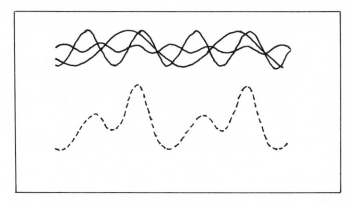

In analytic terms, the theorem says that if *y* is a periodic function that repeats, say, 100 times a second, then *y* has an expansion like

$$y = 7 \text{ sine } 200\pi t + 0.3 \text{ sine } 400\pi t$$
$$+ 0.4 \text{ sine } 600\pi t + \cdots$$

In each term, the time *t* is multiplied by $2\pi$ times the frequency. The first term, with frequency 100, is called the fundamental, or first harmonic; the higher harmonics all have frequencies that are exact multiples of 100. The coefficients 7, 0.3, 0.4, and so on have to be adjusted to suit the particular sound which we have called "*y*." The three dots at the end means that the expansion continues indefinitely; the more terms that are included, the more nearly is the sum equal to *y*.

What if *y* is not periodic—does not repeat itself no matter how long we wait? In that case, we can think of *y* as the limit of a sequence of functions with longer and longer periods—(which means smaller and smaller frequencies). Fourier's theorem would then require a sum which includes *all* frequencies, not just multiples of a given fundamental frequency. The expansion is then called a Fourier integral instead of a Fourier series.

Once we have translated the theorem from physical to mathematical terms, we have a right to ask for a statement and a proof which meet mathematical standards. What precisely do we allow *y* to be as a mathematical function? What precisely do we mean by the sum of an infinite series? These questions, raised by the practical demands of Fourier analysis, have taxed the efforts of every great analyst since Euler and Bernoulli; they are still receiving new answers today.

One new answer, a very practical one, is an efficient and ingenious technique for numerically carrying out Fourier analysis on a digital computer. A famous paper by J. W. Cooley and J. W. Tukey in 1965 exploited the binary notation inherent in today's machine computations to make a radical saving in computing time. By taking maximum advantage of the symmetry properties of sine waves, they reduced the number of operations required to find the Fourier expansion of a function given at $N$ data points from $N^2$ operations to $(2N)$ times the (logarithm of $N$ to the base 2). This reduction was enough to mean that in many applications the effective computational use of Fourier expansions became feasible for the first time. It was reported, for example, that for $N = 8192$, the computations took about five seconds on an IBM 7094; conventional procedures took half an hour.

The origin of Fourier series actually goes back to a problem closely related to the musical interpretation of Fourier analysis with which we began. The problem is that of the motion of a vibrating string.

### Waves on Strings

The "wave equation" which governs the vibration of a string was derived in 1747 by d'Alembert. He also found a solution of the equation, in the form of the sum of two traveling waves, of identical but "arbitrary" form, one moving to the right and one moving to the left. Now, if the string is initially at rest (zero velocity), its future motion is determined completely by its initial displacement from equilibrium. Thus there is one arbitrary function in the problem (the one which gives the initial position of the string before it is released) and there is one arbitrary func-

*Jean Le Rond d'Alembert*
*1717–1783*

tion in d'Alembert's solution (the one which gives the shape of the traveling wave). D'Alembert therefore considered that he had given the general solution of the problem.

However, it is essential to understand that d'Alembert and his contemporaries meant by "function" what nowadays would be called a "formula" or "analytic expression." Euler pointed out that there is no physical reason to require that the initial position of the string is given by a single function. Different parts of the string could very well be described by different formulas (line segments, circular arcs, and so on) as long as they fitted together smoothly. Moreover, the traveling wave solution could be extended to this situation. If the shape of the traveling waves matched the shape of the initial displacement, then Euler claimed the solution was still valid, even though it was not given by a single function but by several, each valid in a different region. The point is that for Euler and d'Alembert every function had a graph, but not every graph represented a single function. Euler argued that any graph (even if not given by a function) should be admitted as a possible initial position of the string. D'Alembert did not accept Euler's physical reasoning.

In 1755 Daniel Bernoulli joined the argument. He found another form of solution for the vibrating string, using "standing waves." A standing wave is a motion of the string in which there are fixed "nodes" which are stationary; between the nodes each segment of the string moves up and down in unison. The "principal mode" is the one without nodes, where the whole string moves together. The "second harmonic" is the name given to the motion with a single node at the mid-point. The "third harmonic" has two equally-spaced nodes, and so on. At any instant, in each of these modes, the string has the form of a sine curve, and at any fixed point on the string the motion in time is given by a cosine function of time. Each "harmonic" thus corresponds to a pure tone of music. Bernoulli's method was to solve the general problem of the vibrating string by summing an infinite number of standing waves. This required that the initial displacement be the sum of an infinite number of sine functions. Physically it meant that any sound produced by the string could be obtained as a sum of pure tones.

Just as d'Alembert had rejected Euler's reasoning, now Euler rejected Bernoulli's. First of all, as Bernoulli acknowledged, Euler himself had already found the standing wave solution in one special case. Euler's objection was to the claim that the standing-wave solution was *general*—applicable to all motions of the string. He wrote,

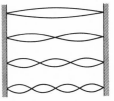

*The first four modes of vibration of a string fixed at both ends.*

> For consider that one has a string which, before release, has a shape which can't be expressed by the equation
>
> $$y = \alpha\sin(\pi x/a) + \beta\sin(2\pi x/a) + \cdots$$
>
> There are none who doubt that the string, after a sudden release, will have a certain movement. It's quite clear that the figure of the string, the instant after release, will also be different from this equation, and even if, after some time, the string conforms to this equation, one cannot deny that before that time, the movement of the string was different from that contained in the consideration of Bernoulli.

Bernoulli's method involved representing the initial position as an infinite sum of sine functions. Such a sum is a concrete analytic expression which Euler would have regarded as a *single function,* and therefore, to his way of thinking, could not represent an initial position composed of several distinct functions joined together. Moreover, it seemed evident to Euler that the sine series couldn't even represent an arbitrary *single* function, for its ingredients are all periodic and symmetric to the origin. How then could it equal a function which lacked these properties?

Bernoulli did not yield his ground; he maintained that since his expansion contained an infinite number of undetermined coefficients, these could be adjusted to match an arbitrary function at infinitely many points. This argument today seems feeble, for equality at an infinite number of points by no means guarantees equality at *every* point. Nevertheless, as it turned out, Bernoulli was closer to the truth than Euler.

Euler returned to the subject of trigonometric series in

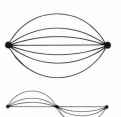

*Standing waves on a string; principal mode and second harmonic.*

259

*A graph composed of two circular arcs and one straight line. To Euler, this graph is not a graph of one function, but of three. To Fourier and Dirichlet, it is the graph of one function which has a Fourier series expansion.*

1777. Now he was considering the case of a function which was known to have a cosine expansion,

$$f(x) = a_0 + a_1 \cos x + a_2 \cos 2x + \cdots$$

and he wanted to find a convenient formula for the coefficients $a_0$, $a_1$, etc. It seems strange that this problem, which we know today can be solved in one line of calculation, had not already been solved by either Bernoulli or Euler. What is even more astonishing, Euler now found the correct formula, but only by a complicated argument involving repeated use of trigonometric identities and two passages to the limit. Once he had arrived at the simple formula for what we now call the "Fourier coefficients," he did notice the easy trick that would have given the answer immediately.

Say that we want the fifth coefficient $a_5$. Write down the assumed expansion of $f$ with unknown coefficients:

$$f = a_0 + a_1 \cos x + \cdots + a_5 \cos 5x + \cdots$$

Multiply both sides by $\cos 5x$ and integrate (i.e., average between the limits $x = 0$ and $x = \pi$).

Now, on the right side we have the integral of an infinite series. By mathematicians of Euler's time, it was taken for granted that one may always evaluate such an expression by integrating each term separately and then adding. But if we integrate each term, we discover that a wonderful thing happens. All the integrals but the fifth equal zero! Since we easily compute $\int_0^\pi a_5 (\cos 5x)^2 dx = \dfrac{\pi a_5}{2}$, we have

$\int_0^\pi f(x) \cos 5x \, dx = \dfrac{\pi a_5}{2}$, and so $a_5 = \dfrac{2}{\pi} \int_0^\pi f(x) \cos 5x \, dx$. Similar arguments of course work for all the other coefficients.

This beautifully simple argument is based completely on the fact that $\int_0^\pi \cos mx \cos nx \, dx = 0$ if $m$ is different from $n$.

(A similar formula holds for sines). This property of the cosines is nowadays described in the phrase, "The cosines are orthogonal over the interval from 0 to $\pi$." The justification for this geometric language (orthogonal means perpendicular) will appear later on in our story.

In order to appreciate Fourier's work, it is essential to understand that Euler believed to the end that only a very special class of functions, given everywhere by a single analytic expression, could be represented by a sine or cosine series. Only in these special cases did he believe his coefficient formula to be valid.

Fourier's use of sines and cosines in studying heat flow was very similar to Bernoulli's method of studying vibrations. Bernoulli's standing wave is a function of two variables (time $t$ and space $x$) which has the very special property that it factors into a function of space times a function of time. For such a product to satisfy the vibrating string equation, the two factors must both be sines or cosines. The boundary conditions (ends fixed, initial velocity zero) and the length of the string then determine that they will be of the form $\sin nx$ and $\cos mt$.

When Fourier derived his equation for heat conduction, he found that it too had special solutions which were factorable into a function of space times a function of time. In this case the function of time is exponential rather than trigonometric, but if the solid whose heat flow we are studying is rectangular, we again obtain trigonometric functions of space.

Suppose, for instance, that we have a block of metal whose surface is maintained at a fixed temperature. Then physical considerations show that the interior temperature distribution at time $t = 0$ is sufficient to determine the interior distribution at all later times. But this initial temperature distribution can be arbitrary. *Fourier asserted nevertheless that it is equal to the sum of a series of sines and cosines.* In this he was repeating Bernoulli's point of view. But whereas Bernoulli had in mind only those functions which are formed analytically by a single expression, Fourier explicitly included functions (temperature distributions) given piecewise by several different formulas. In other words, he was asserting that the distinction between "function" and graph, which had been implicitly recognized by all previous analysts, was nonexistent; just as every "func-

tion" has a graph, so every graph represents a function—
its Fourier series! No wonder Lagrange, the eighteenth
century analyst par excellence, found Fourier's claim hard
to swallow.

### How Fourier Calculated

*Joseph Fourier*
*1768–1830*

Of course an essential step in Fourier's work was to find
the formula for the coefficients in the expansion. Fourier
didn't know Euler had already done this, so he did it over.
And Fourier, like Bernoulli and Euler before him, overlooked the beautifully direct method of orthogonality
which we have just explained. Instead, he went through an
incredible computation, that could serve as a classic example of physical insight leading to the right answer in spite
of flagrantly wrong reasoning.

He started out by expanding each sine function in a
power series (Taylor series), and then rearranging terms,
so that the "arbitrary" function $f$ is now represented by a
power series. This already is objectionable, for the functions Fourier had in mind certainly have no such expansion in general. Nevertheless, Fourier proceeded to find
the coefficients in this nonexistent power-series expansion.
In doing so he used two flagrantly inconsistent assumptions, and arrived at an answer involving division by a divergent infinite product (i.e., an arbitrarily large number).
The only sensible interpretation one could give to this formula for the power series expansion was that all the coefficients vanish—i.e., the "arbitrary" function is identically
zero. "Fourier had no intention whatsoever of drawing
that conclusion, and hence proceeded undismayed with
the analysis of his formula." This is the comment of Rudolf
Langer, in an article to which we are indebted for our syn-
opsis of Fourier's derivation. From this unpromising formula Fourier was able, by dint of still more formal manipulations, ultimately to arrive at the same simple formula that
Euler had obtained correctly, and much more easily, thirty
years earlier.

It is a tribute to the insight of Legendre, Laplace, and
Lagrange that they awarded Fourier the Grand Prize of
the Academy despite the glaring defects in his reasoning.

262

For Fourier's master stroke came *after* he arrived at Euler's formula. At this point he noticed, as Euler had, that the simple formula could have been obtained in one line, by using the orthogonality of the sines. But then he observed further, as no one before him had done, that the final formula for the coefficients, and the derivation by the orthogonality of the sines, remain meaningful for any graph which bounds a definite area—and this, for Fourier, meant any graph at all. He had already computed the Fourier series for a number of special examples. He found numerically in every case that the sum of the first few terms was very close to the actual graph which generated the series. On this basis, he proclaimed that every temperature distribution—or, if you will, every graph, no matter how many separate pieces it consists of—is representable by a series of sines and cosines. It should be clear that while a collection of special examples may carry conviction, it is in no sense a proof as that word was and is understood by mathematicians. "It was, no doubt," says Langer, "partially because of his very disregard for rigor that he was able to take conceptual steps which were inherently impossible to men of more critical genius."

Fourier was right, even though he neither stated nor proved a correct theorem about Fourier series. The tools he used so recklessly give his name a deserved immortality. To make sense out of what he did took a century of effort by men of "more critical genius," and the end is not yet in sight.

### What is a Function?

First of all, what about Euler's seemingly cogent objections of half a century before? How was it possible that a sum of periodic functions (sines and cosines) could equal an arbitrary function which happened *not* to be periodic? Very simply. The arbitrary function is given only on a certain range, say from 0 to $\pi$. Physically, it represents the initial displacement of a string of length $\pi$, or the initial temperature of a rod of length $\pi$. It is only in this range that the physical variables are meaningful, and it is precisely in this range that the Fourier series equals the given function.

*Peter Gustav Lejeune
Dirichlet
1805–1859*

It is irrelevant whether the given function may have a continuation outside this range; if it does, it will *not* in general equal the Fourier series there. In other words, it can perfectly well happen that we have two functions which are identical on a certain range, say from 0 to $\pi$, and unrelated elsewhere. This is a possibility that seems never to have been considered by d'Alembert, Euler, and Lagrange. It not only made possible the systematic use of Fourier series in applied mathematics; it also led to the first careful and critical study of the notion of function, which in all its ramifications is as fruitful as any other idea in science.

It was Dirichlet (1805–1859) who took Fourier's examples and unproved conjectures and turned them into respectable mathematics. The first prerequisite was a clear, explicit definition of a function. Dirichlet gave the definition which to this day is the most often used. A function $y(x)$ is given if we have *any* rule which assigns a definite value $y$ to every $x$ in a certain set of points. "It is not necessary that $y$ be subject to the same rule as regards $x$ throughout the interval," wrote Dirichlet; "indeed, one need not even be able to express the relationship through mathematical operations . . . It doesn't matter if one thinks of this [correspondence] so that different parts are given by different laws or designates it [the correspondence] entirely lawlessly. . . . If a function is specified only for part of an interval, the manner of its continuation for the rest of the interval is entirely arbitrary."

Was this what Fourier meant by "an arbitrary function"? Certainly not in the sense in which Dirichlet interpreted the phrase, "any rule." Consider the following famous example which Dirichlet gave in 1828: $\phi(x)$ is defined to be 1 if $x$ is rational, $\phi(x) = 0$ if $x$ is irrational. Since every interval, no matter how small, contains both rational and irrational points, it would be quite impossible to draw a graph of this function. Thus, with Dirichlet's definition of a function, analysis has overtaken geometry and left it far behind. Whereas the restricted eighteenth-century concept of function was not adequate to describe such easily visualized curves as on page 260 the nineteenth-century concept of an arbitrary function includes creatures beyond anyone's hope of drawing or visualizing.

It is readily evident that one can hardly expect this 0–1 function of Dirichlet to be represented by a Fourier series. Indeed, since the area under such a "curve" is undefined, and since the Euler coefficients are obtained by integrating (i.e., computing an area), Fourier could not have found even a single term of the Fourier series for this example. But of course, the practical-minded physicist Fourier did not have in mind such perverse inventions of pure mathematics as this.

On the positive side, Dirichlet proved, correctly and rigorously, that if a function $f$ has a graph which contains only a finite number of turning points, and is smooth except for a finite number of corners and jumps, then the Fourier series of $f$ actually has a sum whose value at each point is the same as the value of $f$ at that point. (Assuming that at points where $f$ has a jump, it is assigned a value equal to the average of the values on the left and on the right.)

This is the result that used to be presented in old-fashioned courses on engineering mathematics, on the grounds that any function which ever arises in physics would satisfy "Dirichlet's criterion." It is plausible that any curve which can be drawn with chalk or pen satisfies Dirichlet's criterion. Yet such curves are far from adequate to represent all situations of physical or engineering interest.

Let us stress the meaning of Dirichlet's result. The formula $y(x) = b_1 \sin x + b_2 \sin 2x + b_3 \sin 3x + \cdots$ is true in the following sense. If we choose any given value $x_0$ between 0 and $\pi$, then $y(x_0)$ is a number, and the right hand side is a sum of numbers. It is asserted that if enough terms are taken in the series, the sum of numbers is as close as you wish to the value of $y$ at the given point $x_0$. This is *pointwise* convergence, the seemingly simplest, and in reality the most complicated, of many possible notions of convergence. From a purely mathematical viewpoint, Dirichlet's result was not an end but a beginning. What a mathematician wants is a nice, clear-cut answer—a necessary and sufficient condition, as they say in the trade. Dirichlet's criterion is a sufficient condition, but by no means a necessary one.

Bernhard Riemann (1816–1866) saw that further progress required a more general concept of integration, powerful enough to handle functions with infinitely many discontinuities. For Euler's formula gives the Fourier coefficients of $f$ as integrals of $f$ times a sine wave. If the function $f$ is generalized beyond the intuitive notion of a smooth curve, then the integral of $f$ must also be generalized beyond the intuitive notion of area under a curve. Riemann accomplished such a generalization. Using his "Riemann integral," he was able to give examples of functions violating Dirichlet's conditions, yet still satisfying Fourier's theorem.

The search for necessary and sufficient conditions under which Fourier's theorem should be valid has been long and arduous. For physical applications, one certainly wants to admit functions having jumps. This means we allow $f$ to be discontinuous. We certainly want it to be integrable, since the coefficients are computed by an integration. Now, if the values of $f$ are changed at a single point, or several points, this is not enough to affect the value of the integral (which is the average of $f$ over all the uncountably many points between $0$ and $\pi$.) Therefore the Fourier coefficients of $f$ are unchanged. This observation shows that pointwise convergence is not the natural way to study the problem, since there can be points where two functions $f$ and $g$ differ, yet $f$ and $g$ can still have the same Fourier expansion. Indeed, it was the attempt to understand which sets of points were irrelevant for the Fourier series that led Georg Cantor to take the first steps in creating his abstract theory of sets.

A more modest and more reasonable request than convergence at *every* point is to ask that the Fourier series of the function $f$ should equal $f$ except possibly on a set so small it is not noticed by the process of integration. Such sets, defined precisely by H. Lebesgue (1875–1941), are called sets of measure zero, and are used to define a still

266

more powerful notion of the integral than Riemann's. One can think of these sets in the following way: if you pick a point between 0 and 1 at random, your chance of landing in any given interval just equals the length of that interval. If your chance of landing on a given set of points is zero, then that set is said to have measure zero.

*Henri Lebesgue*
*1875–1941*

The length of a point is zero by definition. If we add the lengths of several points, this sum is also zero. Therefore, a set of finitely many points has measure zero. There are also sets of measure zero with infinitely many points. It is even possible for a set to have measure zero and yet to be "everywhere dense"—i.e., have a representative in every interval, however small. In fact, the set of all rational numbers is precisely such an everywhere dense set of measure zero. Thus from Lebesgue's viewpoint, Dirichlet's 0-1 function *does* have a Fourier expansion—and every coefficient is zero, since the function is zero "almost everywhere," as Lebesgue put it. This is the kind of mathematics that makes "practical" people shudder. What use is a Fourier expansion if it gives the wrong answer, not just at a few isolated points, but on an everywhere dense set?

But even if we are willing to accept convergence only "almost everywhere" (i.e., except on a set of measure zero), we may not get it. In 1926 Kolmogorov constructed an integrable function whose Fourier series diverged *everywhere*. So integrability alone certainly is not a basis for even an "almost everywhere" theory.

**Generalized Functions**

A different approach, and one very much in the mainstream of modern analysis, is to take the "orthogonality" property of the sine wave much more seriously. If $f$ is a $\pi$-periodic function whose *square* is integrable, it follows from the orthogonality of sines that $\dfrac{1}{\pi} \int_0^\pi f^2 = b_1^2 + b_2^2 + b_3^2 + \cdots$ where the $b$'s are the coefficients in the sine wave expansion of $f$. (For the proof, evaluate $\int_0^\pi f^2 = \int_0^\pi f \cdot f$ by expanding each factor $f$ in its sine series, multiplying the first series by the second, and integrating term by

267

term. Because of the orthogonality, most of the integrals are zero, and the rest can be evaluated to give the formula.)

The key idea now is to notice that this sum of squares is analogous to that appearing in the Pythagorean theorem of Euclidean geometry.

According to elementary Euclidean geometry, if $P$ is a point with coordinates $(x, y)$ in the plane or $(x, y, z)$ in space, then the vector $OP$ from the origin to $P$ has a length whose square equals

$$\overline{OP}^2 = x^2 + y^2 \quad \text{or}$$
$$\overline{OP}^2 = x^2 + y^2 + z^2 \quad \text{respectively.}$$

This analogy suggests that we think of the function $f$ as a vector in some sort of super-Euclidean space, with rectangular (orthogonal) coordinates $b_1$, $b_2$, $b_3$, etc. Evidently it will be an infinite-dimensional space. Then the "length" of $f$ will have a natural definition as the square root of $\dfrac{1}{\pi} \displaystyle\int_0^\pi f^2$, which is the same as the square root of $b_1^2 + b_2^2 + b_3^2 + \cdots$. The "distance" between two functions $f$ and $g$ would be the "length" of $f - g$.

The space of functions thus defined is called $L_2$, and is the oldest and standard example of the class of abstract spaces known as Hilbert spaces. The 2 in $L_2$ comes from the exponent in the squaring operation. The $L$ reminds us that we must integrate with respect to Lebesgue measure. Now we have a new interpretation for convergence of the Fourier series; we ask that the sum of the first 10,000 terms (or 100,000, or 1,000,000 if necessary) should be close to $f$ in the sense of distance in $L_2$. That is, the difference should give a small number when squared and integrated.

From the Hilbert-space view-point, the subtleties and difficulties of Fourier analysis seem to evaporate like mist. Now the facts are simply proved and simply stated: a function is in $L_2$ (i.e., is square-integrable) if and only if its Fourier series is convergent in the sense of $L_2$. (This fact has gone down in history as the Riesz-Fischer theorem.)

There remained, however, the open question as to how

bad the pointwise behavior of $L_2$ functions could be. In view of Kolmogorov's example of an integrable function whose Fourier series diverged everywhere, there was a great sensation when in 1966 Lennart Carleson proved that if a function is *square* integrable, its Fourier series *converges* pointwise almost everywhere. This includes as a special case the new result that a continuous periodic function has a Fourier series that converges almost everywhere. The theory was rounded off when, also in 1966, Katznelson and Kahane showed that for any set of measure zero there exists a continuous function whose Fourier series diverges on that set.

*This is a graph of a function which is close to zero in the sense of the Hilbert space $L_2$, but not in the ordinary sense of distance between curves. The tall spikes are negligible in $L_2$ because the area they contain is very small.*

It is interesting to observe that this modern development really involves a further evolution of the concept of function. For an element in $L_2$ is not a function, either in Euler's sense of an analytic expression, or in Dirichlet's sense of a rule or mapping associating one set of numbers with another.

It is function-like in the sense that it can be subjected to certain operations normally applied to functions (adding, multiplying, integrating). But since it is regarded as unchanged if its values are altered on an arbitrary set of measure zero, it is certainly not just a rule assigning values at each point in its domain.

As we have seen, the development of Fourier analysis in the nineteenth century achieved logical rigor, but at the price of a certain split between the pure and applied viewpoints. This split still exists, but the thrust of much recent and contemporary work is to reunite these two aspects of Fourier analysis.

First of all, the concept of Hilbert space, abstract as it is, provides the foundation of quantum mechanics. It has therefore been an essential topic in applied mathematics for the last fifty years. Moreover, the major expansion of Fourier analysis, in Norbert Wiener's generalized harmonic analysis, and in Laurent Schwartz's theory of generalized functions, is directly motivated by applications of the most concrete kind. For instance, in electrical engineering one often imagines that a circuit is closed instantaneously. Then the current would jump from a value of zero before

*Andrei Kolmogorov 1903–*

*Norbert Wiener 1894–1964*

*Dirac's delta function is zero except at a single point, where it is infinite. In modern analysis, even this very eccentric "function" can be expressed as an infinite series of cosines.*

the switch is closed to a value of, say, 1, after the switch is closed. Clearly there is no finite rate of change of current at switch-on time. To put it geometrically, the graph of the current is vertical at $t = 0$. Nevertheless, it is very convenient in calculations to use a fictitious rate of change, which is infinite at $t = 0$ (Dirac's delta function). The theory of generalized functions supplies a logical foundation for using such "impulse" functions or pseudofunctions. This theory permits us to differentiate *any* function as many times as we please; the only trouble is that we must allow the resulting object to be, not a genuine function, but a "generalized function." In a historical perspective, the interesting thing is that the concept of function has had to be widened still further, beyond either Dirichlet or Hilbert.

Furthermore, one of the payoffs for this widening is that in a sense we can return to the spirit of Fourier. For when we construct the Fourier expansion of one of these "generalized functions," we obtain a series or integral which is divergent in any of the senses we have considered. Nevertheless, formal manipulations in the style of Euler or Fourier now often become meaningful and reliable in the context of the new theory.

Thus mathematicians have labored for a century and a half to justify some of Fourier's computations. On the other hand, few physicists and engineers ever felt the need for a justification. (After all, a working gadget or a successful experiment speaks for itself.) Still, they do seem to take some comfort from the license which mathematics has now granted them. In recent applied textbooks, the early pages are now lightly sprinkled with references to Laurent Schwartz as if to justify previously "illicit" computations.

### Further Readings. See Bibliography

F. J. Arago; E. T. Bell [1937]; J. W. Dauben; I. Grattan-Guiness; R. E. Langer; G. Weiss

# 6

## TEACHING AND LEARNING

# Confessions of a Prep School Math Teacher

T ED WILLIAMS (pseudonym) is the Chairman of the Mathematics Department in a fine private school in New England. He was interviewed in April, 1978.

Williams, who is in his early forties, teaches mathematics, physics, and general science. He also coaches the boys' baseball team. He says that he prefers to teach mathematics rather than physics because it is hard to keep up with the new developments in physics. Williams has a master's degree in mathematics from an Ivy League school and has had an introductory college course in philosophy. As far as philosophy of science is concerned, he informed the interviewer that some years ago he had read Poincaré's *Science and Hypothesis,* and recently he had read in Minsky's book on perceptrons (but he says he didn't get the point). He also read Bronowski's book that went with the TV series. He has read a bit of the history of mathematics. He says that his school expects him to do too much and he has very little time for reading.

Williams says that the history and philosophy of mathematics just don't come up in his classroom.

In answer to the question of whether mathematics is discovered or invented, he answered with a snap, "There's not much difference between the two. Why waste time trying to figure it out? The thing that is important is that doing math is fun. That's what I try to put across to the kids."

When pressed harder on this question, he said, "Well, I think it's discovered."

272

Asked whether he'd ever thought about the consistency of mathematics, he said, "I've heard about the Russell paradox and all that, but I really don't understand it. I think math is like a sandcastle. It's beautiful, but it's made of sand."

"If it's made of sand, how do you justify its study to your students?"

"I tell them that figures don't lie. You know they don't. No one has come up with any counterexamples to show that they do. But the whole question is irrelevant to me."

In answer to the question as to whether there is a difference between pure and applied mathematics, Williams answered, "Pure math is a game. It's fun to play. We play it for its own sake. It's more fun than applying it. Most of the math that I teach is never used by anyone. Ever. There's no math in fine arts. There's no math in English. There's no math to speak of in banking. But I like pure math. The world of math is nice and clean. Its beautiful clarity is striking. There are no ambiguities."

"But there are applications of math?"

"Of course."

"Why is math applicable?"

"Because nature observes beautiful laws. Physicists didn't get far before math was around."

"Does the number $\pi$ exist apart from people? Would the little green man from Galaxy X–9 know about $\pi$?"

"As one gets older, one is less and less inclined to trouble oneself about this kind of question."

"Is there beauty in mathematics?"

"Oh yes. For example, if you start with a few axioms for a field, you get a whole powerful theory. It's nice to watch a theory grow out of nothing."

Mr. Williams alluded to the fact that his school now had a computer and he was teaching programming.

"What's the purpose of computing?"

"No one in high school asks 'why.' It's there. It's fun."

"Is programming a form of mathematics?"

"No. Programming is thinking. It's not math."

"Is there such a thing as mathematical intuition?"

"Oh, yes. You see it in students. Some are faster than

others. Some have more. Some have less. It can be developed but that takes sweat. Mathematics is patterns. If a person has no visual sense, he is handicapped. If one is 'tuned in,' then one learns fast and the subject is fascinating. Otherwise it is boring. There are many parts of math that bore me. Of course I don't understand them."

"Is there a mystic aspect to math?"

"Math is full of arcane symbols and this is an attraction. If one talks to a 'real' mathematician, one sees he is bright. And he is prying open secrets. And because of this, people know a little more. One is cowed by this larger knowledge."

"Where is mathematical research going?"

"I don't have the foggiest idea."

"How would you sum it all up?"

"As a teacher I am constantly confronted by problem after problem that has nothing to do with math. What I try to do is to sell math to kids on the basis that it's fun. In this way I get through the week."

# The Classic Classroom Crisis of Understanding and Pedagogy

"The grim esotericism, in which even the best of us sometimes fall, the preponderance, in our current writing, of those dreary textbooks which bad teaching concepts have put in place of true synthesis, the curious modesty, which, as soon as we are outside the study, seems to forbid us to expose the honest groping of our methods before a profane public. . . ."

Marc Bloch, *The Historian's Craft*

## 1. Introduction

A certain fraction of time in every college mathematics lecture is devoted to theorem proving. This fraction is very likely to be greater the deeper or the more abstract the material is. One of the objects of proof is to convince the students by reason, by psychology, by intuition, of the truth of certain statements. It often happens in more elementary classes—and this is experienced by all teachers—that some honest and confused student interrupts the proof with the cry: "I don't see why you did what you did, and I don't understand why what you say is so, is so. Moreover, I don't understand how you came to do what you did."

The teacher is confronted with a crisis of understanding. How is it dealt with? Not very well, unfortunately. Perhaps the teacher goes over the sore spot again in slightly different terms or perhaps, anxious to cover a certain amount of material, he brushes the student aside with the remark that understanding will surely come if the student will only go over the material back in his own room.

In the middle of a lecture in a course on the nature of mathematics (of which this book is partially an outcome), such a crisis surfaced. This elicited from me the standard overt reaction. But I caught myself. Instead of brushing the crisis aside, I altered the course of the lectures completely until I had explored the mathematical difficulty and the response to such difficulties along the lines now to be presented.

## 2. The Two-Pancake Problem

The theorem that was under discussion and whose proof elicited the crisis often goes by the name of "the two-pancake problem." The theorem says that the area of two plane pancakes of arbitrary shape can be simultaneously bisected by a single straight-line cut of the knife (see adjoining figure). This interesting theorem is part of elementary real variable theory or elementary topology. It is often given as an application of the properties of continuous functions. One of the things that makes the theorem attractive is its generality. The pancake doesn't have to have

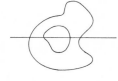

any particular shape, for example circular, square, or elliptical. The pancakes could even have holes or bubbles in them. The price that one pays for this generality is that the theorem has only an existential nature: it tells us that there exists a linear knife-cut that does the bisecting, but doesn't tell us how to find the knife-cut exactly. This could not be done in the absence of numerical information as to the precise shape and location of the pancakes.

The theorem is light-hearted. It has great visual and kinesthetic affect. In one's mind's eye one may imagine making cut after cut in a trial and error process.

I am not aware of any serious application of this theorem, but do not preclude this possibility. There are a number of special cases and generalizations. For example, the pancakes can be in any position in the plane. They may overlap one another. If one is totally contained within the other, then it can be interpreted as an island in a lake. A straight line exists that simultaneously bisects the area of both. Generalizing to three dimensions, one has the famous ham sandwich theorem: a sandwich is formed from a piece of white bread, a piece of dark bread, and a slice of ham. There exists a planar slice of the knife which simultaneously bisects the volume of all three so that each of two people can get equal shares.

### 3. Proof: First Version

I shall now give the proof, more or less as I presented it in class, in the form which led to the crisis. It was modelled on the one found in Chinn and Steenrod. The notion of area and of its continuity properties was taken at the intuitive level. We assume the pancakes are bounded in size. Consider a single pancake. If one starts with a cut that is clear of the pancake ($C_1$), all the area is to one side of the cut. As the knife is moved parallel to itself, less and less area is on that side. Ultimately the knife cut ($C_0$) is again clear of the pancake and there is zero area to that side (see figure). As the cut moves continuously from $C_1$ to $C_0$, the area to one side of it changes continuously and in a constantly decreasing way from 100% at $C_1$ to 0% at $C_0$. Thus,

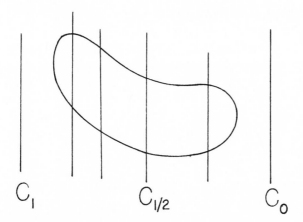

there must exist a unique position $C_{1/2}$ where the area is bisected. This establishes

*Lemma.* Given an arbitrary direction in the plane $\theta$: $0 \leq \theta \leq 180°$, there is a unique knife-cut of this direction (or perpendicular to this direction) which bisects a given pancake.

We are now ready to prove the full theorem.

Draw any circle containing the two pancakes. Use this circle as a reference system. Mark its center and draw diameters through the center with angles $\theta$ where $\theta$ varies from $0°$ to $180°$. For each $\theta$, there is a unique cut perpendicular to the $\theta$ ray and bisecting pancake I. Along the directed ray, let this cut intersect the ray at a distance $p(\theta)$ from the origin. Let $q(\theta)$ be defined similarly for pancake II.

Now consider $r(\theta) = p(\theta) - q(\theta)$. [This was the crisis point: cries of "Why? I don't understand! Say it again! I'm

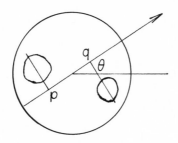

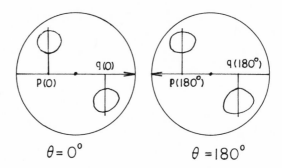

$$\theta = 0°$$    $$\theta = 180°$$

confused!" First response from teacher, "Well, just consider $p(\theta) - q(\theta)$. You'll see it works out! Let me proceed!"]

Consider $p(\theta) - q(\theta)$ as $\theta$ varies from 0° to 180°.

The rays $\theta = 0°$ and $\theta = 180°$ are identical lines but are directed oppositely. On the other hand, the position of the bisectors of I and II are identical, but the measurement of $p(180°)$ is opposite to that of $p(0°)$. Therefore $p(180°) = -p(0°)$. Similarly, $q(180°) = -q(0°)$. Therefore $r(0°) = p(0°) - q(0°)$ and $r(180°) = p(180°) - q(180°)$, so that $r(180°) = -r(0°)$. Now (a) $r(0°) = 0$ or (b) $r(0°) \neq 0$. If (a) $r(0°) = 0$ then $p(0°) = q(0°)$, and this means that the bisectors of I and II coincide, giving us the desired relation with one cut. If (b) $r(0°) \neq 0$ then $r(\theta)$ changes sign as $\theta$ varies from 0° to 180°. This change is continuous; hence there must be a position $\theta$ where $r(\theta) = 0°$. At this point $p(\theta) = q(\theta)$ and this gives us a solution to the problem as the two cuts are really coincident.

## 4. Teacher's Reaction

My first reaction to the crisis was to think: why did it occur where it occurred? At a "mere" definition; and this isn't even the hard part. Well, perhaps this was the straw that broke the camel's back. We had been piling up sophistications. The lemma seemed to go easily enough. Enclosing the pancakes in a reference circle was mysterious. The measurement of directed distances along a ray was troublesome. The notation $p(\theta)$, $q(\theta)$ brought out all the insecurities of inexperience with functional notation. Having gone through the proof as indicated above, I was left with a feeling that at the very best I had bludgeoned my class into accepting the theorem without understanding the proof. It also became clear that, the initial barrier having set in, it would not be cleared up by simply repeating the proof even in excruciatingly fine detail. Another approach was required.

## 5. Documentation of the Discovery Process

It is sometimes said that the way to understanding is the way of the initial discovery. That is, if one knows the way something was originally thought out, then this is a good way for classroom presentation. This is not necessarily the case, for the initial discovery may have been obscure, unnecessarily difficult or buried in a totally different context. It may also happen that a snappy, up-to-date presentation is obscured by being clothed in generalities, and an older presentation has much insight to offer.

Sometimes the surest way to understanding is to reconstruct for oneself a proof along slightly different (or even novel) lines. In this way one confronts squarely the difficulties as well as the brilliant breakthroughs. After the crisis erupted, I went back to my office and constructed (and documented) a slightly different proof which avoided the traumatic step: let $r(\theta) = p(\theta) - q(\theta)$. I knew that the sign-switching argument would have to appear, but I would put it in a slightly different guise.

## 6. Proof: Second Version

(a) Building up preliminary intuition.

1. The area of a circle is bisected by any diameter.

2. Conversely, any area bisector of a circle must be a diameter.

The first is clear, but the second may be slightly less clear. If $AB$ is not a diameter, draw $AOC$ which is a diameter. The presence of the wedge $ABC$ makes it clear that the area on top of $AB$ must be less than a semicircle.

3. The two-circle problem (i.e., simultaneously bisect the area of two circles) has one and only one solution if the centers of the two circles are distinct. For an area bisector of both must be a common diameter. If the centers are

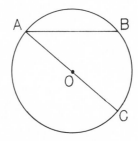

coincident the two-circle problem has infinitely many solutions.

4. *Corollary:* The three-circle problem and hence the three-pancake problem has, in general, no solution. For a common bisector of three circles must be a common diameter. This is impossible unless the three centers are collinear.

This corollary is interesting because it sets limits to how the problem can be generalized. Knowledge of such limits increases the appreciation of the original statement.

(b) The problem thickens.

We keep one circle and allow the other circle to be an odd-shaped pancake, designated by II. Suppose the pancake is completely exterior to the circle. To bisect the circle we need a diameter, so draw a diameter. Put an arrow along the diameter and one perpendicular, creating a coordinate system which distinguishes one side of the diameter from the other. Let $p(\theta)$ be the percentage of the area of pancake II that lies on the arrow side of the diameters. It is clear that as the diameter swings from one side of II to

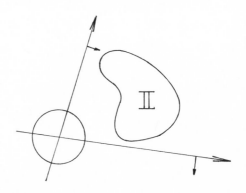

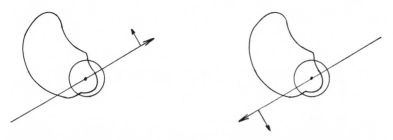

the other $p(\theta)$ goes from 100% to 0%. By continuity, there is a 50% mark.

But what if the circle overlaps the pancake so that we cannot go from 100% to 0%? Well, in that case, simply don't restrict the swing of the diameter, but let it make the full swing $0° \le \theta \le 180°$! The diameters are identical, but the orientation is reversed so that (Aha!) $p(0) + p(180) =$ 100%. Thus, if, e.g., $p(0) = 43\%$, and $p(180) = 57\%$, then there is an intermediate value $\theta$ with $p(\theta) = 50\%$.
(c) The clincher.

Get rid of the circle; deal with two general pancakes. Bisect pancake I (according to the lemma) by a line equipped with an orientation and parallel to direction $\theta$. Designate by $p(\theta)$ the percentage of area of II on the arrow side of the bisector. We have, as before, $p(0) + p(180) = 100\%$, so the conclusion follows in the general case.

This second proof went down, I felt, better than the first. Perhaps it was the previous exposure. Perhaps it was that I felt more secure with it in my own mind—it was something that I had, in fact, thought through myself. At any rate, a higher level of understanding seemed to have been achieved, and the class went on to a discussion of other aspects of the crisis.

### 7. Textbook Presentations

Why are textbook and monograph presentations of mathematics so difficult to follow? The layman might get the idea that a skillful mathematician can sight-read a page of mathematics in the way that Liszt sight-read a page of difficult piano music. This is rarely the case. The absorption of a page of mathematics on the part of the professional is often a slow, tedious, and painstaking process.

The presentation in textbooks is often "backward." The discovery process is eliminated from the description and is not documented. After the theorem and its proof have been worked out, by whatever path and by whatever means, the whole verbal and symbolic presentation is rearranged, polished and reorganized according to the canons of the logico-deductive method. The aesthetics of the craft demands this. Historical precedents—the Greek tradition —demand it. It is also true that the economics of the book business demands maximum information in minimum space. Mathematics tends to achieve this with a vengeance. Brevity is the soul of mathematical brilliance or wit. Fuller explanations are regarded as tedious.

### 8. Authoritarian or Dogmatic Presentations

Mathematical presentations, whether in books or in the classroom, are often perceived as authoritarian and this may arouse resentment on the part of the student. Ideally, mathematical instruction says, "Come, let us reason together." But what comes from the mouth of the lecturer is often, "Look, I tell you this is the way it is." This is proof by coercion. There are several reasons for this to happen. First of all, there is the shortage of time. We must accomplish (or think we must) a certain amount in a semester so that the student is prepared for the next course in mathematics or for Physics 15. Therefore we cannot afford to linger lovingly over any of the difficulties but must rush breathlessly through our set piece.

Then there is a desire on the part of some teachers to appear brilliant. (What I'm telling you is pretty easy and obvious to me, and if you're not getting it, you must really be pretty stupid.)

On the other side of the coin, there may be ignorance or unpreparedness on the part of the teacher which constrains him pretty narrowly to the path laid out in the textbook. Such teachers may have no edge on their subject. Some lack self-confidence as mathematicians and may themselves stand in terror of the authority of the text or the learned monograph. They don't know how to "fiddle around" or, if they do, they are afraid to teach fiddling around.

## 9. Resistance on the Part of Students

What are some of the reasons for resistance, resentment, rejection on the part of students?

First of all there is considerable impatience with the material. Surprisingly, this is often noticeable in the better students. But better students tend to demand instant understanding. Mathematics has always been easy for them. Understanding and intuition have come cheaply. Now as they move into the higher reaches of mathematics, the material is getting difficult. They lack experience. They lack strategies. They don't know how to fiddle around. Understanding is accompanied by pains. It makes little impression to say that the material about to be presented is the end result of centuries of thought on the part of tens or hundreds of brilliant people. The desire for immediate comprehension is very strong and may ultimately be debilitating. (If I don't get it right away, then I never will, and I say to hell with it.)

The key idea is often brilliant but difficult. There may be a psychological unwillingness to accept that there is in the world a brilliance and an understanding which may exceed their own. There may be a sudden revelation that some higher mathematics is beyond them completely and this comes as a shock and a blow to the ego. The resistance may intensify and show up as lack of study, lack of interest, and an unwillingness to attempt a discovery process of one's own.

It is commonly thought that there are "math types" and "nonmath types." No one knows why some people take to math easily, and others with enormous difficulty. For nonmath types resistance may be the honest reaction to innate limitations. Not everyone becomes a piano player or an ice skater. Why should it be otherwise for mathematics?

## 10. At the Core

The flash of insight, the breakthrough, the "aha," symbolizes that something has been brought forth which is genuinely new, a new understanding for the individual, a new concept placed before the larger community. Creativity exists, it occurs every day. It is not spread democratically throughout the populace, but it is plentiful. It is not understood but, within limits, it can be increased or decreased. Within limits it can be taught. But everyone has limits, everyone is baffled and frustrated from more glorious accomplishment; and the proof of this is simply to look around and observe that life and mathematics are full of unsolved problems. This bringing forth of a new element —what is it? An intellectual mutation? A state of grace? A gift of the gods?

There is currently much study and experimentation aimed at splitting open the core of insight. There is even an effort to automate it by computer, to make it more plentiful, to turn our period into one of the great periods of history.

Yet, it is clear that one learns by example and precept, by sitting at the feet of the masters and imitating what they do; and it is equally clear that the masters are able to transmit something of their strategy and insight. Let us examine some concrete experience.

### Further Readings. See Bibliography

Chinn and Steenrod, J. Hadamard

# Pólya's Craft of Discovery

"My mind was struck by a flash of lightning in which its desire was fulfilled."

Dante, *Paradiso,* Canto XXXIII
Quoted by G. Pólya

GEORGE PÓLYA (1888–) has had a scientific career extending more than seven decades. A brilliant mathematician who has made fundamental contributions in many fields, Pólya has also been a brilliant teacher, a teacher's teacher, and an expositor. Pólya believes that there is a craft of discovery. He believes that the ability to discover and the ability to invent can be enhanced by skillful teaching which alerts the student to the principles of discovery and which gives him an opportunity to practise these principles.

*George Pólya*
*1888–*

In a series of remarkable books of great richness, the first of which was published in 1945, Pólya has crystallized these principles of discovery and invention out of his vast experience, and has shared them with us both in precept and in example. These books are a treasure-trove of strategy, know-how, rules of thumb, good advice, anecdote, mathematical history, together with problem after problem at all levels and all of unusual mathematical interest. Pólya places a global plan for "How to Solve It" in the endpapers of his book of that name:

HOW TO SOLVE IT
First: you have to *understand* the problem.
Second: find the connection between the data and the unknown. You may be obliged to consider auxiliary problems if an immediate connection cannot be found. You should obtain eventually a *plan* of the solution.
Third: *Carry out* your plan.
Fourth: *Examine* the solution obtained.

These precepts are then broken down to "molecular" level on the opposite endpaper. There, individual strategies are suggested which might be called into play at appropriate moments, such as

> If you cannot solve the proposed problem, look around for an appropriate related problem
> Work backwards
> Work forwards
> Narrow the condition
> Widen the condition
> Seek a counterexample
> Guess and test
> Divide and conquer
> Change the conceptual mode

Each of these heuristic principles is amplified by numerous appropriate examples.

Subsequent investigators have carried Pólya's ideas forward in a number of ways. A. H. Schoenfeld has made an interesting tabulation of the most frequently used heuristic principles in college-level mathematics. We have appended it here.

## FREQUENTLY USED HEURISTICS*

### Analysis

1) DRAW A DIAGRAM if at all possible.
2) EXAMINE SPECIAL CASES:
   a) Choose special values to exemplify the problem and get a "feel" for it.
   b) Examine limiting cases to explore the range of possibilities.
   c) Set any integer parameters equal to 1, 2, 3, . . . , in sequence, and look for an inductive pattern.
3) TRY TO SIMPLIFY THE PROBLEM by
   a) exploiting symmetry, or
   b) "Without Loss of Generality" arguments (including scaling)

*From: A. H. Schoenfeld*

## Exploration

1) CONSIDER ESSENTIALLY EQUIVALENT PROBLEMS:
    a) Replacing conditions by equivalent ones.
    b) Re-combining the elements of the problem in different ways.
    c) Introduce auxiliary elements.
    d) Re-formulate the problem by
        i) change of perspective or notation
        ii) considering argument by contradiction or contrapositive
        iii) assuming you have a solution, and determining its properties

2) CONSIDER SLIGHTLY MODIFIED PROBLEMS:
    a) Choose subgoals (obtain partial fulfillment of the conditions)
    b) Relax a condition and then try to re-impose it.
    c) Decompose the domain of the problem and work on it case by case.

3) CONSIDER BROADLY MODIFIED PROBLEMS:
    a) Construct an analogous problem with fewer variables.
    b) Hold all but one variable fixed to determine that variable's impact.
    c) Try to exploit any related problems which have similar
        i) form
        ii) "givens"
        iii) conclusions.

Remember: when dealing with easier related problems, you should try to exploit both the RESULT and the METHOD OF SOLUTION on the given problem.

## Verifying your solution

1) DOES YOUR SOLUTION PASS THESE SPECIFIC TESTS:
    a) Does it use all the pertinent data?
    b) Does it conform to reasonable estimates or predictions?
    c) Does it withstand tests of symmetry, dimension analysis, or scaling?

2) DOES IT PASS THESE GENERAL TESTS?
    a) Can it be obtained differently?
    b) Can it be substantiated by special cases?
    c) Can it be reduced to known results?
    d) Can it be used to generate something you know?

To give the flavor of Pólya's thinking and writing in a very beautiful but subtle case, a case that involves a change in the conceptual mode, I shall quote at length from his *Mathematical Discovery* (vol. II, pp. 54 ff):

**Example**

I take the liberty of trying a little experiment with the reader. I shall state a simple but not too commonplace theorem of geometry, and then I shall try to reconstruct the sequence of ideas that led to its proof. I shall proceed slowly, very slowly, revealing one clue after the other, and revealing each clue gradually. I think that before I have finished the whole story, the reader will seize the main idea (unless there is some special hampering circumstance). But this main idea is rather unexpected, and so the reader may experience the pleasure of a little discovery.

A. *If three circles having the same radius pass through a point, the circle through their other three points of intersection also has the same radius.*

This is the theorem that we have to prove. The statement is short and clear, but does not show the details distinctly enough. If we *draw a figure* (Fig. 10.1) and *introduce suitable notation,* we arrive at the following more explicit restatement:

B: *Three circles k, l, m have the same radius r and pass through the same point O. Moreover, l and m intersect in the point A, m and k in B, k and l in C. Then the circle e through A, B, C has also the radius r.*

Figure 10.1 exhibits the four circles *k, l, m,* and *e* and their four points of intersection *A, B, C,* and *O.* The figure is apt to be unsatisfactory, however, for it is not simple, and it is still incomplete; something seems to be missing; we failed to take into account something essential, it seems.

We are dealing with circles. What is a circle? A circle is determined by center and radius; all its points have the same distance, measured by the length of the radius, from the center. We failed to introduce the common radius *r,* and so we failed to *take into account an essential part of the hypothesis.* Let us, therefore, introduce the centers, *K* of *k, L* of *l,* and *M* of *m.* Where should we exhibit the radius *r*? There

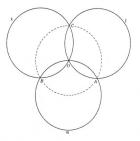

*Fig. 10.1. Three circles through one point.*

seems to be no reason to treat any one of the three given circles $k$, $l$, and $m$ or any one of the three points of intersection $A$, $B$, and $C$ better than the others. We are prompted to connect all three centers with all the points of intersection of the respective circle: $K$ with $B$, $C$, and $O$, and so forth.

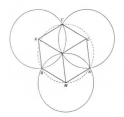

Fig. 10.2. Too crowded.

The resulting figure (Fig. 10.2) is disconcertingly crowded. There are so many lines, straight and circular, that we have much trouble in "seeing" the figure satisfactorily; it "will not stand still." It resembles certain drawings in old-fashioned magazines. The drawing is ambiguous on purpose; it presents a certain figure if you look at it in the usual way, but if you turn it to a certain position and look at it in a certain peculiar way, suddenly another figure flashes on you, suggesting some more or less witty comment on the first. Can you recognize in our puzzling figure, overladen with straight lines and circles, a second figure that makes sense?

......................................................................................

We may hit in a flash on the right figure hidden in our overladen drawing, or we may recognize it gradually. We may be led to it by the effort to solve the proposed problem, or by some secondary, unessential circumstance. For instance, when we are about to redraw our unsatisfactory figure, we may observe that the *whole* figure is determined by its *rectilinear part* (Fig. 10.3).

This observation seems to be significant. It certainly simplifies the geometric picture, and it possibly improves the logical situation. It leads us to restate our theorem in the following form.

C. *If the nine segments*

$$K\,O, \qquad K\,C, \qquad K\,B,$$
$$L\,C, \qquad L\,O, \qquad L\,A,$$
$$M\,B, \qquad M\,A, \qquad M\,O,$$

*are all equal to r, there exists a point E such that the three segments*

$$E\,A, \qquad E\,B, \qquad E\,C$$

*are also equal to r.*

This statement directs our attention to Fig. 10.3. This figure is attractive; it reminds us of something familiar. (Of what?)

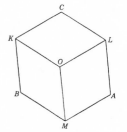

Fig. 10.3. It reminds you—of what?

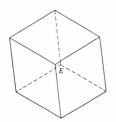

*Fig. 10.4. Of
course!*

Of course, certain quadrilaterals in Fig. 10.3, such as *OLAM* have, by hypothesis, four equal sides, they are rhombi. A rhombus is a familiar object; having recognized it, we can "see" the figure better. (Of what does the *whole* figure remind us?)

Opposite sides of a rhombus are parallel. Insisting on this remark, we realize that the 9 segments of Fig. 10.3 are of three kinds; segments of the same kind, such as *AL, MO,* and *BK,* are parallel to each other. (Of what does the figure remind us *now?*)

We should not forget the conclusion that we are required to attain. Let us assume that the conclusion is true. Introducing into the figure the center *E* or the circle *e,* and its three radii ending in *A, B,* and *C,* we obtain (supposedly) still more rhombi, still more parallel segments; see Fig. 10.4. (Of what does the whole figure remind us *now?*)

..................................................................................................

Of course, Fig. 10.4 is the projection of the 12 edges of a parallelepiped having the particularity that the projection of all edges are of equal length.

Figure 10.3 is the projection of a "nontransparent" parallelepiped; we see only 3 faces, 7 vertices, and 9 edges; 3 faces, 1 vertex, and 3 edges are invisible in this figure. Figure 10.3 is just a part of Fig. 10.4, but this part defines the whole figure. If the parallelepiped and the direction of projection are so chosen that the projections of the 9 edges represented in Fig. 10.3 are all equal to *r* (as they should be, by hypothesis), the projections of the 3 remaining edges must be equal to *r*. These 3 lines of length *r* are issued from the projection of the 8th, the invisible vertex, and this projection *E* is the center of a circle passing through the points *A, B,* and *C,* the radius of which is *r*.

Our theorem is proved, and proved by a surprising, artistic conception of a plane figure as the projection of a solid.

(The proof uses notions of solid geometry. I hope that this is not a great wrong, but if so it is easily redressed. Now that we can characterize the situation of the center *E* so simply, it is easy to examine the lengths *EA, EB,* and *EC* independently of any solid geometry. Yet we shall not insist on this point here.)

This is very beautiful, but one wonders. Is this the "light that breaks forth like the morning," the flash in which desire is fulfilled? Or is it merely the wisdom of the Monday morning quarterback? Do these ideas work out in the classroom? Followups of attempts to reduce Pólya's program to practical pedagogics are difficult to interpret. There is more to teaching, apparently, than a good idea from a master.

### Further Readings. See Bibliography

I. Goldstein and S. Papert; E. B. Hunt; A. Koestler [1964]; J. Kestin; G. Pólya [1945], [1954], [1962]; A. H. Schoenfeld; J. R. Slagle

# The Creation of New Mathematics:
## An Application of the Lakatos Heuristic

I N *PROOFS AND REFUTATIONS*, Imre Lakatos presents a picture of the "logic of mathematical discovery." A teacher and his class are studying the famous Euler-Descartes formula for polyhedra

$$V - E + F = 2.$$

In this formula $V$ is the number of vertices of a polyhedron, $E$, the number of its edges, and $F$, the number of its faces. Among the familiar polyhedra, these quantities take the following values:

*Leonhard Euler*
*1707–1783*

|  | V | E | F |
|---|---|---|---|
| tetrahedron | 4 | 6 | 4 |
| (Egyptian) pyramid | 5 | 8 | 5 |
| cube | 8 | 12 | 6 |
| octahedron | 6 | 12 | 8 |

(See also Chapter 7, Lakatos and the Philosophy of Dubitability.)

The teacher presents the traditional proof in which the polyhedron is stretched on the plane. This "proof" is immediately followed by a barrage of counterexamples presented by the students. Under the impact of these counterexamples, the statement of the theorem is modified, the proof is corrected and elaborated. New counterexamples are produced, new adjustments are made.

This development is presented by Lakatos as a model for the development of mathematical knowledge in general.

The Lakatos heuristic example of proofs and refutations which was formulated for the mathematical culture at large can of course be applied by the individual in his attempt to create new mathematics. The writer has used the method with moderate success in his classes. The initial shock of presenting students not with a fixed problem to be cracked, but with an open-ended situation of potential discovery, must and can be overcome. The better students then experience a sense of exhilaration and freedom in which they are in control of the material.

I shall illustrate the method with a little example from the elementary theory of numbers.

*Simplified Lakatos model for the heuristics of mathematical discovery*

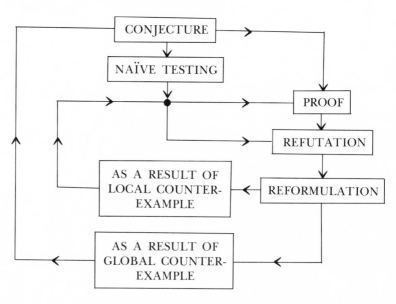

I begin with an initial statement which I shall call the "seed." The seed statement should be an interesting one, quite simple. The object of the exercise is for the student to water the seed so that it grows into a sturdy plant. I usually present the class with a variety of seeds and they select them for watering, depending upon their experience.

## Act I

*Seed:* "If a number ends in 2, it is divisible by 2."

*Examples:* 42 ends in and is divisible by 2. 172 ends in and is divisible by 2.

*Proof:* A number is even if and only if it ends in 0, 2, 4, 6, 8. All even numbers are divisible by 2. In particular, those that end in 2 are divisible by 2.

*Proof* (more sophisticated): If the number, in digit form, is $ab \ldots c2$, then it is clearly of the form $(ab \ldots c0) + 2$, hence of the form $10Q + 2 = 2(5Q + 1)$.

*Conjectural Leap:* If a number ends in $N$ it is divisible by $N$.

*Comment:* Be bold and make the obvious generalization. The heavens won't fall in if it turns out to be false.

*Example:* If a number ends in 5 it is divisible by 5. Sure: 15, 25, 128095, etc. But, alas,

*Counterexample:* If a number ends in 4 is it divisible by 4? Is 14 divisible by 4? No. Too bad.

*Objection:* But some numbers that end in 4 *are* divisible by 4: 24. Some numbers ending in 9 *are* divisible by 9: 99.

*Recap of Experience:* The numbers 1, 2, . . . , 9 seem to be divisible into two categories. *Category I:* The digits $N$ such that when a number ends in $N$ it is divisible by $N$, always. *Category II:* The digits $N$ such that when a number ends in $N$ it is divisible by $N$ only occasionally.
*Category I:* 1, 2, 5.
*Category II:* 3, 4, 6, 7, 8, 9.

*Point of Order:* What about numbers that end in 0? Are they divisible by 0? No. But they are divisible by 10. Hmm.

We may have to watch this. This phenomenon doesn't fit in with the form of the seed.

**Definition:** Let's call the numbers in Category I "magic numbers." They have a delightful property.

**Tentative Theorem:** The numbers 1, 2 and 5 are magic numbers. They are the only magic numbers.

**Counterexample:** What about the number 25? Isn't it magic? If a number ends in 25 it is divisible by 25. For example: 225, or 625.

**Objection:** We thought you were talking about single-digit numbers.

**Rebuttal:** Well, we were originally. But the 25 phenomenon is interesting. Let's open up the original inquiry a bit.

**Reformulation:** Let $N$ now represent not necessarily a single digit but a whole group of digits like 23, 41, 505, etc. Make the definition that $N$ is magic if a number that ends with the digit group $N$ is divisible by $N$. Does this extended definition make sense?

**Examples:** Yes it does. 25 is magic. 10 is magic. 20 is magic. 30 is magic.

**Counterexample:** 30 isn't magic. 130 is not divisible by 30. Come to think of it, how do you know 25 is magic?

**Theorem:** 25 is a magic number.

**Proof:** If a number ends in 25 it is of the digital form $abc \ldots e25 = abc \ldots e00 + 25$, hence of the form $100Q + 25 = 25(4Q + 1)$.

**Reformulation of Goal:** Find all the magic numbers.

**Accumulation of Experience:** 1, 2, 5, 10, 25, 50, 100, 250, 500, 1000 are all magic numbers.

**Observation:** All the magic numbers we have been able to find seem to be products of 2's and 5's. Certainly the ones in the above list are.

***Conjecture:*** Any number $N$ of the form $N = 2^p \cdot 5^q$ where $p \geq 0$, $q \geq 0$ is a magic number.

***Comment:*** Seems reasonable. What have we got to lose?

***Counterexample:*** Take $p = 3$, $q = 1$. Then $N = 2^3 \cdot 5 = 40$. Is a number that ends in 40 always divisible by 40? No. E.g., 140.

***Reformulation:*** How about the other way around, though? All the magic numbers we have found are of the form $2^p \cdot 5^q$. Perhaps all magic numbers are of that form.

***Objection:*** Isn't that what you just proposed?

***Rebuttal:*** No, what was proposed was the other way around: a number of the form $2^p \cdot 5^q$ is magic. See the difference?

***Theorem:*** If $N$ is a magic number then $N = 2^p \cdot 5^q$.

***Proof:*** Let a number end in $N$ (recall: in this statement $N$ is acting as a group of digits.) Then the number looks like $abc \ldots eN$, digitwise. We would like to split it up, as before. Therefore, let $N$ have $d(N)$ digits. Then the number $abc \ldots eN$ is really $abc \ldots e00 \ldots 0 + N$ where there are $d(N)$ 0's at the end. Therefore the number is of the form $Q \cdot 10^{d(N)} + N$. (Try this out when $d(N) = 2, 3$, etc.) All numbers that end with $N$ are of this form. Conversely, if $Q$ is any number whatever then the number $Q \cdot 10^{d(N)} + N$ ends with $N$. Now if $N$ is magic it always divides $Q \cdot 10^{d(N)} + N$. Since $N$ divides $N$, it must always divide $Q \cdot 10^{d(N)}$ for all $Q$. But $Q$ might be the simple number 1, for example. Therefore $N$ must divide $10^{d(N)}$. Since $10^{d(N)} = 2^{d(N)} \cdot 5^{d(N)}$ is a prime factorization, it follows that $N$ must itself factor down to a certain number of 2's and 5's.

***Current Position:*** We now know that a magic number is one of the form $N = 2^p \cdot 5^q$ for some integers $p, q \geq 0$. We would like to turn it around. Then we should have a necessary and sufficient condition for magicality.

***Refocussing of Experience:*** Since we know that all magic numbers are of the form $N = 2^p \cdot 5^q$, the problem comes down to: what must be asserted about $p$ and $q$ to make the resulting $N$ magic?

*Conjecture:*  $p \leq q$?

*Counterexample:*  $p = 0$, $q = 4$, $N = 2^0 \cdot 5^4 = 625$. Is 625 magic? No: 1625 doesn't divide by 625.

*Conjecture:*  $p = q$?

*Objection:*  Then $N = 2^p 5^p = 10^p$ or 1, 10, 100, . . . . O.K. But there are other magic numbers.

*Conjecture:*  $p \geq q$?

*Counterexample:*  $p = 3$, $q = 1$, $N = 2^3 \cdot 5^1 = 40$. This is not magic.

*Observation:*  Hmm. Something subtle at work here. This brings down the curtain on Act I. The process goes on for those with sufficient interest and strength.

### Act II

(In this act, the heuristic line is severely abbreviated in the write-up.)

*Strategy Conference:*  Let's go back to the proof of the necessity of the form $N = 2^p \cdot 5^q$. We found that if $N$ is magic it divides $10^{d(N)}$. Recall that $d(N)$ stands for the number of digits in the group of digits $N$. Perhaps this is sufficient as well? Aha! A breakthrough?

*Theorem:*  $N$ is magic if and only if it divides $10^{d(N)}$.

*Proof:*  The necessity has already been proved. If a number ends in $N$, then, as we know, it is of the form $Q \cdot 10^{d(N)} + N$. But $N$ divides $N$ and $N$ is assumed to divide $10^{d(N)}$. Therefore it surely divides $Q \cdot 10^{d(N)} + N$.

*Aesthetic Objection:*  While it is true that we now have a necessary and sufficient condition for magicality, this condition is on $N$ itself and not on its factored form $2^p \cdot 5^q$.

*Conference:*  When does $N = 2^p \cdot 5^q$ divide $10^{d(N)}$? Well, $10^{d(N)} = 2^{d(N)} \cdot 5^{d(N)}$, so that obviously a necessary and sufficient condition for this is $p \leq d(N)$, $q \leq d(N)$. But this is equivalent to max($p$, $q$) $\leq d(N)$. We still have the blasted $d(N)$ to contend with. We don't want it. We'd like a condition on $N$ itself, or possibly on $p$ and $q$. How can we convert max($p$, $q$) $\leq d(N) = d(2^p \cdot 5^q)$ into a more convenient form? As we know, $p = q$ is O.K. Let's see this written in the new form: $p = $ max($p$, $p$) $\leq d(2^p \cdot 5^p) = d(10^p)$. Now the number of digits in $10^p$ is $p + 1$. So this is saying $p \leq p + 1$ which is O.K. What if, in the general case, we "even out" the powers of 2 and the powers of 5? Write $q = p + h$ where $h > 0$. (Aha!)

***Objection:*** What if $p > q$ so that $q = p + h$ is impossible with $h > 0$?

***Rebuttal:*** Treat that later.

***Conference:*** max($p$, $p + h$) $\leq d(2^p \cdot 5^{p+h}) = d(2^p \cdot 5^p \cdot 5^h) = d(10^p \cdot 5^h)$. Now since $h > 0$, max($p$, $p + h$) $= p + h$. Also, the number of digits in $10^p \cdot Q$ where $Q$ is any number $= p + $ number of digits in $Q$. Therefore $p + h \leq p + d(5^h)$ or: $h \leq d(5^h)$.

***Query:*** When is it true that $h > 0$ and $h \leq d(5^h)$?

***Experimentation:*** $h = 1$: $1 \leq d(5^1)$ O.K. $h = 2$: $2 \leq d(5^2)$ O.K. $h = 3$: $3 \leq d(5^3)$ O.K. $h = 4$: $4 \leq d(5^4) = d(625) = 3$. No good. $h = 5$: $5 \leq d(5^5) = d(3125) = 4$. No good.

***Conjecture:*** $h \leq d(5^h)$ if and only if $h = 1, 2, 3$.

***Proof:*** Omitted.

***Reprise:*** What about $p > q$?

***Conference:*** Set $p = q + h$, $h > 0$. $q + h = $ max($q + h$, $q$) $\leq d(2^{q+h} \cdot 5^q) = d(10^q \cdot 2^h) = q + d(2^h)$, or $h \leq d(2^h)$. When is $h \leq d(2^h)$?

***Experimentation:*** $h = 1 : 1 \leq d(2^1)$ O.K. $h = 2 : 2 \leq d(2^2)$ No good.

***Conjecture:*** $h \leq d(2^h)$ if and only if $h = 1$.

***Proof:*** Omitted.

297

***Theorem:*** *N* is magic if and only if it equals a power of ten times 1, 2, 5, 25, or 125.

***Proof:*** Omitted.

In anticipation of further developments we might like to write this theorem in a different way.

***Theorem:*** *N* is magic if and only if $N = 2^p \cdot 5^q$, where $0 \le q - p + 1 \le 4$.

***Proof:*** Omitted.

Act III might begin by asking what would happen if we wrote our numbers in some base other than 10. What about a prime base, or a base equal to a power of a prime?

**Further Readings. See Bibliography**

M. Gardner, U. Grenander; I. Lakatos, [1976].

# Comparative Aesthetics

WHAT ARE THE ELEMENTS that make for creativity? Is it a deep analytic ability deriving from ease of combinatorial or geometric visualization—a mind as restless as a swarm of honeybees in a garden, flitting from fact to fact, perception to perception and making connections, aided by a prodigious memory—a mystic intuition of how the universe speaks mathematics—a mind that operates logically like a computer, creating implications by the thousands until an appropriate configuration emerges?

Or is it some extralogical principle at work, a grasp and a use of metaphysical principles as a guide? Or, as Henri Poincaré thought, a deep appreciation of mathematical aesthetics?

There is hardly a science of mathematical aesthetics. But we can look at an instance and discuss it at length. We can get close to the reason for Poincaré's assessment.

I shall take a famous mathematical theorem, one with a high aesthetic component, and present two different proofs of it. The theorem is the famous result of Pythagoras that $\sqrt{2}$ is not a fraction.

The first proof is the traditional proof.

***Proof I.*** One supposes that $\sqrt{2} = p/q$ where $p$ and $q$ are integers. This equation is really an abbreviation of $2 = p^2/q^2$. One supposes that $p$ and $q$ are in lowest terms, i.e. they have no common factor. (For if they do, strike it out.) Now $2 = p^2/q^2$ implies $p^2 = 2q^2$. Therefore $p^2$ is an even number. Therefore $p$ is even (for if it were odd, $p^2$ would be odd since odd × odd = odd). If $p$ is even it is of the form $p = 2r$, so that we have $(2r)^2 = 2q^2$ or $4r^2 = 2q^2$ or $q^2 = 2r^2$. Thus, as before, $q^2$ is even so that $q$ must be even. Now we are in a logical bind, for we have proved that $p$ and $q$ are both even, having previously asserted that they have no common factor. Thus, the equation $\sqrt{2} = p/q$ must be rejected if $p$, $q$ are integers.

The second proof is not traditional and is argued a bit more loosely.

***Proof II.*** As before, suppose that $p^2 = 2q^2$. Every integer can be factored into primes, and we suppose this has been done for $p$ and $q$. Thus in $p^2$ there are a certain number of primes doubled up (because of $p^2 = p \cdot p$). And in $q^2$ there are a certain number of doubled-up primes. But (aha!) in $2 \cdot q^2$ there is a 2 that has no partner. Contradiction.

I have no doubt that nine professional mathematicians out of ten would say that Proof II exhibits a higher level of aesthetic delight. Why? Because it is shorter? (Actually, we have elided some formal details.) Because, in comparison, Proof I with its emphasis on logical inexorableness seems heavy and plodding? I think the answer lies in the fact that Proof II seems to reveal the heart of the matter, while Proof I conceals it, starting with a false hypothesis and ending with a contradiction. Proof I seems to be a wiseguy

**The Vegreville Alberta Easter Egg.** *Ron Resch, computer scientist and artist, created this monumental polyhedral egg. The Egg stands 31 ft. tall, is 18 ft. wide, weighs 5000 lb. and has 3512 visible facets. It is made of 524 star-shaped pieces of $\frac{1}{16}$ in. anodized aluminum and 2208 triangular pieces of $\frac{1}{8}$ in. aluminum.*

*Courtesy: the artist.*

300

argument; Proof II exposes the "real" reason. In this way, the aesthetic component is related to the purer vision.

**Further Readings. See Bibliography**

S. A. Papert, [1978].

# Nonanalytic Aspects of Mathematics

## Conscious and Unconscious Mathematics

If we accept the common belief that the natural universe is governed by mathematical laws, then we understand that the universe and all within it are perpetually mathematizing—carrying out mathematical operations. If we are fanciful, we can think of each particle or each aggregate as the residence of a mathematical demon whose function it is to ride herd and say: "Mind the inverse square law. Mind the differential equations." Such a demon would also reside in human beings, for they, too, are constantly mathematizing without conscious thought or effort. They are mathematizing when they cross the street in fierce traffic, thereby solving mechanistic-probabilistic extremal problems of the utmost complexity. They are mathematizing when their bodies constantly react to transient conditions and seek regulatory equilibrium. A flower seed is mathematizing when it produces petals with six-fold symmetry.

Let us call the mathematizing that is inherent in the universe "unconscious" mathematics. Unconscious mathematizing goes on despite what anyone thinks; it cannot be prevented or shut off. It is natural, it is automatic. It does not require a brain or special computing devices. It requires no intellectual force or effort. In a sense, the flower or the planet is its own computer.

301

In opposition to unconscious mathematics we may distinguish "conscious" mathematics. This seems to be limited to humans and possibly to some of the higher animals. Conscious mathematics is what one normally thinks of as mathematics. It is acquired largely by special training. It seems to take place in the brain. One has special awareness of its going on or not going on. It is often tied to a symbolic and abstract language. It often is assisted by pencil and paper, mathematical instruments, or reference books.

But conscious mathematics does not always proceed through abstract symbols. It may operate through a "number sense" or a "space sense" or a "kinesthetic sense." For example, the problem "will this object fit into that box?" is answerable with high reliability on the basis of a mere glance. What lies behind these special senses is often not clear. Whether they represent stored experiences, analog-type solutions performed on the spot, inspired but partly random guesses, nonetheless the fact remains that this type of judgement can often be arrived at quickly and correctly. Although one is conscious of the problem, one is only partly conscious of the means by which the solution is brought about. Reflection after the fact often reveals a mixture of independent and overlapping operations. There is therefore no sharp dividing line between conscious and unconscious mathematizing.

## Analog and Analytical Mathematics

It is convenient to divide conscious mathematics into two categories. The first, possibly more primitive, will be called "analog-experimental" or analog, for short. The second category will be called "analytic." Analog mathematizing is sometimes easy, can be accomplished rapidly, and may make use of none, or very few, of the abstract symbolic structures of "school" mathematics. It can be done to some extent by almost everyone who operates in a world of spatial relationships and everyday technology. Although sometimes it can be easy and almost effortless, sometimes it can be very difficult, as, for example, trying to understand the arrangement and relationship of the parts of a machine, or trying to get an intuitive feeling for a complex system. Results may be expressed not in words but in "understanding," "intuition," or "feeling."

In analytic mathematics, the symbolic material predominates. It is almost always hard to do. It is time consuming. It is fatiguing. It requires special training. It may require constant verification by the whole mathematical culture to assure reliability. Analytic mathematics is performed only by very few people. Analytic mathematics is elitist and self-critical. The practitioners of its higher manifestations form a "talentocracy." The great virtue of analytic mathematics arises from this, that while it may be impossible to verify another's intuitions, it is possible, though often difficult, to verify his proofs.

Insofar as the words analog and analytic are commonplace words that are used in many specific contexts in science, and since we intend them to have a special meaning in the course of this essay, we shall illustrate our intent with a number of examples. We begin with a very ancient problem which had a religious basis.

**Problem:** When is the time of the summer solstice, or of the new moon, or of some other important astronomical event?

*Analog solution:*
  (i) Wait till it happens. Relay the happening to those concerned by messengers from the point of first detection.
 (ii) Build some kind of physical device to detect important astronomical measures. Numerous astronomical "computers" are thought to have been used from prehistoric times in both the Old and New Worlds to detect the solstices and important lunar or stellar alignments, which are often of great importance for agricultural or religious reasons.

*Analytic solution:* Formulate a theory of astronomical periodicities and build it into a calendric structure.

**Problem:** How much liquid is in this beaker?

*Analog solution:* Pour the liquid into a graduated measure and read off the volume directly.

303

*Analytic solution:*  Apply the formula for the volume of a conical frustrum. Measure the relevant linear dimensions and then compute.

**Problem:**  What route should a bus take between downtown Providence and downtown Boston in order to maximize profits for the company?

*Analog solution:*  Lay out a half dozen plausible routes. Collect time-cost-patronage data from the bus runs and adopt the maximizing solution.

*Analytical solution:*  Make a model of the mileage, toll, and traffic conditions. Solve the model in closed form, if possible. If not, run it on a computer.

*Analytical-existential solution:*  Demonstrate that on the basis of certain general assumptions, the calculus of variations assures us that a solution to the problem exists.

**Problem:**  Given a function of two variables $f(x, y)$ defined over a square in the $x - y$ plane. It is desired to formulate a computer strategy which will give a plot of the contour lines of the function ($f(x, y) = $ constant).

*Analytic solution:*  Starting at some point $(x_0, y_0)$ compute $c = f(x_0, y_0)$. By means of inverse interpolation, find nearby points $(x_1, y_1)$, $(x_2, y_2)$, . . . for which $f(x_i, y_i) = c$. Connect these points. Iterate.

*Analog-like solution:*  Place a fine grid over the square and think of the final picture as produced in a raster-scan fashion. Compute the function on the grid points and divide the range of values with say, 20 values: $v_1, \ldots, v_{20}$. Selecting a value $v_i$, draw in each small square either (a) nothing or (b) a straight line segment in the event that the four corner values are compatible with $v_i$. Iterate on $i$.

## Contrasting Analog vs. Analytical Solutions

In some problems both analog and analytical solutions may be available. It may also happen that one is available and not the other, or that both are lacking. Neither type is to be preferred a priori over the other as regards accuracy or ease of performance. If both types of solutions are available, then the agreement of the two solutions is highly desirable. This may constitute the crucial experiment for a physical theory.

An attack on a problem is often a mixture of the two approaches. In the real world, an analytical solution, no matter how good, must always be fine tuned when a real system is to be modeled or constructed. Therefore, in engineering, the analytical solution is generally taken as a point of departure, and, we hope, a good first approximation.

The analog solution appears to be closer to the unconscious mathematizing that goes on in the universe. Analog solutions probably predominate in the world of technology —but this is a pure conjecture.

## The Hierarchy of Intellectual Values

When it comes however to the intellectual value that is set on these two modes, it is clear what the ordering is. Although an analog solution may be clever, based on sophisticated and subtle instrumentation, it does not carry the accolade of the purely intellectual solution.

The intellect looks after its own. What is consciously more difficult is the more praiseworthy. The level of intellectual acclaim is proportional to the apparent complexity of the abstract symbolization. Up to a point, of course; for the house built by intellect may crash to the ground when confronted by experimental reality. Scientific education is often directed not to the solution of specific problems, but toward the carrying on of a discourse at the highest possible intellectual level.

The hierarchy we have just described is that of the practicing mathematician. From the point of view of an engineer, say, an analytical solution is of little interest unless it leads to a functioning device, which corresponds to an analog solution to a problem. A well-designed, highly developed device can show the economy of means and the elegance of thought that characterizes the best science and mathematics, but this elegance is seldom recognized by theoretically oriented scientists. The ingenuity and artistry that go to make a functioning airplane or a reliable, efficient computer are seldom appreciated until one tries to do something along these lines oneself.

## Similar Hierarchies Exist in the Nonscientific World

Nor is this tendency toward stressing the intellect limited to science. It occurs, for example, in the art world. At the very lowest level is the commercial artist. Somewhat higher is the artist who paints portraits on demand. At the highest level stands the "fine artist" who is supposed to respond to the abstract and unfettered promptings of the intellect and spirit. The work of art is often accompanied by an explication de texte whose abstraction may rival the deepest productions of mathematics.

## Mathematical Proof and its Hierarchy of Values

The definition-theorem-proof approach to mathematics has become almost the sole paradigm of mathematical exposition and advanced instruction. Of course, this is not the way mathematics is created, propagated, or even understood. The logical analysis of mathematics, which reduces a proof to an (in principle) mechanizable procedure, is a hypothetical possibility, which is never realized in full. Mathematics is a human activity, and the formal-logical account of mathematics is only a fiction; mathematics itself is to be found in the actual practice of mathematicians.

An interesting phenomenon should be noted in connection with difficulties of proof comprehension. A mathematical theorem is called "deep" if its proof is difficult. Some of the elements that contribute to depth are nonintuitiveness of statement or of argument, novelty of ideas, complexity or length of proof-material measured from some origin which itself is not deep. The opposite of deep is "trivial" and this word is often used in the sense of a put-down. However, it does not follow that what is trivial is uninteresting, unuseful, or unimportant.

Now despite this hierarchical ordering, what is deep is in a sense undesirable, for there is a constant effort towards simplification, towards the finding of alternative ways of looking at the matter which trivialize what is deep. We all

306

feel better when we have moved from the analytic toward the analog portion of the experiential spectrum.

## Cognitive Style

An obvious statement about human thought is that people vary dramatically in what might be called their "cognitive style," that is, their primary mode of thinking.

This was well known to nineteenth-century psychologists. Galton, in 1880, asked a wide range of people to "describe the image in their mind's eye of their breakfast table on a given morning." He found that some subjects could form vivid and precise pictures while others could form only blurry images or, in some cases, no image at all. William James reported that people varied greatly in the sense modality they primarily used to think in, most people being auditory or visual. There was a smaller number, however, who were powerfully influenced by the sense of touch or of kinesthesis (movement), even in what is usually called abstract thought.

Such a wide range of ways of thinking should cause no problems. Indeed, we might regard with pleasure the diversity of methods of thinking about the world that our species shows, and value all of them highly as valid ways of approaching problems.

Unfortunately, tolerance is a rare virtue, and a common response to different ways of thinking is to deny, first, that they are possible, and, second, that they are valuable.

William James remarks, "A person whose visual imagination is strong finds it hard to understand how those who are without the faculty can think at all."

Conversely, some of those who think mostly in words are literally incapable of imagining nonlinguistic thought. W. V. O. Quine says that ". . . memories mostly are traces not of past sensation but of past conceptualizations or verbalizations." Max Muller wrote, "How do we know that there is a sky and that it is blue? Should we know of a sky if we had no name for it?" He then argues that thought without language is impossible. Abelard said, "Language is generated by the intellect and generates intellect." The Chandogya Upanishad states, "The essence of man is speech."

*Willard V. O. Quine
1908–*

307

The Gospel of John starts, "In the beginning was the word. . . ."

However, many have held opposing views. Aristotle said we often think and remember with images. Bishop Berkeley held that words are an impediment to thought. Many philosophers and theologians view concepts and words as dangerously misleading "word play." The Lankavatara Sutra is typical, "Disciples should be on their guard against the seductions of words and sentences and their illusive meanings, for by them, the ignorant and dullwitted become entangled and helpless as an elephant floundering around in deep mud. Words and sentences . . . cannot express highest reality. . . . The ignorant and simple minded declare that meaning is not otherwise than words, that as words are, so is meaning. . . . Truth is beyond letters and words and books." The Tao Te Ching (LXXXI) says, "True words are not fine sounding; fine-sounding words are not true. The good man does not prove by argument and he who proves by argument is not good. . . ." A Biblical quotation in this tradition says, "The letter killeth, but the spirit giveth life. . . ."

## Cognitive Style in Mathematics

In his book, Hadamard tried to find out how famous mathematicians and scientists actually thought while doing their work. Of those he contacted in an informal survey, he wrote "Practically all of them . . . avoid not only the use of mental words, but also . . . the mental use of algebraic or precise signs . . . they use vague images." (p. 84) and ". . . the mental pictures of the mathematicians whose answers I have received are most frequently visual, but they may also be of another kind—for example kinetic." (p. 85)

*Albert Einstein*
*1879–1955*

Albert Einstein wrote to Hadamard that "the words or the language, as they are written or spoken, do not seem to play any role in my mechanism of thought. . . . The physical entities which seem to serve as elements in thought are certain signs and more or less clear images which can be 'voluntarily' reproduced and combined. . . . The above mentioned elements are, in my case, of visual and some of muscular type. Conventional words or other signs have to be sought for laboriously only in a secondary stage . . ." (p. 142) Several recent studies on the way in which nonmathematical adults perform simple arithmetic seem to suggest the same is true for non-mathematicians as well.

### An Example of Cognitive Style in Combinatorial Geometry

We have already described and contrasted analog solutions and analytical solutions. Insofar as mathematical discovery may have large components of one or the other, we are dealing here with differences in cognitive style.

Here is a striking example where an analytic proof might be very difficult while an analog-like proof makes the whole business transparent.

***Gomory's Theorem.*** Remove one white and one black square from an ordinary checkerboard. The reduced board can always be covered with 31 dominos of size 2 × 1.

***Analog Proof.*** Convert the checkerboard into a labyrinth as in the accompanying figure. No matter which black square "A" and which white square "B" are deleted, the board can be covered by threading through the labyrinth with a caterpillar tractor chain of dominos which break off at "A" and "B." (See Honsburger, p. 66.)

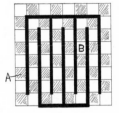

The imagery of a moving caterpillar tread is sufficient to enable one to grasp the solution at a glance. Note the powerful kinesthetic and action-oriented mode of proof. It would be difficult for the normal (i.e., somewhat visual) reader to work through this proof and not feel a sense of movement.

We are not aware of an analytical solution to the problem. Of course one could tighten up the above solution by counting blacks and whites to provide a more formal proof.

Here is a geometrical problem where both types of solution are available.

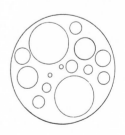

***Theorem.*** *It is impossible to fill up a circle C with a finite number of nonoverlapping smaller circles contained in C.*

***Analog Solution.*** This is visually obvious.

***Analytical Solution.*** For a neat proof based on notions of linear independence, see Davis, 1965. The analog solution is so apparent that to insist on more is a piece of mathematical pedantry.

This leads us to the notorious

***Jordan Curve Theorem.*** A simple closed curve in the plane separates the plane into two regions, one finite and one infinite.

***Analog Solution.*** This is visually obvious.

***Analytical Solution.*** Very difficult, the difficulty deriving from the fact that an excessive degree of analytical generality has been introduced into the problem.

### Mathematical Imagery

Hadamard described the semiconscious stream of thought which may accompany the process of conscious mathematizing. This kind of thing surely exists although it is very difficult to describe and document.

A few more words along this line describing my own experiences might therefore be appropriate.

The semiconscious stream of thought—which might be referred to as mathematical imagery—does not seem to relate directly to the analytical work attempted. It feels to be more analog, almost visual, sometimes even musical. It accompanies and occasionally helps the dominant stream of thought. It frequently seems irrelevant, a mere hovering background presence.

Some years ago I spent considerable time working in the theory of functions of a complex variable. This theory has a considerable geometric underlay. In fact, it can be developed independently from a geometric (Riemannian) or from an analytic (Weierstrassian) point of view. The geometrical illustrations in textbooks often feature spheres, maps, surfaces of an unusual kind, configurations with circles, overlapping chains of circles, etc. As I was working along with the analytic material, I found it was accompanied by the recollection or the mixed debris of dozens of pictures of this type that I had seen in various books, together with inchoate but repetitious nonmathematical thought and musical themes.

I worked out, more or less, a body of material which I set down in abbreviated form. Something then came up in my calendar which prevented me from pursuing this material for several years. I hardly looked at it in the interval. At the end of this period, time again became available, and I decided to go back to the material and see whether I could work it into a book.

At the beginning I was completely cold. It required several weeks of work and review to warm up the material. After the time I found to my surprise that what appeared to be the original mathematical imagery and the melody returned, and I pursued the task to a successful completion.

### The Proper Goal of Mathematical Applications is for the Mathematics to Become Automated

Analytical mathematics is hard to do and is inaccurately performed if done quickly. We do not expect that an astronaut landing on the moon will be laying out his actions as the result of computing with a table of logarithms and trigonometric functions in real time. Ideally, as in unmanned flight, the whole system might be automated. Although there is a considerable mathematical underlay to landing manoeuvres, we expect that the astronaut will

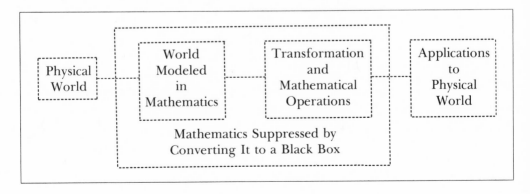

respond to instrument or computer readouts or verbal instructions which are surrogate readouts. The analytical mathematics must be suppressed or bypassed and replaced by analog mathematics.

One sees this over and over again in applied mathematics. The more complete and successful is an application, the more automatic, programmed, rote, it must become.

### An Example From Computer Graphics

A striking example of the suppression of the mathematical underlay occurred in the last ten to fifteen years in the area of computer art and animation.

Computer art can be traced back to mechanical instruments that produced cyclic motions of various sorts, hypocycloids, Lissajou's figures, "spirographic" figures, etc. These were easily produced on oscillators or analog circuitry. The idea was to draw figures which were visually pleasing or exciting. There was a constant identification of the figure with the mathematical equation.

In the first decade or so of computer art, the mathematical presence was quite noticeable. In addition to having artistic sensibilities, practioners had to know computer programming, graphical programming, a certain amount of basic mathematics such as analytic geometry, elementary transformations, and interpolation schemes. Gradually, higher and higher level languages were written for computer art. The mode of operation became less analytic, more linguistic, and more analog. As this occurred the mathematical substructure was either built in, suppressed, or bypassed.

An excellent instance of this suppression is the PAINT program developed at the University of Utah and New York Institute of Technology. As a response to a desire to do commercial animation by computer, a very high level language was developed which could be learned easily and used by commercial animators without a knowledge of mathematics. Working in color, the artist is able to select a palette and create brushes of various widths and spatter characteristics. He creates shapes working with a stylus on a computer sketch pad. Numerous "menu items" allow him to fill with color, to exchange colors continuously in real time, to replicate, to transform (zoom), to phantom, to animate, via linear and nonlinear interpolation.

Thus the artist inputs to the computer his own muscle movements of wrist and arm and shoulder as well as a set of "menu options." Since these options are also controlled by the stylus, the whole process proceeds in imitation of the manner of painting in conventional media.

## The Degradation of the Geometric Consciousness

It has often been remarked over the past century and a half that there has been a steady and progressive degradation of the geometric and kinesthetic elements of mathematical instruction and research. During this period the formal, the symbolic, the verbal, and the analytic elements have prospered greatly.

What are some of the reasons for this decline? A number of explanations come to mind:

(1) The tremendous impact of Descartes' *La Géométrie*, wherein geometry was reduced to algebra.

(2) The impact in the late nineteenth century of Felix Klein's program of unifying geometries by group theory.

(3) The collapse, in the early nineteenth century, of the view derived largely from limited sense experience that the geometry of Euclid has a priori truth for the universe, that it is *the* model for physical space.

(4) The incompleteness of the logical structure of classical Euclidean geometry as discovered in the nineteenth century and as corrected by Hilbert and others.

(5) The limitations of two or three physical dimensions which form the natural backdrop for visual geometry.

(6) The discovery of non-Euclidean geometries. This is related to the limitations of the visual ground field over which visual geometry is built, as opposed to the great generality that is possible when geometry is algebraized and abstracted (non-Euclidean geometries, complex geometries, finite geometries, linear algebra, metric spaces, etc.).

(7) The limitations of the eye in its perception of mathematical "truths" (continuous, nondifferentiable functions; optical illusions; suggestive, but misleading special cases).

313

An excellent exposition of the counter-intuitive nature of analytical mathematics when it attempts to extend the visual field can be found in the article by Hahn in volume III of *The World of Mathematics.* It has become traditional to regard these "pathological examples" as pointing to failures of the visual intuition. *But they can equally well be interpreted as examples of the inadequacies of the analytical modeling of the visual process.*

### Right Hemisphere and Left Hemisphere

There is an intriguing, but speculative similarity between the two approaches to mathematics we have described and current work on the functions of the two cerebral hemispheres. Although this work is still in its beginnings, it seems clear that the right and left hemispheres are specialized to do somewhat different tasks. (For a review of this rapidly growing field, see part one of Schmitt and Worden, and for an informal discussion, see Gardner, *The Shattered Mind.*)

It has been known for over a hundred years that in virtually all right-handed humans and in about half of the left-handed, the parts of the brain associated with speech are primarily located in the left cerebral hemisphere. This appears to be an innate biological specialization, and a slight anatomical asymmetry has been shown to exist between the hemispheres in both newborn infants and adults. Damage to certain areas of the left hemisphere will cause characteristic types of difficulties with speech, while damage to the right hemisphere in the same location will not. To oversimplify a complex issue, the left hemisphere in most humans is primarily concerned with language-based behavior and with the cognitive skills we might crudely characterize as analytical or logical. It has become apparent recently that the right hemisphere is far superior to the left in most visual and spatial abilities, discriminations by touch, and in some nonverbal aspects of hearing, for example, music.

A great deal of information about hemispheric specialization has come from careful study of a small number of neurosurgical patients who had to have the two hemispheres disconnected as a last resort against life-threatening epilepsy. Sperry has summarized many of the results of research on these patients:

> Repeated examination during the past 10 years has consistently confirmed the strong lateralization and dominance for speech, writing, and calculation in the disconnected left hemisphere in these right handed patients. . . . Though predominantly mute . . . the minor hemisphere is nevertheless clearly the superior cerebral member for certain types of tasks. . . . Largely they involve the apprehension and processing of spatial patterns, relations, and transformations. They seem to be holistic and unitary rather than analytic and fragmentary . . . and to involve concrete perceptual insight rather than abstract, symbolic, sequential reasoning. (p. 11)

We should always remember that the two hemispheres combine to give a whole brain. Even in speech, the left hemisphere function, the right hemisphere plays an important role, and the hemispheres work in harmony in a normal individual.

The melodic aspects of speech—rhythm, pitch, and intonation—seem related to the right hemisphere. Gardner provides an apt description of other functions:

> With individuals who have disease of the right hemisphere, the abilities to express oneself in language and to understand . . . others are deceptively normal . . . [however] these patients are strangely cut off from all but the verbal messages of others. . . . They are reminiscent of language machines. . . . appreciative of neither the subtle nuances or non-linguistic contexts in which the message was issued . . ." (p. 434)

Gardner then goes on, in a way somewhat unflattering to the public image of mathematicians:

> Here the patient (with right hemisphere damage) exemplifies the behavior . . . associated with the brilliant young mathematician or computer scientist. This highly rational individual is ever alert to an inconsistency in what is being said, always seeking to formulate ideas in the most airtight way; but in neither case does he display any humor about

315

his own situation, nor . . . the many subtle intuitive inter-
personal facets which form so central a part of human in-
tercourse. One feels rather that the answers are being
typed out at high speed on computer printout paper. (p.
435)

The anecdotes we have given earlier, and our own experiences, indicate that mathematics makes use of the talents
that are found in both hemispheres, rather than being restricted to the linguistic, analytic specialties of the left
hemishere. The nonverbal, spatial, and holistic aspects of
thought are prominent in what most good mathematicians
actually do though perhaps not so much in what they say
they do.

It is a reasonable conclusion that a mathematical culture
that specifically downgrades the spatial, visual, kinesthetic,
and nonverbal aspects of thought does not fully use all the
capacities of the brain.

The de-emphasis of the analog elements of mathematics
represents the closing off of one channel of mathematical
consciousness and experience. Surely, it would be better to
develop and use all the special talents and abilities of our
brains, rather than to suppress some by education and professional prejudice. We suggest that in mathematics it
would be better for the contributions of the two halves of
the brain to cooperate, complement, and enhance each
other, rather than for them to conflict and interfere.

### Further Readings. See Bibliography

T. Banchoff and C. Strauss; P. J. Davis [1974]; P. J. Davis and J. Anderson; H. Gardner; I. Goldstein and S. Papert; J. Hadamard; R. Honsberger; W. James [1962]; M. Kline [1970]; K. Knowlton; Wikuyk;
E. Michener; R. S. Moyer and T. K. Landauer [1967], [1973];
S. Papert [1971]; M. Polányi [1960]; R. D. Resch; F. Restle; F. O.
Schmitt and F. G. Worden; R. W. Sperry; C. M. Strauss; R. Thom
[1971].

# 7

## FROM CERTAINTY TO FALLIBILITY

# Platonism, Formalism, Constructivism

IF YOU DO mathematics every day, it seems the most natural thing in the world. If you stop to think about what you are doing and what it means, it seems one of the most mysterious. How are we able to tell about things no one has ever seen, and understand them better than the solid objects of daily life? Why is Euclidean geometry still correct, while Aristotelian physics is dead long since? What do we know in mathematics, and how do we know it?

In any discussion of the foundations of mathematics, three standard dogmas are presented: Platonism, formalism, and constructivism.

According to Platonism, mathematical objects are real. Their existence is an objective fact, quite independent of our knowledge of them. Infinite sets, uncountably infinite sets, infinite-dimensional manifolds, space-filling curves —all the members of the mathematical zoo are definite objects, with definite properties, some known, many unknown. These objects are, of course, not physical or material. They exist outside the space and time of physical existence. They are immutable—they were not created, and they will not change or disappear. Any meaningful question about a mathematical object has a definite answer, whether we are able to determine it or not. According to Platonism, a mathematician is an empirical scientist like a geologist; he cannot invent anything, because it is all there already. All he can do is discover.

Two whole-hearted Platonists are René Thom and Kurt Gödel. Thom writes (1971),

318

Everything considered, mathematicians should have the courage of their most profound convictions and thus affirm that mathematical forms indeed have an existence that is independent of the mind considering them. . . . Yet, at any given moment, mathematicians have only an incomplete and fragmentary view of this world of ideas.

And here is Gödel,

Despite their remoteness from sense experience, we do have something like a perception also of the objects of set theory, as is seen from the fact that the axioms force themselves upon us as being true. I don't see any reason why we should have less confidence in this kind of perception, i.e., in mathematical intuition, than in sense perception. . . . They, too, may represent an aspect of objective reality.

Thom's world of ideas is geometric, whereas Gödel's is the set-theoretic universe. On the other side, there is Abraham Robinson,

I cannot imagine that I shall ever return to the creed of the true Platonist, who sees the world of the actual infinite spread out before him and believes that he can comprehend the incomprehensible. (A. Robinson, 1969)

According to formalism, on the other hand, there are *no* mathematical objects. Mathematics just consists of axioms, definitions and theorems—in other words, formulas. In an extreme view, there are rules by which one derives one formula from another, but the formulas are not *about* anything; they are just strings of symbols. Of course the formalist knows that mathematical formulas are sometimes applied to physical problems. When a formula is given a physical interpretation, it acquires a meaning, and may be true or false. But this truth or falsity has to do with the particular physical interpretation. As a purely mathematical formula, it has no meaning and no truth value.

An example that demonstrates the difference between formalist and Platonist comes out in the continuum hypothesis of Cantor. Cantor conjectured that there is no infinite cardinal number which is greater than $\aleph_0$ (the cardinality of the integers) and smaller than $C$ (the cardinality of the real numbers). K. Gödel and P. J. Cohen showed that, on the basis of the axioms of formal set theory, the continuum hypothesis can neither be proved (Gödel, 1937) nor disproved (Cohen, 1964). (See Chapter 5, Non-Cantorian Set Theory.) To a Platonist, this means that our axioms are incomplete as a description of the set of real numbers. They are not strong enough to tell us the whole truth. The continuum hypothesis is either true or false, but we don't understand the set of real numbers well enough to find the answer.

To the formalist, on the other hand, the Platonist's interpretation makes no sense, because there *is* no real number system, except as we choose to create it by laying down axioms to describe it. Of course we are free to change this axiom system if we desire to do so. Such a change can be for convenience or usefulness or some other criterion we choose to introduce; it cannot be a matter of a better correspondence with reality, because there is no reality there.

Formalists and Platonists are at opposite sides on the question of existence and reality; but they have no quarrel with each other on what principles of reasoning should be permissible in mathematical practice. Opposed to both of them are the constructivists. The constructivists regard as genuine mathematics only what can be obtained by a finite construction. The set of real numbers, or any other infinite set, cannot be so obtained. Consequently, the constructivist regards Cantor's hypothesis as meaningless talk. Any answer at all would be sheer waste of breath.

# The Philosophical Plight of the Working Mathematician

MOST WRITERS on the subject seem to agree that the typical working mathematician is a Platonist on weekdays and a formalist on Sundays. That is, when he is doing mathematics he is convinced that he is dealing with an objective reality whose properties he is attempting to determine. But then, when challenged to give a philosophical account of this reality, he finds it easiest to pretend that he does not believe in it after all.

We quote two well-known authors:

> On foundations we believe in the reality of mathematics, but of course when philosophers attack us with their paradoxes we rush to hide behind formalism and say, "Mathematics is just a combination of meaningless symbols," and then we bring out Chapters 1 and 2 on set theory. Finally we are left in peace to go back to our mathematics and do it as we have always done, with the feeling each mathematician has that he is working with something real. This sensation is probably an illusion, but is very convenient. That is Bourbaki's attitude toward foundations. (J. A. Dieudonné, 1970, p. 145.)

> To the average mathematician who merely wants to know his work is accurately based, the most appealing choice is to avoid difficulties by means of Hilbert's program. Here one regards mathematics as a formal game and one is only concerned with the question of consistency. . . . The Realist [i.e., Platonist] position is probably the one which most mathematicians would prefer to take. It is not until he becomes aware of some of the difficulties in set theory that he would even begin to question it. If these difficulties particularly upset him, he will rush to the

shelter of Formalism, while his normal position will be somewhere between the two, trying to enjoy the best of two worlds. (P. J. Cohen, "Axiomatic Set Theory," ed. D. Scott.)

In these quotations from Dieudonné and Cohen we use the term "formalism" to mean the philosophical position that much or all of pure mathematics is a meaningless game. It should be obvious that to reject formalism as a philosophy of mathematics by no means implies any critique of mathematical logic. On the contrary, logicians, whose own mathematical activity *is* the study of formal systems, are in the best position to appreciate the enormous difference between mathematics as it is done and mathematics as it is schematized in the notion of a formal mathematical system.

According to Monk, the mathematical world is populated with 65% Platonists, 30% formalists, and 5% constructivists. Our own impression is that the Cohen-Dieudonné picture is closer to the truth. The typical mathematician is both a Platonist and a formalist—a secret Platonist with a formalist mask that he puts on when the occasion calls for it. Constructivists are a rare breed, whose status in the mathematical world sometimes seems to be that of tolerated heretics surrounded by orthodox members of an established church.

### Further Readings. See Bibliography

J. Dieudonné [1970]; M. Dummett; K. Gödel; L. Henkin; J. D. Monk; A. Robinson; R. Thom; D. Scott

# The Euclid Myth

T HE TEXTBOOK PICTURE of the philosophy of mathematics is a strangely fragmentary one. The reader gets the impression that the whole subject appeared for the first time in the late nineteenth century, in reponse to contradictions in Cantor's set theory. At that time there was talk of a "crisis in the foundations." To repair the foundations, three schools appeared on the scene, and spent some thirty or forty years quarreling with each other. It turned out that none of the three could really do much about the foundations, and the story ends in mid-air some forty years ago, with Whitehead and Russell having abandoned logicism, Hilbert's formalism defeated by Gödel's theorem, and Brouwer left to preach constructivism in Amsterdam, disregarded by all the rest of the mathematical world.

This episode in the history of mathematics actually is a remarkable story. Certainly it was a critical period in the philosophy of mathematics. But, by a striking shift in the meaning of words, the fact that foundationism was at a certain critical period the dominant trend in the philosophy of mathematics has led to the virtual identification of the philosophy of mathematics with the study of foundations. Once this identification is made, we are left with a peculiar impression: the philosophy of mathematics was an active field for only forty years. It was awakened by the contradictions in set theory, and after a while went back to sleep.

In reality there has always been a philosophical background, more or less explicit, to mathematical thinking. The foundationist period was one in which leading mathematicians were overtly concerned with philosophical issues, and engaged in public controversy about them. To make sense out of what happened during that period, one should look at what went before and after.

There are two strands of history that should be followed. One is in philosophy of mathematics; the other is in mathematics itself. For the crisis was a manifestation of a long-standing discrepancy between the traditional ideal of mathematics, which we can call the Euclid Myth, and the reality of mathematics, the actual practice of mathematical activity at any particular time. Bishop Berkeley recognized this discrepancy in 1734, in his book *The Analyst*. The book had a long subtitle, *A Discourse Addressed to an Infidel Mathematician, Wherein it is Examined Whether the Object, Principles and Inferences of the Modern Analysis are More Distinctly Conceived, or More Evidently Deduced, than Religious Mysteries and Points of Faith. "First cast out the beam of thine own Eye; and*

*then shalt thou see clearly to cast the mote out of thy brother's Eye."*
(The infidel was Edmund Halley.)

Berkeley exposed the obscurities and inconsistencies of differential calculus, as explained in his time by Newton, Leibnitz, and their followers. That is to say, he showed how far the calculus fell short of fitting the idea of mathematics according to the Euclid myth.

What is the Euclid myth? It is the belief that the books of Euclid contain truths about the universe which are clear and indubitable. Starting from self-evident truths, and proceeding by rigorous proof, Euclid arrives at knowledge which is certain, objective, and eternal. Even now, it seems that most educated people believe in the Euclid myth. Up to the middle or late nineteenth century, the myth was unchallenged. Everyone believed it. It has been the major support for metaphysical philosophy, that is, for philosophy which sought to establish some a priori certainty about the nature of the universe.

The roots of the philosophy of mathematics, as of mathematics itself, are in classical Greece. For the Greeks, mathematics meant geometry, and the philosophy of mathematics in Plato and Aristotle is the philosophy of geometry.

For Plato, the mission of philosophy was to discover true knowledge behind the veil of opinion and appearance, the change and illusion of the temporal world. In this task, mathematics had a central place, for mathematical knowledge was the outstanding example of knowledge independent of sense experience, knowledge of eternal and necessary truths.

In Plato's book the *Meno,* Socrates questions a slave boy, and leads him to discover that the area of the large square (see figure) is twice that of the square *ABCD,* whose diagonal is the side of the large square. How does the slave boy know this? Socrates argues that the boy did not learn it in this mortal life, so his knowledge must be recollection from life before birth. To Plato, this example shows that there is such a thing as true knowledge, knowledge of the eternal. Plato argues that

*Aristotle*
*384–322* B.C.

*Plato*
*c. 427* B.C.

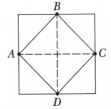

*The square on the diagonal* AC *is twice as large as the square on the side* AB. *Plato used this example to show that we do possess knowledge which is certain and timeless.*

325

1. We know truths of geometry which we have not learned through education or experience.

2. This knowledge is an example of the changeless, universal truths which, indeed, we can perceive and recognize.

3. Thus there must exist a realm of absolute, unchanging truth, the source and the basis for our knowledge of the Good.

Plato's conception of geometry was a key element in his conception of the world. Geometry played a similar role for the rationalist philosophers, Spinoza, Descartes, and Leibnitz. Like Plato, the rationalists regarded the faculty of Reason as an innate trait of the human mind, by which truths could be perceived a priori, independent of observation. For example, I may be mistaken in thinking that I am sitting at my writing desk composing this sentence, and I surely may be mistaken in thinking that the sun will rise tomorrow, but by no means can I be mistaken in my knowledge that the angle sum in a triangle equals a straight angle. (Spinoza's favorite example of an indubitably true statement was this theorem of Euclid, which, incidentally, is proved to be false in non-Euclidean geometry.)

Reason was the faculty that allowed man to know the Good and to know the Divine. The existence of this faculty was best seen in mathematics. Mathematics started with self-evident truths, and proceeded by careful reasoning to discover hidden truths. The truths of geometry had as subject matter ideal forms whose existence was evident to the mind. To question their existence would have been a sign of ignorance or insanity.

Mathematics and religion were the preeminent examples of knowledge obtained by Reason. The knowledge of the Good in Plato was transmuted into knowledge of God in the thought of the Renaissance rationalists.

The service of rationalism to science was that it denied the supremacy of authority, in particular religious authority, while yet maintaining the truth of religion. This philosophy gave science room to grow without being strangled as a rebel. It claimed for Reason—in particular for

326

science—the right to independence from authority—in particular, the authority of the Church. Yet this independence of reason was not too dangerous to authority, since philosophers declared that science was but the study of God. "The heavens proclaim the glory of God and the firmament showeth His handiwork."

The existence of mathematical objects in a realm of ideas independent of human minds was no difficulty for Newton or Leibnitz; as Christians, they took for granted the existence of a Divine Mind. In that context, the existence of ideal objects such as numbers and geometric forms is no problem. The problem is rather to account for the existence of nonideal, material objects. After rationalism succeeded in replacing medieval scholasticism, it was challenged in turn by materialism and empiricism; by Locke and Hobbes in Britain and by the Encyclopedists in France. In the contest between rationalism and empiricism, it was the advance of natural science on the basis of the experimental method that gave empiricism its decisive victory. The conventional wisdom in science became the belief in the material universe as the fundamental reality, with experiment and observation as the only legitimate means of obtaining knowledge.

The empiricists held that all knowledge, except mathematical knowledge, comes from observation. They usually did not try to explain how mathematical knowledge is obtained. An exception was John Stuart Mill. He proposed an empiricist theory of mathematical knowledge—that mathematics is a natural science no different from the others. For instance, we know that $3 + 4 = 7$ because we observe that, joining a pile of three buttons to a pile of four buttons, we get a pile of seven buttons. In Frege's *Foundations of Arithmetic* great sport is made of Mill's crudities, and it is only in Frege's critique that Mill's mathematical philosophy is discussed nowadays.*

In the philosophical controversy, first between rationalism and scholasticism, later between rationalism and the

* A recent attempt to return to an empiricist philosophy of mathematics is the book by Lehman (1979), which is influenced by the realism of Hilary Putnam.

new radical currents of empiricism and materialism, the sanctity of geometry remained unchallenged. Philosophers disputed whether we proceed from Reason (which humans possess as a gift from the Divine) to discover the properties of the physical world, or whether we have only our bodily senses with which to discover the properties of physical objects and of their Creator. In these battles, both sides took it for granted that geometrical knowledge is not problematical, even if all other knowledge is. Hume exempted only books of mathematics and of natural science from his celebrated instruction to "commit it to the flames." Even he did not perceive a problem in defining the status of mathematical knowledge.

For the rationalists, mathematics was the best example to confirm their view of the world. For empiricists it was an embarrassing counterexample which had to be ignored or somehow explained away. If, as seemed obvious, mathematics contains knowledge independent of sense perception, then empiricism is inadequate as an explanation of all human knowledge. This embarrassment is still with us; it is a reason for our difficulties with the philosophy of mathematics.

We are likely to forget that the modern scientific outlook gained ascendancy only in the last century. By the time of Russell and Whitehead, only logic and mathematics could still be claimed as nonempirical knowledge that is obtained directly by Reason.

Mathematics always had a special place in the battle between rationalism and empiricism. The mathematician-in-the-street, with his common-sense belief in mathematics as knowledge, is the last vestige of rationalism.

From the viewpoint customary among scientists nowadays, the prevalence of Platonism as an informal or tacit working philosophy is a remarkable anomaly. The accepted assumptions in science are now, and have been for many years, those of materialism with respect to ontology, and empiricism with respect to epistemology. That is to say, the world is all one stuff, which is called "matter" and which is studied by physics; if matter gets itself into sufficiently complicated configurations, it becomes subject to more special sciences with their own methodology, such as chemistry, geology, and biology. We learn about the world by looking at it and thinking about what we see. Until we look, we have nothing to think about.

Yet in mathematics we have knowledge of things that we have never observed and can never observe. At least, this is the naive point of view we assume when we are not trying to be philosophical.

By the end of the eighteenth century, the culmination of classical philosophy came with Kant, whose work attempted to unify the two conflicting traditions of rationalism and empiricism. Kant's metaphysics is a continuation of the Platonic heritage, of the search for certainty and timelessness in human knowledge. Kant wanted to rebut Hume's critique of the possibility of certainty in human knowledge. He made a sharp distinction between the noumena, the things in themselves, which we can never know, and the phenomena, the appearances, which are all that our senses can tell us about. But for all his careful skepticism, Kant's principal concern was still knowledge a priori —knowledge which was timeless and independent of experience. He distinguished two types of a priori knowledge. The "analytic a priori" is what we know to be true by logical analysis, by the very meaning of the terms being used. Kant, like the rationalists, believed that we also possess another kind of a priori knowledge, that is not simply logical truism. This is the "synthetic a priori." Our intuitions of time and space, according to Kant, are such knowledge. He explains their a priori nature by asserting that these intuitions are inherent properties of the human mind. Our knowledge of time is systematized in arithmetic, which is based on the intuition of *succession*. Our knowledge of space is systematized in geometry. For Kant, as for Plato, there is only one geometry—the same one that today we call Euclidean, to distinguish it from the many other systems of concepts which are also called geometries. The truths of geometry and arithmetic are forced on us by the way our minds work; this explains why they are supposedly true for everyone, independent of experience. The intuitions of time and space, on which arithmetic and geometry are based, are objective in the sense that they are universally valid for all human minds. No claim is made for existence outside the human mind, yet the Euclid myth remains as a central element in Kantian philosophy.

The Kantian dogma of the a priori remained a dominant influence on the philosophy of mathematics well into the twentieth century. All three foundationist schools sought to save for mathematics the special role that Kant ascribed to it.

# Foundations, Found and Lost

U NTIL WELL INTO the nineteenth century, the Euclid myth was as well established among mathematicians as among philosophers. Geometry was regarded by everybody, including mathematicians, as the firmest, most reliable branch of knowledge. Mathematical analysis—calculus and its extensions and ramifications—derived its meaning and legitimacy from its link with geometry. We need not use the term "Euclidean geometry" because the use of the qualifier became necessary and meaningful only after the possibility of more than one geometry had been recognized. Until that recognition, geometry was simply geometry—the study of the properties of space. These existed absolutely and independently, were objectively given, and were the supreme example of properties of the universe which were exact, eternal, and knowable with certainty by the human mind.

In the nineteenth century, several disasters took place. One disaster was the discovery of non-Euclidean geometries, which showed that there was more than one thinkable geometry.

A greater disaster was the development of analysis so that it overtook geometrical intuition, as in the discovery of space-filling curves and continuous nowhere-differentiable curves. These stunning surprises exposed the vulnerability of the one solid foundation—geometrical intuition—on which mathematics had been thought to rest. The loss of certainty in geometry was philosophically intolerable, because it implied the loss of all certainty in human knowledge. Geometry had served, from the time of Plato, as the supreme exemplar of the possibility of certainty in human knowledge.

*Richard Dedekind*
*1831–1916*

The mathematicians of the nineteenth century met the challenge. Led by Dedekind and Weierstrass, they turned from geometry to arithmetic as the foundation for mathematics. To do this, it was necessary to give a construction of the linear continuum, i.e., the real number system, to show how it could be built up from the integers, 1, 2, 3, . . . . Three different methods of doing this were proposed by Dedekind, Cantor, and Weierstrass. In all three methods one had to use some infinite set of rational numbers in order to define or construct a real number. Thus, in the effort to reduce analysis and geometry to arithmetic, one was led to introduce infinite sets into the foundations of mathematics.

*Karl Weierstrass*
*1815–1897*

The theory of sets was developed by Cantor as a new and fundamental branch of mathematics in its own right. It seemed that the idea of a set—an arbitrary collection of distinct objects—was so simple and fundamental that it could be the building block out of which all of mathematics could be constructed. Even arithmetic could be downgraded (or upgraded) from a fundamental to a secondary structure, for Frege showed how the natural numbers could be constructed from nothing—i.e., from the empty set—by using operations of set theory.

Set theory at first seemed to be almost the same as logic. The set-theory relation of inclusion, *A* is a subset of *B*, can always be rewritten as the logical relation of implication, "If *A*, then *B*." So it seemed possible that set-theory-logic could serve as the foundation for all of mathematics. "Logic," as understood in this context, refers to the fundamental laws of reason, the bedrock of the universe. The law of contradiction and the rules of implication are re-

331

*Alfred North Whitehead*
*1861–1947*

*Bertrand Russell*
*1872–1970*

*Gottlob Frege*
*1848–1925*

garded as objective and indubitable. To show that all of mathematics is just an elaboration of the laws of logic would have been to justify Platonism, by passing on to the rest of mathematics the indubitability of logic itself. This was the "logicist program," pursued by Russell and Whitehead in their *Principia Mathematica.*

Since all mathematics can be reduced to set theory, all one need consider is the foundation of set theory. However, it was Russell himself who discovered that the seemingly transparent notion of set contained unexpected traps.

The controversies of the late nineteenth and early twentieth century came about because of the discovery of contradictions in set theory. A special word—"antinomies"—was used as a euphemism for contradictions of this type.

The paradoxes arise from the belief that any reasonable predicate—any verbal description that seemed to make sense—could be used to define a set, the set of things that shared the stated property.

The most famous example of such a set was discovered by Russell himself. To state the Russell paradox, we define an "*R*-set" as "a set which includes itself." (An example is "the set of all objects describable in exactly eleven English words.") Now consider another set *M*: the set whose members are all possible sets *except* the *R*-sets. Is *M* an *R*-set? No. Is *M* not an *R*-set? Again, no. Moral: the definition of *M*, which seemed harmless though a bit tricky, is self-contradictory.

Russell sent his example in a letter to Gottlob Frege. Frege was about to publish a monumental work in which arithmetic was reconstructed on the foundation of set theory in the intuitive form. Frege added a postscript to his treatise—"A scientist can hardly meet with anything more undesirable than to have the foundations give way just as the work is finished. In this position I was put by a letter from Mr. Bertrand Russell, as the work was nearly through the press."

The Russell paradox and the other antinomies showed that intuitive logic, far from being more secure than classical mathematics, was actually much riskier, for it could lead

332

to contradictions in a way that never happens in arithmetic or geometry.

This was the "crisis in foundations," the central issue in the famous controversies of the first quarter of this century. Three principal remedies were proposed.

The program of "logicism," the school of Frege and Russell, was to find a reformulation of set theory which could avoid the Russell paradox and thereby save the Frege-Russell-Whitehead project of establishing mathematics upon logic as a foundation.

The work on this program played a major role in the development of logic. But it was a failure in terms of its original intention. By the time set theory had been patched up to exclude the paradoxes, it was a complicated structure which one could hardly identify with logic in the philosophical sense of "the rules for correct reasoning." So it became untenable to argue that mathematics is nothing but logic—that mathematics is one vast tautology. Russell wrote,

> I wanted certainty in the kind of way in which people want religious faith. I thought that certainty is more likely to be found in mathematics than elsewhere. But I discovered that many mathematical demonstrations, which my teachers expected me to accept, were full of fallacies, and that, if certainty were indeed discoverable in mathematics, it would be in a new field of mathematics, with more solid foundations than those that had hitherto been thought secure. But as the work proceeded, I was continually reminded of the fable about the elephant and the tortoise. Having constructed an elephant upon which the mathematical world could rest, I found the elephant tottering, and proceeded to construct a tortoise to keep the elephant from falling. But the tortoise was no more secure than the elephant, and after some twenty years of very arduous toil, I came to the conclusion that there was nothing more that I could do in the way of making mathematical knowledge indubitable. (Bertrand Russell, "Portraits from Memory.")

After the logicist, the next major school was the constructivist. This originated with the Dutch topologist L. E. J. Brouwer about 1908. Brouwer's position was that the

*Russell and Whitehead were pioneers in a program to reduce mathematics to logic. Here, after 362 pages, the arithmetic proposition 1 + 1 = 2 is established.*

*From "Principia Mathematica", Cambridge U. Press, 1910.*

$*54 \cdot 42$. $\vdash :: \alpha \epsilon 2 . \supset :. \beta \subset \alpha . \quad ! \beta . \beta \neq \alpha . \equiv . \beta \epsilon \iota " \alpha$

Dem.

$\vdash . *54 \cdot 4 . \quad \supset \vdash :: \alpha = \iota ' x \cup \iota ' y . \supset :.$

$\qquad\qquad \beta \subset \alpha . \exists ! \beta . \equiv : \beta = \Lambda . v . \beta = \iota ' x . v . \beta = \iota ' y . v . \beta = \alpha : \exists ! \beta :$

$[*24 \cdot 53 \cdot 56 . *51 \cdot 161] \qquad \equiv : \beta = \iota ' x . v . \beta = \iota ' y . v . \beta = \alpha \qquad (1)$

$\vdash . *54 \cdot 25 . \text{Transp} . *52 \cdot 22 . \supset \vdash : x \neq y . \supset . \iota ' x \cup \iota ' y \neq \iota ' x . \iota ' x \cup \iota ' y \neq \iota ' y :$

$[*13 \cdot 12] \qquad \supset \vdash : \alpha = \iota ' x \cup \iota ' y . x \neq y . \supset . \alpha \neq \iota ' x . \alpha \neq \iota ' y \qquad (2)$

$\vdash . (1) . (2) . \supset \vdash :: \alpha = \iota ' x \cup \iota ' y . x \neq y . \supset :.$

$\qquad\qquad \beta \subset \alpha . \exists ! \beta . \beta \neq \alpha . \equiv : \beta = \iota ' x . v . \beta = \iota ' y :$

$[*51 \cdot 235] \qquad\qquad \equiv : (\exists z) . z \epsilon \alpha . \beta = \iota ' z :$

$[*37 \cdot 6] \qquad\qquad \equiv : \beta \epsilon \iota " \alpha \qquad (3)$

$\vdash . (3) . *11 \cdot 11 \cdot 35 . *54 \cdot 101 . \supset \vdash . \text{Prop}$

$*54 \cdot 43$. $\vdash :. \alpha , \beta \epsilon 1 . \supset : \alpha \cap \beta = \Lambda . \equiv . \alpha \cup \beta \epsilon 2$

Dem.

$\vdash . *54 \cdot 26 . \supset \vdash :. \alpha = \iota ' x . \beta = \iota ' y . \supset : \alpha \cup \beta \epsilon 2 . \quad \equiv . x \neq y .$

$[*51 \cdot 231] \qquad\qquad \equiv . \iota ' x \cap \iota ' y = \Lambda .$

$[*13 \cdot 12] \qquad\qquad \equiv . \alpha \cap \beta = \Lambda \qquad (1)$

$\vdash . (1) . *11 \cdot 11 \cdot 35 . \supset$

$\qquad\qquad \vdash :. (\exists x , y) . \alpha = \iota ' x . \beta = \iota ' y . \supset : \alpha \cup \beta \epsilon 2 . \equiv . \alpha \cap \beta = \Lambda \qquad (2)$

$\vdash . (2) . *11 \cdot 54 . *52 \cdot 1 . \supset \vdash . \text{Prop}$

From this proposition it will follow, when arithmetical addition has been defined, that $1 + 1 = 2$.

natural numbers are given to us by a fundamental intuition, which is the starting point for all mathematics. He demanded that all mathematics should be based *constructively* on the natural numbers. That is to say, mathematical objects may not be considered meaningful, may not be said to exist, unless they are given by a construction, in finitely many steps, starting from the natural numbers. It is not sufficient to show that the assumption of nonexistence would lead to a contradiction.

For the constructivists, many of the standard proofs in classical mathematics are invalid. In some cases they are able to supply a constructive proof. But in other cases they show that a constructive proof is impossible: theorems which are considered to be well-established in classical mathematics are actually declared to be false for constructivist mathematics.

An important example is the "law of trichotomy": *Every real number is either zero, positive, or negative.*

When the real numbers are constructed set-theoretically, according to the recipe of Dedekind or Cantor, for example, the law of trichotomy can be proved as a theorem. It plays a fundamental part in all of calculus and analysis.

However, Brouwer gave an example of a real number such that we are unable to prove constructively that it is zero, positive, or negative. (For details, see the next chapter, "$\pi$ and $\hat{\pi}$".) From Brouwer's point of view, this is a counterexample and shows that the law of trichotomy is false.

In fact, the classical proof of the law of trichotomy uses proof by contradiction (the law of the excluded middle) and so is not a valid proof by Brouwer's criteria.

Although many prominent mathematicians had expressed misgivings and disagreements with nonconstructive methods and free use of infinite sets, Brouwer's call for restructuring analysis from the ground up seemed to most mathematicians unreasonable, indeed fanatical.

Hilbert was particularly alarmed. "What Weyl and Brouwer do comes to the same thing as to follow in the footsteps of Kronecker! They seek to save mathematics by throwing overboard all that which is troublesome. . . . They would chop up and mangle the science. If we would follow such a reform as the one they suggest, we would run the risk of losing a great part of our most valuable treasure!" (*Hilbert*, C. Reid, p. 155)

Hilbert undertook to defend mathematics from Brouwer's critique by giving a *mathematical proof* of the consistency of classical mathematics. Furthermore, he proposed to do so by arguments of a purely finitistic, combinatorial type—arguments that Brouwer himself could not reject.

This program involved three steps.

(1) Introduce a formal language and formal rules of inference, sufficient so that every "correct proof" of a classical theorem could be represented by a formal derivation, starting from axioms, with each step mechanically checkable. This had already been accomplished in large part by Frege, Russell and Whitehead.

(2) Develop a theory of the combinatorial properties of

*David Hilbert*
*1862–1943*

this formal language, regarded as a finite set of symbols subject to permutation and rearrangements as provided by the rules of inference, now regarded as rules for transforming formulas. This theory was called "meta-mathematics."

(3) Prove by purely finite arguments that a contradiction, for example $1 = 0$, cannot be derived within this system.

In this way, mathematics would be given a secure foundation—in the sense of a guarantee of consistency.

This kind of foundation is not at all the same as a foundation based on a theory known to be *true,* as geometry had been believed to be true, or at least impossible to doubt, as it is supposed to be impossible to doubt the law of contradiction in elementary logic.

The formalist foundation of Hilbert, like the logicist foundation, offered certainty and reliability at a price. As the logicist interpretation tried to make mathematics safe by turning it into a tautology, the formalist interpretation tried to make it safe by turning it into a meaningless game. The "proof-theoretic program" comes into action only after mathematics has been coded in a formal language and its proofs written in a way checkable by machine. As to the *meaning* of the symbols, that becomes something extra-mathematical.

Hilbert's writings and conversation display full conviction that mathematical problems are questions about real objects, and have meaningful answers which are true in the same sense that any statement about reality is true. If he was prepared to advocate a formalist interpretation of mathematics, this was the price he considered necessary for the sake of obtaining certainty.

The goal of my theory is to establish once and for all the certitude of mathematical methods. . . . The present state of affairs where we run up against the paradoxes is intolerable. Just think, the definitions and deductive methods which everyone learns, teaches and uses in mathematics, the paragon of truth and certitude, lead to absurdities! If mathematical thinking is defective, where are we to

find truth and certitude? (D. Hilbert, "On the Infinite,"
in *Philosophy of Mathematics* by Benacerraf and Putnam.)

As it happened, certainty was not to be had, even at this
price. In 1930 Gödel's incompleteness theorems showed
that the Hilbert program was unattainable—that any consistent formal system strong enough to contain elementary
arithmetic would be unable to prove its own consistency.
The search for secure foundations has never recovered
from this defeat.

Hilbert's program rested on two unexamined premises;
first, the Kantian premise that *something* in mathematics—
at least the purely "finitary part"—is a solid foundation, is
indubitable; and second, the formalist premise, that a sol-
idly founded theory about formal sentences could validate
the mathematical activity of real life, wherein formalization
even as a hypothetical possibility is present only in the remote background if at all.

The first premise was shared by the constructivists; the
second, of course, was rejected by them.

The program of formalization amounts to a mapping of
set theory and analysis into a part of itself—namely, into
finite combinatorics. At best, then, one would be left with
the claim that all of mathematics is consistent if the "finitistic" principle to be allowed in "metamathematics," as Hilbert's mathematics *about* mathematics used to be called, is
itself reliable. Again, one is looking for the last tortoise
under the last elephant.

The bottom tortoise or elephant is in fact the Kantian
synthetic a priori, the intuition. Although Hilbert does not
explicitly refer to Kant, his conviction that mathematics
can and must provide truth and certainty "or where else
are we to find it?" is in the Platonic heritage as transmitted
through the rationalists to Kant, and thereby to the intellectual milieu of nineteenth-century western Europe. In
this respect, he is as much a Kantian as Brouwer, whose
label of constructivism openly avows his Kantian heritage.

To Brouwer, the Hilbert program was misconceived at
Step 1, because it rested on the identification of mathematics itself with the formulas which are used to represent or

express it. But it was only by this transition to languages and formulas that Hilbert was able to envision even the possibility of a *mathematical* justification of mathematics.

Brouwer, who like Hilbert took it for granted that mathematics could and should be established on a "sound" and "firm" foundation, took the other road, of insisting that mathematics must start from the intuitively given, the finite, and must contain only what is obtained in a constructive way from this intuitively given starting point. Intuition here means the intuition of *counting* and that alone. For both Brouwer and Hilbert, the acceptance of geometric intuition as a basic or fundamental "given" on a par with arithmetic, would have seemed utterly retrograde and unacceptable *within the context of foundational discussions.* At the same time, for Brouwer as for Hilbert, the use of geometrical intuition in his "regular" (nonfoundational) mathematical research was a matter of course. Brouwer no more felt obligated to sacrifice his research in topology to his intuitionist dogma than Hilbert felt obligated in his work to deal with formulas rather than meanings. For both of them, the split between their ordinary mathematical practice and their theories of foundations did not seem to call for an explanation or apology. It is reported that Brouwer in his later years was prepared to sacrifice his research in topology to his intuitionist dogma.

### Further Readings. See Bibliography

D. Hilbert; S. Kleene; C. Reid.

# The Formalist
# Philosophy of
# Mathematics

I N THE MID-TWENTIETH century, formalism became the predominant philosophical attitude in textbooks and other "official" writing on mathematics.
Constructivism remained a heresy with only a few adherents. Platonism was and is believed by (nearly) all mathematicians. But, like an underground religion, it is observed in private and rarely mentioned in public.

Contemporary formalism is descended from Hilbert's
formalism, but it is not the same thing. Hilbert believed in
the reality of finite mathematics. He invented metamathematics in order to justify the mathematics of the infinite.
This realism-of-the-finite with formalism-for-the-infinite is
still advocated by some writers. But more often the formalist doesn't bother with this distinction. For him, mathematics, from arithmetic on up, is just a game of logical deduction.

The formalist defines mathematics as the science of rigorous proof. In other fields some theory may be advocated
on the basis of experience or plausibility, but in mathematics, he says, either we have a proof or we have nothing.

Any logical proof must have a starting point. So a mathematician must start with some undefined terms, and some
unproved statements about these terms. These are called
"assumptions" or "axioms." For example, in plane geometry we have the undefined terms "point" and "line" and the
axiom "Through any two distinct points passes exactly one
straight line." The formalist points out that the logical import of this statement does not depend on any mental picture we may associate with it. Only tradition prevents us
from using other words than point and line—"Through
any two distinct bleeps passes exactly one neep."

If we give some interpretation to the terms point and

line then these axioms may become true or false. Presumably there is some interpretation in which they are true; otherwise, it would be senseless to be interested in them. However, so far as pure mathematics is concerned, the interpretation we give to the axioms is irrelevant. We are concerned only with valid logical deductions from them.

Results deduced in this way are called theorems. One cannot assert that a theorem is true, any more than one can assert that the axioms are true. As statements in pure mathematics, they are neither true nor false, since they talk about undefined terms. All we can say in mathematics is that the theorem follows logically from the axioms. Thus the statements of mathematical theorems have no content at all; they are not *about* anything. On the other hand, according to the formalist, they are free of any possible doubt or error, because the process of rigorous proof and deduction leaves no gaps or loopholes.

In brief, to the formalist mathematics is the science of formal deductions, from axioms to theorems. Its primitive terms are undefined. Its statements have no content until they are supplied with an interpretation. For example, we can interpret statements in geometry in terms of distances between physical locations.

In some textbooks the formalist viewpoint is stated as a simple matter of fact, and the uncritical reader or student may accept it as the authorized or "official" view. It is not a simple matter of fact, but rather a complex matter of interpretation. The reader has the right to a skeptical attitude, and to expect evidence to justify this view.

Indeed, brief reflection shows that the formalist view is not plausible according to ordinary mathematical experience. Every elementary school teacher talks about "facts of arithmetic" or "facts of geometry," and in high school the Pythagorean theorem and the prime factorization theorem are taught as true statements about right triangles or about integers. In the official view, any talk of facts or truths is incorrect.

One argument for the official view comes from the history of geometry, as a response to the dethronement of Euclidean geometry.

For Euclid, the axioms of geometry were not assumptions but "self-evident truths." The formalist view results, in part, from the rejection of the idea that one can start from "self-evident truths."

In our discussion of non-Euclidean geometry in Chapter 5, we saw how the attempt to prove Euclid's fifth postulate (the postulate of parallels, which was not as "self-evident" as the other four postulates) led to the discovery of non-Euclidean geometry in which the parallel postulate is assumed to be false.

Now, can we claim that Euclid's parallel postulate and its negation are *both* true? The formalist concludes that if we want to keep our freedom as mathematicians to study both Euclidean and non-Euclidean geometry, it is necessary to give up the notion that either is true. It is enough if each is consistent.

As a matter of fact, Euclidean and non-Euclidean geometry appear to conflict only if we believe in an objective physical space which obeys a single set of laws and which both theories attempt to describe. If we give up this belief, then Euclidean and non-Euclidean geometry are no longer rival candidates for a solution of the same problem, but just two different mathematical theories. The parallel postulate is true for the Euclidean straight line, false for the non-Euclidean. But are the theorems of geometry meaningful even apart from physical interpretations? May we still use the words "true" and "false" about statements in pure geometry? The Platonist would say yes, since mathematical objects exist in their own world apart from the world of physical application. The formalist, on the other hand, says no, the statements can't be true or false because they aren't about anything and don't mean anything.

The formalist makes a distinction between geometry as a deductive structure and geometry as a descriptive science. Only the first is regarded as mathematical. The use of pictures or diagrams, or even mental imagery, all are non-mathematical. In principle, they should be unnecessary. Consequently, he regards them as inappropriate in a mathematics text, perhaps even in a mathematics class.

Why do we give *this* particular definition, and not some

341

other one? Why *these* axioms and not some others? Such questions, to the formalist, are premathematical. If they are admitted at all to his text or his course, it will be in parentheses, and in brief.

What examples or applications can be made from the general theory he has developed? This is also not strictly relevant, and can be left for parenthetical remarks, or to be worked out as a problem.

From the formalist point of view, we haven't really started doing mathematics until we have stated some hypotheses and begun a proof. Once we have reached our conclusions, the mathematics is over. Anything more we have to say about it is, in a sense, superfluous. We measure how much we have accomplished in a class by how much we have *proved* in our lectures. The question of what was learned and understood by the audience is another matter —not a mathematical question.

One reason for the dominance of formalism was its connection with logical positivism. This was the dominant trend in the philosophy of science during the 1940s and 1950s. Its after-effects linger on, if only because nothing definitive has appeared to replace it. The "Vienna school" of logical positivists advocated the goal of a unified science, coded in a formal logical calculus and with a single deductive method. Formalization was held up as the goal for all the sciences. Formalization meant choice of a basic vocabulary of terms, statement of fundamental laws using these terms, and logical development of a theory from the fundamental laws. The example that was followed was that of classical and quantum mechanics.

In order to relate the formal theory to experimental data, each science has to have its rules of interpretation, which are not part of the formal theory. For example, in classical mechanics there are rules for physical measurements of the basic quantities (mass, length, time). Quantum mechanics has its own rules by which the term "observable" in the formal theory is related to experimental measurements. In this scheme of things, mathematics appears as the tool for formulating and developing the theory. The fundamental laws are mathematical formulas. In

342

mechanics, they are differential equations. The theory is developed by deriving the consequences of these laws, using mathematical reasoning.

Mathematics itself is seen, not as a science, but as a language for other sciences. It is not a science because it has no subject matter. It has no observed data to which one can apply rules of interpretation. By the philosophical categories that logical positivism admits, mathematics seems to be *only* a formal structure. So logical positivism in the philosophy of science leads to formalism in the philosophy of mathematics.

As a philosophy of mathematics, formalism is not compatible with the mode of thought of working mathematicians. But this was not a problem to positivist philosophers of science. Since their main orientation was to theoretical physics, they could look upon mathematics simply as a tool, not as a living and growing subject in itself. From the viewpoint of a user, it is possible, and sometimes even convenient, to identify mathematics itself with its axiomatic presentation in textbooks. From the viewpoint of the producer, the axiomatic presentation is secondary. It is only a refinement that is provided after the primary work, the process of mathematical discovery, has been carried out. This fact can be ignored by a physicist, and still more by a philosopher of physics, whose notions of mathematics come mainly from logic and the philosophy of mathematics, not from participation in the development of mathematics itself.

Logical positivism is no longer popular in the philosophy of science. A historical-critical point of view, derived largely from Karl Popper's work, is now available as an alternative. But this has had little effect on the philosophy of mathematics.

The heritage of Russell, Frege, and Wittgenstein has left a school of analytic philosophy which holds that the central problem of philosophy is the analysis of meaning, and logic is its essential tool. Since mathematics is the branch of knowledge whose logical structure is best understood, it is held that the philosophy of mathematics is the most advanced branch of philosophy and a model for other parts of philosophy. As the dominant style of Anglo-American philosophy, analytic philosophy tends to perpetuate identification of the philosophy of mathematics with logic and the study of formal systems.

From this standpoint, a problem of principal concern to the mathematician becomes totally invisible. This is the problem of giving a philosophical account of the actual development of mathematics, of preformal mathematics, the mathematics of the classroom and seminar, including an examination of how this preformal mathematics relates to and is affected by formalization.

The most influential example of formalism as a style in mathematical exposition was the writing of the group known collectively as Nicolas Bourbaki. Under this pseudonym, a series of basic graduate texts in set theory, algebra and analysis was produced which had a tremendous influence all over the world in the 1950s and 1960s.

The formalist style gradually penetrated downward into undergraduate mathematics teaching and, finally, in the name of "the new math," even invaded kindergarten, with preschool texts of set theory. A game of formal logic called "WFF and Proof" was invented to teach grade-school children how to recognize a "well-formed formula" (WFF) according to formal logic.

In recent years, a reaction against formalism has been growing. In recent mathematical research, there is a turn toward the concrete and the applicable. In texts and treatises, there is more respect for examples, less strictness in formal exposition. The formalist philosophy of mathematics is the intellectual source of the formalist style of mathematical work. The signs seem to indicate that the formalist philosophy may soon lose its privileged status.

**Further Readings. See Bibliography**

H. B. Curry, [1951]; A. Robinson [1964], [1969].

# Lakatos and the Philosophy of Dubitability

*Imre Lakatos*
*1922–1973*

**F**OUNDATIONISM, i.e., attempts to establish a basis for mathematical indubitability, has dominated the philosophy of mathematics in the twentieth century. A radically different alternative was offered in the remarkable work of Imre Lakatos to which we now turn. It was an outgrowth of new trends in the philosophy of science.

In science, the search for "foundations" leads to the traditional problem of "inductive logic:" how to derive general laws from particular experiments and observations. In 1934 there was a revolution in the philosophy of science when Karl Popper proposed that it is neither possible nor necessary to justify the laws of science by justifying inductive reasoning. Popper asserted that scientific theories are not derived inductively from the facts; rather, they are invented as hypotheses, speculations, even guesses, and are then subjected to experimental tests where critics attempt to refute them. A theory is entitled to be considered scientific, said Popper, only if it is in principle capable of being tested and risking refutation. Once a theory survives such tests, it acquires a degree of credibility, and may be considered to be tentatively established; but it is never *proved*. A scientific theory may be objectively true, but we can never know it to be so with certainty.

Although Popper's ideas have been criticized and are now sometimes regarded as one-sided and incomplete, his criticism of the inductivist dogma has made a fundamental change in the way people think about scientific knowledge.

While Popper and other recent thinkers transformed the philosophy of science, the philosophy of mathematics remained relatively stagnant. We are still in the aftermath of the great foundationist controversies of the early twentieth century. Formalism, intuitionism, and logicism, each left its trace in the form of a certain mathematical research program that ultimately made its own contribution to the body of mathematics itself. As *philosophical* programs, as attempts to establish a secure foundation for mathematical knowledge, all have run their course and petered out or dried up. Yet there remains, as a residue, an unstated consensus that the philosophy of mathematics *is* research on the foundations of mathematics. If I find research in foundations uninteresting or irrelevant, I conclude that I'm simply not interested in philosophy (thereby depriving myself of any chance of confronting and clarifying my own uncertainties about the meaning, nature, purpose or significance of mathematical research).

Lakatos entered the arena as a mathematically literate philosopher and a follower of Popper's theory of scientific knowledge. He was a graduate in mathematics, physics, and philosophy at Debrecen (1944), a survivor of the Nazis whose mother and grandmother died at Auschwitz. (He was born a Lipschitz, changed to Imre Molnar in 1944 for safety under the Germans, then to Imre Lakatos when he came back into possession of some shirts bearing the monogram I.L.) After the war he was an active communist, and for a time a high official of the Ministry of Education, but in 1950 he was arrested, and served three years in prison. After his release, he found employment, with the help of Rényi, as a translator of mathematical works into Hungarian; one of the books he translated was Pólya's *How to Solve It*. After the 1956 uprising he fled Hungary, arriving ultimately in England where he came under Popper's influence and started work on a doctorate in philosophy. Popper and Pólya are the joint godfathers of Lakatos' work; it was at Pólya's suggestions that he took as his theme the history of the Euler-Descartes formula: $V - E + F = 2$. (See Chapter 6, The Creation of New Mathematics.)

Instead of presenting symbols and rules of combination, he presents human beings, a teacher and his students. Instead of presenting a system built up from first principles, he presents a clash of views, arguments and counterarguments. Instead of mathematics skeletalized and fossilized,

346

he presents mathematics growing from a problem and a conjecture, with a theory taking shape before our eyes, in the heat of debate and disagreement, doubt giving way to certainty and then to renewed doubt.

*Proofs and Refutations* is the name of Lakatos' masterpiece. For fifteen years it was a sort of underground classic among mathematicians, known only to those few intrepid souls who ventured into the bound volumes of the *British Journal for Philosophy of Science,* where it appeared in 1963 as a series of four articles. Finally in 1976 it was published in book form by the Oxford University Press, three years after Lakatos' death of a brain tumor at the age of 51.

*Proofs and Refutations* uses history as the text on which to base its sermon: mathematics, too, like the natural sciences, is fallible, not indubitable; it too grows by the criticism and correction of theories which are never entirely free of ambiguity or the possibility of error or oversight. Starting from a problem or a conjecture, there is a simultaneous search for proofs and counterexamples. New proofs explain old counterexamples, new counterexamples undermine old proofs. To Lakatos, "proof" in this context of informal mathematics does not mean a mechanical procedure which carries truth in an unbreakable chain from assumptions to conclusions. Rather, it means explanations, justifications, elaborations which make the conjecture more plausible, more convincing, while it is being made more detailed and accurate under the pressure of counterexamples.

Each step of the proof is itself subject to criticism, which may be mere skepticism or may be the production of a counterexample to a particular argument. A counterexample which challenges one step in the argument is called by Lakatos a "local counterexample"; a counterexample which challenges, not the argument, but the conclusion itself, he calls a global counterexample.

Thus Lakatos applied his epistemological analysis, not to formalized mathematics, but to *informal* mathematics, mathematics in process of growth and discovery, which is of course mathematics as it is known to mathematicians and students of mathematics. Formalized mathematics, to

347

which most philosophizing has been devoted in recent years, is in fact hardly to be found anywhere on earth or in heaven outside the texts and journals of symbolic logic.

In form, *Proofs and Refutations* is a classroom dialogue, a continuation of one in Pólya's *Induction and Analogy in Mathematics.* The teacher presents the traditional proof, due to Cauchy, of the Euler formula, in which the edges of a polyhedron are stretched out to form a network on the plane, and then successively reduced to a single triangle. No sooner is the proof complete than the class produces a whole menagerie of counterexamples. The battle is on. What did the proof prove? What do we know in mathematics, and how do we know it? The discussion proceeds to ever-deeper levels of sophistication, both mathematical and logical. There are always several different points of view in contest, and many about-faces when one character changes his point of view and adopts a position which has just been abandoned by his antagonist.

In counterpoint with these dialectical fireworks, the footnotes provide the genuine, documented history of the Euler-Descartes conjecture, in amazing detail and complexity. The main text is in part a "rational reconstruction" of the actual history; or perhaps it would be better to say, as Lakatos once did, that the actual history is a parody of its rational reconstruction.

*Proofs and Refutations* is an overwhelming work. The effect of its polemical brilliance, its complexity of argument and self-conscious sophistication, its sheer weight of historical learning, is to dazzle the reader.

It would be fair to say that in *Proofs and Refutations* Lakatos *argues* that dogmatic philosophies of mathematics (logicist or formalist) are unacceptable, and he *shows* that a Popperian philosophy of mathematics is possible. However, he does not actually carry out the program of reconstructing the philosophy of mathematics with a fallibilist epistemology.

In the main text of *Proofs and Refutations,* we hear the author's characters, but not the author himself; he shows us mathematics as he sees it, but he does not make explicit the full import of what he is showing us. Or rather, he states its import only in the critical sense, especially in an all-out tooth-and-nail attack on formalism. But what is its import in the positive sense?

First of all, we need to know what mathematics is *about.* The Platonist (in particular, the logical Platonist such as Frege or the early Russell) would say it is about objectively existing ideal entities, which a certain intellectual faculty permits us to perceive or intuit directly, just as our five senses permit us to perceive physical objects. But few modern readers, and certainly not Lakatos, are prepared to contemplate seriously the existence, objectively, timelessly, and spacelessly, of all entities contained in modern set theory, let alone in future theories yet to be revealed. The formalist, on the other hand, says mathematics isn't about anything, it just *is.* A mathematical formula is just a formula, and our belief that it has content is an illusion that need not be defended or justified. This position is tenable only if one forgets that informal mathematics *is* mathematics. Formalization is only an abstract possibility which no one would want or be able actually to carry out.

Lakatos holds that informal mathematics is a science in the sense of Popper, that it grows by a process of successive criticism and refinement of theories and the advancement of new and competing theories (*not* by the deductive pattern of formalized mathematics.) But in natural science Popper's doctrine depends on the objective existence of the world of nature. Singular spatio-temporal statements such as "The voltmeter showed a reading of 3.2" provide the tests whereby scientific theories are criticized and sometimes refuted. To use the Popperian jargon, these "basic statements" are the "potential falsifiers."

If informal mathematics is on a par with natural science, we must locate its "objects." What are the data, the "basic statements" of the subject, which provide potential falsifiers to proposed informal mathematical theories? This question is not even posed in *Proofs and Refutations,* yet it is the main question which must be dealt with if one wants to go further in constructing a fallibilist or nondogmatic epistemology of mathematics.

We will never know if Lakatos could have solved this

Descartes (1635) and
Euler (1752) stated that
$V - E + F = 2$ for all
polyhedra. Imre Lakatos
focused attention on the
subsequent comedy of
errors wherein mathemat-
ics discovered successive
invalidating polyhedral
monsters and attempted
to patch up the theory.
Has the final word now
been said?

$V$ = Number of Verti-
ces
$E$ = Number of Edges
$F$ = Number of Faces

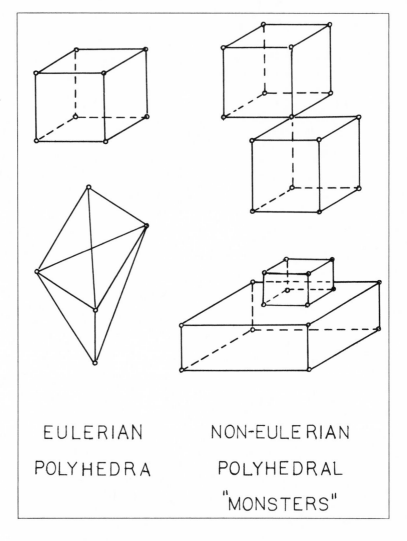

EULERIAN

POLYHEDRA

NON-EULERIAN

POLYHEDRAL

"MONSTERS"

problem. After he wrote *Proofs and Refutations,* he turned
away from philosophy of mathematics. He became a prominent polemicist in controversies about philosophy of science involving such authors as Carnap, Popper, Thomas
Kuhn, Polányi, Toulmin, and Feyerabend. No doubt he intended to return to mathematics; but at the time of his sudden death in February, 1974 he had not done so.

A partial answer is contained in one of the articles in volume 2 of his posthumous collected papers. This article, *"A*

*renaissance of empiricism in the philosophy of mathematics?"*, starts out with an impressive collection of quotations, from a dozen or so eminent mathematicians and logicians, of both the logicist and formalist stripe, all showing that the search for secure foundations has been given up. All agree that there is no reason to believe in mathematics, except that it seems to work. Von Neumann says that at least it is no worse than modern physics, in which many people also seem to believe. Having cut the ground from under his opponents by showing that his "heretical" view is really not opposed to that of the mathematical establishment, Lakatos goes on to draw the contrast between "Euclidean" theories such as the traditional foundationist philosophies of mathematics, and "quasi-empiricist" theories which regard mathematics as intrinsically conjectural and fallible. He points out that his theory is quasi-empiricist (not empiricist pure and simple) because the potential falsifiers or basic statements of mathematics, unlike those of natural science, are certainly not singular spatiotemporal statements (i.e. such statements as "the reading on the voltmeter was 3.2"). He gives his own answer in two parts. First of all, for formalized mathematical theories, the potential falsifiers are informal theories. In other words, if it is a question of accepting or rejecting a proposed set of axioms for set theory, we make our decision according to how well the formal system reproduces or conforms to the informal mathematical theory which we had in mind in the first place. Of course, Lakatos is well aware that we may also decide to modify our informal theory, and that the decision of which road to take may be a complex and controversial one.

At this point, he is face to face with the main problem. What are the "objects" of *informal* mathematical theories? When we talk about numbers, triangles, or betting odds apart from any system of axioms and definitions, what kinds of entities are we talking about? There are many possible answers, some going back to Aristotle and Plato, all having difficulties and long histories of attempts to evade the difficulties. The fallibilist position should lead to a new critique of the old answers, and perhaps to a new answer that would bring the philosophy of mathematics into the mainstream of contemporary philosophy of science. But Lakatos was not prepared to commit himself. He wrote, "The answer will scarcely be a monolithic one. Careful historico-critical case-studies will probably lead to a sophisticated and composite solution." A reasonable viewpoint, certainly, but a disappointing one.

The introduction to *Proofs and Refutations* is a blistering attack on formalism, which Lakatos defines as that school

> which tends to identify mathematics with its formal axiomatic abstraction and the philosophy of mathematics with metamathematics.
>
> Formalism disconnects the history of mathematics from the philosophy of mathematics. . . . Formalism denies the status of mathematics to most of what has been commonly understood to be mathematics, and can say nothing about its growth. . . .
>
> Under the present dominance of formalism, one is tempted to paraphase Kant: the history of mathematics, lacking the guidance of philosophy, has become blind, while the philosophy of mathematics, turning its back on the most intriguing phenomena in the history of mathematics, has become empty. . . . The formalist philosophy of mathematics has very deep roots. It is the latest link in the long chain of dogmatist philosophies of mathematics. For more than 2,000 years there has been an argument between dogmatists and sceptics. In this great debate, mathematics has been the proud fortress of dogmatism. . . . A challenge is now overdue.

However, Lakatos did not claim that his own work was making the overdue challenge. He wrote,

> The core of this case-study will challenge mathematical formalism, but will not challenge directly the ultimate positions of mathematical dogmatism. Its modest aim is to elaborate the point that informal, quasi-empirical, mathematics does not grow through a monotonous increase of the number of indubitably established theorems, but through the incessant improvement of guesses by speculation and criticism, by the logic of proofs and refutation.

Lakatos quickly became a major figure internationally in the philosophy of science, but (except for a fairly complete summary which appeared in Mathematical Reviews) I am not aware of any published critique or response to *Proofs and Refutations* until it was republished posthumously in book form by Cambridge University Press in 1976.

The first critique appeared in the book itself, in the form of footnotes and comments added by the editors, John Worrall and Elie Zahar. One finds their additions in footnotes on pages 56, 100, 138, and 146, and in two pages of new dialog in pages 125–26. The critique on page 138 is the most forthright.

On that page, Lakatos wrote that to revise the infallibilist philosophy of mathematics "one had to give up the idea that our deductive inferential intuition is infallible."

The editors note,

> This passage seems to us mistaken and we have no doubt that Lakatos, who came to have the highest regard for formal deductive logic, would himself have changed it. First-order logic has arrived at a characterisation of the validity of an inference which (relative to a characterisation of the "logical" terms of a language) does make valid inference essentially infallible.

This same point is made in the other footnotes and additional material. Lakatos "underplays a little the achievements of the mathematical 'rigorists'." The goal of rigorous correct proofs *is* attainable. "There is no serious sense in which such proofs are fallible."

The editors evidently think it is a matter of great importance to correct Lakatos whenever he calls into question the existence of a final solution of the problem of mathematical rigor. Noting their comments, the unwary reader might well believe that present-day mathematical practice has really reached the stage where there is no room for error in the decision whether a proof is valid or not. They assert that a modern formal deductive proof is infallible, so that the only source of doubt as to the truth of the conclusions is doubt as to the truth of the premises. If we regard the theorem, not as a statement of its conclusions, but as a conditional statement, of the form, "If the hypotheses are true, then the conclusion is true," then, in this conditional form, say Worrall and Zahar, the achievements of first-order logic make its truth indubitable. To this extent, they say, Lakatos' fallibilism is incorrect.

In my opinion, Lakatos is right; Zahar and Worrall are wrong. What is more surprising, their objection is rooted in the very error which Lakatos attacked so vehemently in his introduction—the error of identifying mathematics itself (what real mathematicians really do in real life) with its model or representation in metamathematics, or, if you prefer, first-order logic.

Worrall and Zahar are asserting that a formal derivation in first order logic is not fallible in any serious sense. But they fail to make clear that such derivations are purely hypothetical activities (except for "toy" problems that might be played with as exercises in a course in logic).

The actual situation is this. On the one side, we have real mathematics, with proofs which are established by "consensus of the qualified." A real proof is not checkable by a machine, or even by any mathematician not privy to the gestalt, the mode of thought of the particular field of mathematics in which the proof is located. Even to the "qualified reader," there are normally differences of opinion as to whether a real proof (i.e., one that is actually spoken or written down) is complete or correct. These doubts are resolved by communication and explanation, never by transcribing the proof into first-order predicate calculus. Once a proof is "accepted," the results of the proof are regarded as true (with very high probability). It may take generations to detect an error in a proof. If a theorem is widely known and used, its proof frequently studied, if alternative proofs are invented, if it has known applications and generalizations and is analogous to known results in related areas, then it comes to be regarded as "rock bottom." In this way, of course, all of arithmetic and Euclidean geometry are rock bottom.

On the other side, to be distinguished from real mathematics, we have "metamathematics" or "first-order logic." As an activity, this is indeed part of real mathematics. But as to its content, it portrays a structure of proofs which are indeed infallible "in principle." We are thereby able to study mathematically the consequences of an imagined ability to construct infallible proofs; we can, for example, give constructivist variations on the rules of proof, and see what are the consequences of such variations.

How does the availability of this picture of mathematics affect our understanding and practice of real mathematics? Worrall and Zahar, in their critique of Lakatos, seem to be saying that the problem of fallibility in real proofs (which is what Lakatos is talking about) has been conclusively settled by the presence of a notion of infallible proof within metamathematics. (This term of Hilbert's is now old-fashioned, if not obsolete, as a name for proof theory, but it is still convenient in this discussion as a name for the study of the formal systems model of mathematics.) One wonders how they would justify such a claim.

Recently a well-known analyst, having lunch with a group of fellow mathematicians, recalled that in his graduate student days he was once reading the book *Logic for Mathematicians*, by Paul Rosenbloom. His major professor (a very well-known analyst) told him to get rid of the book. "It will be time to read that stuff when you are too old and tired to do real mathematics," he said, or words to that effect. The other mathematicians, listening to the story, were amused at this display of narrow-mindedness. But no one was surprised or shocked; in fact, it was generally agreed that the old professor was right, in the sense that studying logic would certainly not help an analyst and might well interfere with him.

Today, this would no longer be the case, for, as a part of mathematics, logic offers theories which can serve as tools for the analyst or algebraist, as for example, in nonstandard analysis. But this has no relation to justifying proofs by translating them into formulas of first-order logic.

Most likely, Worrall and Zahar would say that a real proof is merely an abbreviated or incomplete formal proof. This has a plausible sound, but it raises several difficulties. In real mathematical practice, we make distinctions between a complete (informal) proof and an incomplete proof. (In a complete informal proof, every step in the argument is convincing to the intended reader.) As formal proofs, *both* are incomplete. So it is hard to see what we are being told when we are told that a real mathematical proof is an abbreviation of a formal derivation, since the same could be said of an incomplete, unacceptable proof.

Spokesmen for formalism (to use Lakatos' term for people like Zahar and Worrall of the footnotes) never explain in just what sense formal systems are a model of mathematics. Is it in the normative sense—mathematics *should* be like a formal system? Or is it in the descriptive sense—mathematics *is* like a formal system?

If one is guided by the rhetoric found in prefaces to logic texts, logic claims no normative role. It is content to study its model of mathematics in the same way a theoretical physicist studies the wave equation as a model for the propagation of sound. The wave equation is a subject of study within pure mathematics. If we want to connect this study to the physical phenomena of sound propagation, we must have rules of interpretation. How do we actually observe or measure the physical variables which are described by the mathematical equation. And then, most important, how close is the agreement between our physical observations and our theoretical prediction? Under what circumstances is the wave equation an accurate description of the physics?

In a recent and very interesting expository paper, Solomon Feferman says that the aim of a logical theory is "to model the reasoning of an idealized platonistic or an idealized constructivistic mathematician." He points out that, although an analogy is sometimes made between the use by logicians of formal systems to study mathematical reasoning and the use by physicists of differential equations to study physical problems, the analogy breaks down inasmuch as there is no analog of the experimental method of physics by which the logicians' models are tested against experience.

He writes, "we have no such tests of logical theories. Rather, it is primarily a matter of individual judgment how well these square with ordinary experience. The accumulation of favorable judgment by many individuals is of course significant."

There is no talk about infallibility. Feferman goes on to say, "Although the significance of the logical work is thus not conclusive, I hope to convince the reader that there is much of interest which is known or being investigated." He reports on various results of the following type: a formal system *A*, seemingly weaker (having fewer "permitted moves" in the rules of inference) than a second formal system *B* is actually just as strong as *B*. For example, if *A* is a "constructivist formal system" and *B* one that violates some of the constructivist restrictions, it follows that anything proved in *B* is actually true constructively. (However, the proof of such a result in logic would *not* be a constructive proof. Would a constructivist find it illuminating? Perhaps it might encourage him, in some cases, to look for a constructive proof of something proved in *B*.)

Such modest claims for logic certainly are not controversial at all. The impact of *Proofs and Refutations* is that it presents a philosophical picture of mathematics utterly at variance with the picture presented by logic and meta-mathematics. What is more, when these two pictures are placed side by side, there is no question as to which one seems more true to life.

Feferman writes,

> The mathematician at work relies on surprisingly vague in-tuitions and proceeds by fumbling fits and starts with all too frequent reversals. Clearly logic as it stands fails to give a direct account of either the historical growth of mathe-matics or the day-to-day experience of its practitioners. It is also clear that the search for ultimate foundations via for-mal systems has failed to arrive at any convincing conclu-sion.

Feferman has serious reservations about Lakatos' work. He argues that Lakatos' scheme of proofs and refutations is not adequate to explain the growth of all branches of mathematics. Other principles, such as the drive toward the unification of diverse topics, seem to provide the best explanation for the development of abstract group theory

357

or of point-set topology. But Lakatos did not claim to give a complete and all inclusive explanation of how mathematics develops. His purpose, stated clearly in his introduction, was to show the inadequacy of formalism by presenting an alternate picture, a picture of mathematics living and growing, and not fossilized in formal axioms.

In this respect, Feferman is fully appreciative of Lakatos' achievement. He writes,

> Many of those who are interested in the practice, teaching and/or history of mathematics will respond with larger sympathy to Lakatos' program. It fits well with the increasingly critical and anti-authoritarian temper of these times. Personally, I have found much to agree with both in his general approach and in his detailed analysis.

This is a reasonable and encouraging beginning. It leads us to hope for a new, enlightening dialogue that might lead to progress on the fundamental problem, the problem of truth and meaning in mathematics, the problem of the nature of mathematical knowledge.

YOU DON'T UNDERSTAND what a cube is if you can only visualize it head on. It helps to see it from many different angles. It helps even more to pick it up, feel its corners and edges as you see them, watch what happens as you turn it around. It helps if you build a cube, construct it out of stiff wire bent and twisted together, or mold it out of soft clay, or cut it out of steel on a milling machine.

You can learn about a hypercube by looking at pictures of it or by handling it at the console of an interactive graphics system. (See pp. 403–405.) As you turn it around and see how one picture transforms into another, you learn to think about a hypercube as one single thing.

In an analogous way, mathematics is one single thing. The Platonist, formalist and constructivist views of it are believed because each corresponds to a certain view of it, a view from a certain angle, or an examination with a particular instrument of observation.

Our problem is to find an understanding of the thing itself, to fit together the partial views—each of which is wrong if taken by itself, just because it is incomplete and one-sided. Since they are pictures of the same thing, they are compatible. Their seeming incompatibility is created by our looking at them with an inappropriate preconception.

For example, the different pictures of a hypercube are mutually contradictory if we think of it as a three-dimensional object. In four-dimensional space, the different three-dimensional projections fit together.

Or to bring it one step lower down, different two-dimensional views of an ordinary solid cube look like pictures of two different objects, until we develop the three-dimensional insight or "intuition" that permits us to transform one into the other.

There are many different ways of looking at mathematics. In the twentieth century most of the systematic writing on mathematics from a philosophical viewpoint has been in the foundationist traditions.

If one asks what is mathematics, it is easy to take the formal-systems model as an answer, although it is not hard to find criticisms of the formal-systems model by mathematicians who were well aware how little it accorded with their own practice. But since the time of Frege hardly any philosopher of note has discussed mathematics in terms different from those of foundationism, of formal logic. The best corrective is to be confronted with a totally different model. This is what Lakatos has given us in *Proofs and Refutations*.

## Further Readings. See Bibliography

S. Feferman; I. Hacking; R. Hersh [1978]; I. Lakatos [1962], [1967], [1976], [1978]

# 8
## MATHEMATICAL REALITY

Let us look at some specific examples of mathematical work and see what philosophical lessons we can draw from them. We will find that the activity of mathematical research *forces* a recognition of the objectivity of mathematical truth. The "Platonism" of the working mathematician is not really a belief in Plato's myth; it is just an awareness of the refractory nature, the stubbornness of mathematical facts. They are what *they* are, not what *we* wish them to be.

At the same time, we will see that our knowledge of these mathematical truths is attained by various methods, heuristic and "rigorous." The heuristic method may be utterly convincing; the rigorous method may leave us with nagging doubt.

# The Riemann Hypothesis

W E TAKE AS OUR first example the most re-
vered and uncontroversial branch of pure
mathematics—number theory.
Within number theory, we take as our case
study the problem of the distribution of the primes, which
we have presented earlier in Chapter 5. The attraction of
this problem is that we are able to *see* what is going on long
before we can *prove* it. For instance, the table on page 213
(taken from Zagier, *Math. Intelligencer* #0), shows that for $x$
less than 10,000,000,000 the number of primes less than or
equal to $x$, when multiplied by log $x$, fall in a near-perfect
straight line when graphed against $x$.

When one is confronted with such evidence as this, it is
impossible not to be impressed by the weight of the argument. Exactly as in Popper's theory of scientific knowledge,
one formulates a "bold conjecture"—very precise and in-
formative, and therefore not likely to be true "by accident,"
so to speak. Then one subjects this conjecture to the test—
by a numerical calculation, rather than by a physical experiment. The test fails to refute the conjecture. The conjecture thereby becomes greatly strengthened—proved, so
to speak, in the sense of natural science though certainly
not in the sense of deductive mathematics.

A more refined piece of natural scientific research into
prime numbers was reported in a paper by I. J. Good and
R. F. Churchhouse in 1967. They are interested in the Riemann zeta-function, which we have defined and discussed
in Chapter 5. The Riemann hypothesis concerns the "roots"

363

of the zeta function—the complex numbers $z$ at which the zeta function equals zero. Riemann conjectured that these roots all have real part $= \frac{1}{2}$. Geometrically, they lie on the line "real part of $z = \frac{1}{2}$" —i.e., a line parallel to the imaginary axis and $\frac{1}{2}$ unit to the right of it.

Now, this conjecture of Riemann is by universal agreement *the outstanding unsolved problem* in mathematics. One proof of the "prime number theorem" depends on the fact (which *has* been proved) that all the zeros are somewhere between the imaginary axis and the line $x = 1$. To prove that they all lie exactly on $x = \frac{1}{2}$ would imply even more precise conclusions about the distribution of prime numbers. It was a major triumph of G. H. Hardy to prove that there are infinitely many zeros of the zeta-function on the line $x = \frac{1}{2}$. We still do not know if *all* of them are there.

It has been verified by calculations that the first 70,000,000 complex zeros of the zeta function are on $x = \frac{1}{2}$. But, Good and Churchhouse say,

> this is not a very good reason for believing that the hypothesis is true. For in the theory of the zeta function, and in the closely allied theory of the distribution of prime numbers, the iterated logarithm log log $x$ is often involved in asymptotic formulae, and this function increases extremely slowly. The first zero off the line $R(s) = \frac{1}{2}$, if there is one, might have an imaginary part whose iterated logarithm is, say, as large as 10, and, if so, it might never be practicable to find this zero by calculation.

(If log log $x = 10$, then $x$ is approximately $10^{10,000}$.)

If this seems far-fetched, they mention another well-verified conjecture—known to be true in the first billion cases—which Littlewood proved is false *eventually*. Nevertheless, Good and Churchhouse write that the aim of their own work is to suggest a "reason" (their quotation marks) for believing Riemann's hypothesis.

Their work involves something called the Möbius function, which is written $\mu(x)$ (pronounced "mu of $x$"). To calculate $\mu(x)$, factor $x$ into primes. If there is a repeated prime factor, as in $12 = 1 \cdot 2 \cdot 2 \cdot 3$ or $25 = 5 \cdot 5$, then $\mu(x)$

is defined to be zero. If all factors are distinct, count them. If there is an even number of factors, we set $\mu(x) = 1$; if there is an odd number, set $\mu(x) = -1$. For instance, $6 = 2 \cdot 3$ has an even number of factors, so $\mu(6) = 1$. On the other hand, $70 = 2 \cdot 5 \cdot 7$ so $\mu(70) = -1$.

Now add up the values of $\mu(n)$ for all $n$ less than or equal to $N$. This sum of $+1$'s and $-1$'s is a function of $N$, and it is called $M(N)$. It was proved a long time ago that the Riemann conjecture is equivalent to the following conjecture: $M(N)$ grows no faster than a constant multiple of $N^{1/2+\epsilon}$ as $N$ goes to infinity (here $\epsilon$ is arbitrary but greater than 0). Either conjecture implies the other; both, of course, are still unproven.

Good and Churchhouse give a "good reason" for believing the Riemann hypothesis by giving a "good reason" (not a proof!) that $M(N)$ has the required rate of growth.

Their "good reason" involves thinking of the values of the Möbius function as if they were random variables.

Why is this a good reason? The Möbius function is completely deterministic; once a number $n$ is chosen, then there is no ambiguity at all as to whether it has any repeated factors—or, if it has no repeated factors, whether the number of factors is even or odd.

On the other hand, if we make a table of the values of the Möbius function, it "looks" random, in the sense that it seems to be utterly chaotic, with no discernible pattern or regularity, except for the fact that $\mu$ is "just as likely" to equal 1 or $-1$.

What is the chance that $n$ has no repeated factor—i.e., that $\mu(n) \neq 0$? This will happen if $n$ is not a multiple of 4 or a multiple of 9 or a multiple of 25 or any other square of a prime. Now, the probability that a number chosen at random is not a multiple of 4 is $\frac{3}{4}$, the probability that it is not a multiple of 9 is $\frac{8}{9}$, the probability that it is not a multiple of 25 is $\frac{24}{25}$, and so on. Moreover, these conditions are all independent—knowing that $n$ is not a multiple of 4 tells us nothing about whether it is a multiple of 9. So according to the basic probabilistic law that the probability of occurrence of two independent events is the product of their separate probabilities, we conclude that the probability that $\mu(n)$ does not equal zero is the product

$$\tfrac{3}{4} \cdot \tfrac{8}{9} \cdot \tfrac{24}{25} \cdot \tfrac{48}{49} \cdot \cdot \cdot \cdot \cdot$$

Even though this product has an infinite number of factors, it can be evaluated analytically, and it is known that it is equal to $6/\pi^2$.

Therefore, the probability that $\mu(n) = 1$ is $3/\pi^2$, and the probability that $\mu(n) = -1$ is the same. The "expected value" of $\mu$ is, of course, zero; on the average, the $+1$'s and the $-1$'s should just about cancel.

Now suppose we choose a very large number of integers at random and independently. Then, for each of these choices, we would have $\mu = 0$ with probability $1 - 6/\pi^2$, $\mu = 1$ with probability $3/\pi^2$, and $\mu = -1$ with probability $3/\pi^2$. If we should then add up all the values of $\mu$, we would get a number which might be very large, if most of our choices happened to have $\mu = 1$, say. On the other hand, it would be unlikely that our choices gave $\mu = 1$ very much more often than $\mu = -1$. In fact, a theorem in probability (Hausdorff's inequality) says that, if we pick $N$ numbers in this way, then, with probability 1, the sum grows no faster than a constant times $N^{1/2+\epsilon}$ as $N$ goes to infinity.

This conclusion is exactly what we need to prove the Riemann conjecture! However, we have changed the terms in our summation. For the Riemann conjecture, we should have added the values of $\mu$ for the numbers from 1 to $N$. Instead, we took $N$ numbers at random.

What justifies this? It is justified by our feeling or impression that the table of values of $\mu$ is "chaotic," "random," "unpredictable." By that token, the first $N$ values of $\mu$ are nothing special, they are a "random sample."

If we grant this much, then if follows that the Riemann hypothesis is true *with probability one*. This conclusion seems at the same time both compelling and nonsensical. Compelling because of the striking way in which probabilistic reasoning gives *precisely* the needed rate of growth for $M(N)$; nonsensical because the truth of the Riemann hypothesis is surely not a random variable which may hold only "with probability one."

The author of the authoritative work on the zeta function, H. M. Edwards, calls this type of heuristic reasoning "quite absurd." (Edwards refers, not to Good and Churchhouse, but to a 1931 paper of Denjoy which uses similar but less detailed probabilistic arguments.)

To check their probabilistic reasoning, Good and Churchhouse did some numerical work. They tabulated the values of the sum of $\mu(n)$ for $n$ ranging over intervals of length 1,000. They found statistically excellent confirmation of their random model.

In a separate calculation, they found that the total number of zeros of $\mu(n)$ for $n$ between 0 and 33,000,000 is 12,938,407. The "expected number" is $33,000,000 \cdot (1 - 6/\pi^2)$, which works out to 12,938,405.6. They call this "an astonishingly close fit, better than we deserved." A nonrigorous argument has predicted a mathematical result to 8 place accuracy.

In physics or chemistry, experimental agreement with theory to 8 place accuracy would be regarded as a very strong confirmation of the theory. Here, also, it is impossible to believe that such agreement is accidental. The principle by which the calculation was made *must* be right.

When we respond in this way to heuristic evidence, we are in a certain sense committed to the realist or Platonist philosophy. We are asserting that the regularity which has been predicted and confirmed is not illusory—that there is *something there* which is lawful and regular.

It is easy to make up an example of a sequence of statements which are true for $n = 1, 2$, up to 1,000,000,000,000 and false from then on (for instance, the statement "$n$ is not divisible by both $2^{12}$ and $5^{12}$"). So the fact that a conjecture about the natural numbers is true for the first 2,000,000,000 cases certainly does not prove it will be true for the 2,000,000,001st case. But for conjectures such as those about the distribution of primes, no one believes that the behavior we observe in our sample will suddenly change to something radically different in another sample, taken farther out toward infinity.

Only with some confidence in the orderliness or "rationality" of the number system is it possible to do successful research. "God is subtle, but not malicious," said Einstein. This faith, which a physicist needs in order to believe he can understand the universe, is also needed by a mathematician trying to understand his mental universe of number and form. Perhaps this is what Dieudonné means when he calls realism "convenient." It is more than convenient; it is indispensable.

The point to notice in this discussion is that none of it makes any sense from a constructivist *or* formalist point of view. The constructivist says that the Riemann hypothesis will become true or false only when a constructive proof one way or the other is given. It makes no sense to discuss whether it is *already* true or false, apart from any proof. The formalist says that the Riemann hypothesis makes no sense except as a conjecture that a certain statement can be derived from certain axioms. Again, there is no acceptance of truth or falsity in mathematics apart from what is proved or disproved.

It is interesting to ask, in a context such as this, why we still feel the need for a proof, or what additional conviction would be carried if a proof should be forthcoming which was, say, 200 or 300 pages long, full of arduous calculations where even the most persistent may sometimes lose their way.

It seems clear that we want a proof because we are convinced that all the properties of the natural numbers *can* be deduced from a single set of axioms, and if something is true and we *can't* deduce it in this way, this is a sign of a lack of understanding on our part. We believe, in other words, that a proof would be a way of understanding *why* the Riemann conjecture is true; which is something more than just knowing from convincing heuristic reasoning that it *is* true.

But then a proof which is so complex and nonperspicuous that it sheds no light on the matter would fail to serve this purpose.

Why would we still want a proof, even a hopelessly complex and nonperspicuous one? Suppose a proof were published which took 500 pages to write. How would it be decided that the proof was correct? Suppose it were so decided by a sufficient number of experts. Would we be overjoyed because we would now know definitely that Riemann's conjecture is true?

Perhaps, though, there is another purpose to proof—as a testing ground for the stamina and ingenuity of the mathematician. We admire the conqueror of Everest, not because the top of Everest is a place we want to be, but just because it is so hard to get there.

## Further Readings. See Bibliography

H. Edwards; I. J. Good and R. F. Churchhouse; E. Grosswald; M. Kac; G. Pólya [1954]; D. Zagier

# $\pi$ and $\hat{\pi}$

WE HAVE SEEN that in number theory, there may be heuristic evidence so strong that it carries conviction even without rigorous proof. This is a piece of mathematical experience that philosophy has to accommodate. It is true that number theory is far from typical in this respect.

In most areas of mathematics, one deals with much more complicated objects than in number theory. It is often very difficult or impossible to test a conjecture on concrete examples. It may be a major achievement even to exhibit a single nontrivial example of the structure one is considering; or the assertion one wishes to prove may be of a nature difficult or impossible to check by computation, even in particular examples. This would be the case, for example, in set theory and functional analysis.

Nevertheless, even in these areas the presumption of an objective reality about which one seeks to ascertain the truth is an unavoidable one for the researcher or the student.

Rather than argue the case by means of an example from one of the more abstract fields of research, we will present here a famous example of Brouwer. It has to do with the "law of trichotomy" for the real number system: every real number is either positive, negative, or zero. Brouwer asserted that his example is a *counterexample* to the law of trichotomy. He gives a real number which, he asserts, is neither positive, negative, nor zero. Most mathematicians, when presented with the example, violently reject Brouwer's conclusion. His number, they say, *is* either zero, negative or positive. We simply don't know what it is.

In presenting the example here, we will also be giving a little more detail to our account of constructivism. Our principal motive, however, is to further expose the Platonistic thinking that is embodied in the very structure of the real number system as understood by ordinary (nonconstructivist) mathematics. Since it would be hard to think of any branch of mathematics that does not depend crucially on the real numbers, this will show that Platonism is intimately associated with most of the practice of mathematics as it is done today.

To give Brouwer's counterexample, we start with $\pi$, and then use its decimal expansion to define a second, related real number, which we call $\hat{\pi}$ (read: "pi-hat"). Our definition of $\hat{\pi}$ involves a good deal of arbitrariness—there are many other constructions that would give the same essential result. Instead of $\pi$, we could start with $\sqrt{2}$, or any other familiar irrational number. All that is required is that (1) as with $\pi$, we have a definite calculating procedure ("algorithm") which gives the decimal expansion to as many terms as we like; and (2) there is some property of this decimal expansion—for instance, the appearance in it of a row of 100 successive zeros—which, so far as we know, is "accidental." That is to say, we know no reason why this property is either excluded or required by the definition of $\pi$. To determine whether there are anywhere in the expansion of $\pi$ a row of 100 successive zeros, we have no procedure except to actually generate the expansion of $\pi$ and take a look. So far as $\pi$ has been calculated at the present date, there is no such row. If we should generate the first billion digits and find therein a row of 100 zeros, then of course the matter would be settled (provided we are totally confident in the correctness of our computation). On the

PI = 3.+

```
1415926535 8979323846 2643383279 5028841971 6939937510 5820974944 5923078164 0628620899 8628034825 3421170679
8214808651 3282306647 0938446095 5058223172 5359408128 4811174502 8410270193 8521105559 6446229489 5493038196
4428810975 6659334461 2847564823 3786783165 2712019091 4564856692 3460348610 4543266482 1339360726 0249141273
7245870066 0631558817 4881520920 9628292540 9171536436 7892590360 0113305305 4882046652 1384146951 9415116094
3305727036 5759591953 0921861173 8193261179 3105118548 0744623799 6274956735 1885752724 8912279381 8301194912
9833673362 4406566430 8602139494 6395224737 1907021798 6094370277 0539217176 2931767523 8467481846 7669405132
0005681271 4526356082 7785771342 7577896091 7363717872 1468440901 2249534301 4654958537 1050792279 6892589235
4201995611 2129021960 8640344181 5981362977 4771309960 5187072113 4999999837 2978049951 0597317328 1609631859
5024459455 3469083026 4252230825 3344685035 2619311881 7101000313 7838752886 5875332083 8142061717 7669147303
5982534904 2875546873 1159562863 8823537875 9375195778 1857780532 1712268066 1300192787 6611195909 2164201989

3809525720 1065485863 2788659361 5338182796 8230301952 0353018529 6899577362 2599413891 2497217752 8347913151
5574857242 4541506959 5082953311 6861727855 8890750983 8175463746 4939319255 0604009277 0167113900 9848824012
8583616035 6370766010 4710181942 9555961989 4676783744 9448255379 7747268471 0404753464 6208046684 2590694912
9331367702 8989152104 7521620569 6602405803 8150193511 2533824300 3558764024 7496473263 9141992726 0426992279
6782354781 6360093417 2164121992 4586315030 2861829745 5570674983 8505494588 5869269956 9092721079 7509302955
3211653449 8720275596 0236480665 4991198818 3479775356 6369807426 5425278625 5181841757 4672890977 7727938000
8164706001 6145249192 1732172147 7235014144 1973568548 1613611573 5255213347 5741849468 4385233239 0739414333
4547762416 8625189835 6948556209 9219222184 2725502542 5688767179 0494601653 4668049886 2723279178 6085784383
8279679766 8145410095 3883786360 9506800642 2512520511 7392984896 0841284886 2694560424 1965285022 2106611863
0674427862 2039194945 0471237130 8696095636 4371917287 4677646575 7396241389 0865832645 9958133904 7802759009

9465764078 9512694683 9835259570 9825822620 5224894077 2671947826 8482601476 9909026401 3639443745 5305068203
4962524517 4939965143 1429809190 6592509372 2169646151 5709858387 4105978859 5977297549 8930161753 9284681382
6868386894 2774155991 8559252459 5395943104 9972524680 8459872736 4469584865 3836736222 6260991246 0805124388
4390451244 1365497627 8079771569 1435997700 1296110894 4169486855 5848406353 4220722258 2848864815 8456028506
0168427394 5226746767 8895252138 5225499546 6672782398 6456596116 3548862305 7745649803 5593634568 1743241125
1507606947 9451096596 0940252288 7971089314 5669136867 2287489405 6010150330 8617928680 9208747609 1782493858
9009714909 6759852613 6554978189 3129784821 6829989487 2265880485 7564014270 4775551323 7964145152 3746234364
5428584447 9526586782 1051141354 7357395231 1342716610 2135969536 2314429524 8493718711 0145765403 5902799344
0374200731 0578539062 1983874478 0847848968 3321445713 8687519435 0643021845 3191048481 0053706146 8067491927
8191197939 9520614196 6342875444 0643745123 7181921799 9839101591 9561814675 1426912397 4894090718 6494231961

5679452080 9514655022 5231603881 9301420937 6213785595 6638937787 0830390697 9207734672 2182562599 6615014215
0306803844 7734549202 6054146659 2520149744 2850732518 6660021324 3408819071 0486353174 6496514539 0579626856
1005508106 6587969981 6357473638 4052571459 1028970641 4011097120 6280439039 7595156771 5770042033 7869936007
2305587631 7635942187 3125147120 5329281918 2618612586 7321579198 4148488291 6447060957 5270695722 0917567116
7229109816 9091528017 3506712748 5832228718 3520935396 5725121083 5791513698 8209144421 0067510334 6711031412
6711136990 8658516398 3150197016 5151168517 1437657618 3515565088 4909989859 9823873455 2833163550 7647918535
8932261854 8963213293 3089857064 2046752590 7091548141 6549859461 6371802709 8199430992 4488957571 2828905923
2332609729 9712084433 5732654893 8239119325 9746366730 5836041428 1388303203 8249037589 8524374417 0291327656
1809377344 4030707469 2112019130 2033038019 7621101100 4492932151 6084244485 9637669838 9522868478 3123552658
2131449576 8572624334 4189303968 6426243410 7732269780 2807318915 4411010446 8232527162 0105265227 2111660396

6655730925 4711055785 3763466820 6531098965 2691862056 4769312570 5863566201 8558100729 3606598764 8611791045
3348850346 1136576867 5324944166 8039626579 7877185560 8455296541 2665408530 6143444318 5867697515 5661406800
7002378776 5913440171 2749470420 5622305389 9456131407 1127000407 8547332699 3908145466 4645880797 2708266830
6343285878 5698305235 8089330657 5740679545 7163775254 2021149557 6158140025 0126228594 1302164715 5097925923
0990796547 3761255176 5675018517 7829666454 7791154711 2996148903 0463994713 2962107340 4375189573 5961458901
9389713111 7904297828 5647503203 1986915140 2870808599 0480109412 1472213179 4764777262 2414254854 5403321571
8530614228 8137585043 0633217518 2979866223 7172159160 7716692547 4873898665 4949450114 6540628433 6639379003
9769265672 1463853067 3609657120 9180763832 7166416274 8888007869 2560290228 4721040317 2118608204 1900042296
6171196377 9213375751 1495950156 6049631862 9472654736 4252308177 0367515906 7350235072 8354056704 0386743513
6222247715 8915049530 9844489333 0963408780 7693259939 7805419341 4473774418 4263129860 8099888687 4132604721
```

other hand, if there should fail to be 100 zeros in the expansion we have computed, we are left no wiser than before; we know nothing about the second billion digits. Even if there is a sequence of 100 zeros in our calculated expansion, we could change the question to 1,000 successive 9's (for example) and still have an open question. The point is that there are now, and always will be, simple questions about $\pi$ of this kind to which we never expect to have an answer.

Let $P$ denote the statement, "In the decimal expansion of $\pi$, there eventually occurs a row of 100 successive zeros." Let $\bar{P}$ denote the contrary, "In the decimal expansion of $\pi$,

*$\pi$ to 5,000 Decimals. From Daniel Shanks and John W. Wrench, Jr.*

*Courtesy: Mathematics of Computation, Vol. XVI, No. 77, January 1962*

there nowhere appears a row of 100 successive zeros." Is the statement "Either $P$ or $\overline{P}$" a true statement?

Most mathematicians would answer yes. In fact, the "law of the excluded middle" requires a "yes"; we are simply asking whether $P$ is either true or false, and the law of the excluded middle says that every statement is true or false.

The constructivist disagrees. He argues that the law of the excluded middle does not apply in this case. For he regards "the expansion of $\pi$" as a mythical beast. The belief that either $P$ or $\overline{P}$ is true comes from a mistaken conception of the expansion of $\pi$ *already existing* as a completed object. But this is false. All that exists, or that we know how to construct, is a finite part of this expansion.

The argument may seem a bit theological. Why does it matter?

It does matter. For the mathematician to give up his Platonic belief in the existence of the expansion of $\pi$, in the truth of either $P$ or $\overline{P}$, would require a restructuring of all of mathematical analysis. This is illustrated by the example of the law of trichotomy. We define a number $\hat{\pi}$ by giving a rule according to which the first 1,000, the first million, or the first hundred billion digits of the decimal expansion of $\hat{\pi}$ are to be calculated. That is all that is meant by "defining" a real number.

$\hat{\pi}$ is going to look very much like $\pi$. In fact, it *is* the same as $\pi$ in the first 100, the first 1,000, even the first 10,000 decimal places. Our rule is: expand $\pi$ until we find a row of 100 successive zeros (or until we have passed the desired precision for $\hat{\pi}$, whichever comes first). Up to this first run of 100 successive zeros, the expansion of $\hat{\pi}$ is to be identical with that of $\pi$. Suppose the first run of 100 successive zeros starts in the $n$'th digit. If $n$ is odd, let $\hat{\pi}$ terminate in its $n$'th digit. If $n$ is even, let $\hat{\pi}$ have a 1 in the $n + 1$'st digit, and then terminate.

Notice that we do not know at present, and probably never will know, *if there is any such number as n.* If we never find a sequence of 100 successive zeros in $\pi$, then we will never have a value for $n$. Nevertheless, our recipe for constructing $\hat{\pi}$ is perfectly definite; we know it to as many decimal places as we know $\pi$. Also, we know that $\hat{\pi} = \pi$ if and only if $\pi$ does not contain a sequence of 100 zeros. If $\pi$ does contain such a sequence, and it starts at an even number in the expansion, then $\hat{\pi}$ is greater than $\pi$. If it starts at an odd number, $\hat{\pi}$ is less than $\pi$.

Now let's compute, not $\pi$, but the difference $\hat{\pi} - \pi$. Call this difference $Q$. Is $Q$ positive, negative, or zero?

If we try to find out by setting a computing machine to calculate the expansion of $\pi$, we will not receive an answer until we find a row of 100 successive zeros. If our machine runs for 1,000 years, and we have not found a run of 100 zeros, we still will not know whether $Q$ is positive, negative, or zero. What is more, we will have no reason to think we have made any progress, or that we are any closer to the answer than when we started.

In a situation such as this, what significance can we attach to the basic law of standard mathematics, the so-called "law of trichotomy"—"every number is either zero, positive or negative"? We are saying, it is clear, that $Q$ *is* either positive, negative, or zero, *regardless* of the fact that we can never know which. The law of trichotomy, taken literally at face value, asserts that one of these three statements *must* be true, quite aside from whether there is any way, even in principle, to determine which.

The constructivists' argument is that none of the three is true. $Q$ *will* be zero, positive or negative at such time as someone determines which of the three is the case; until then, it is none of the three. Thus, mathematical truth is time-dependent, and is *subjective,* although it does not depend on the consciousness of any particular live mathematician.

The main thrust of their critique is that any conclusion based on the compound statement "either $Q > 0$, $Q = 0$, or $Q < 0$" is unjustifiable. More generally, any conclusion based on reasoning about an infinite set is defective if it relies on the principle that every statement is either true or false—the law of the excluded middle. As the example shows, a statement may well be neither true nor false, in the constructive sense. That is to say, no one may have any means of showing that it is either true or false.

The standard mathematician finds the argument not

convincing but annoying. He has no intention of giving up the law of trichotomy for a more precise version which would be provable constructively. Neither does he want to acknowledge that his practice and teaching of mathematics is dependent on a Platonist ontology. He neither defends his Platonism nor reconsiders it. He adopts the ostrich strategy—pretend nothing has happened.

In recent years, a major effort to reconstruct analysis along constructivist lines has been carried out by Errett Bishop, who was well known for major work in classical analysis before he turned to constructivism. Bishop has attracted a small band of followers. He argues, as Brouwer did, that much of standard mathematics is a meaningless game; but he goes far beyond Brouwer in showing by example how it can be made over in a constructively meaningful way.

Most mathematicians respond to his work with indifference or hostility. The nonconstructivist majority should be able to do better. We should be able to state our own philosophical view as clearly as the constructivists state theirs. We have a right to prefer our own viewpoint, but we ought to recognize honestly what it is.

The account of constructivism given here is the conventional one, stated from the viewpoint of ordinary or classical mathematics.

This means that it is unacceptable from the viewpoint of the constructivist. From his point of view, classical mathematics is a jumble of myth and reality. He prefers to do without the myth. From his point of view, it is classical mathematics that appears as an aberration; constructivism is just the refusal to participate in the acceptance of a myth.

Gabriel Stolzenberg has written a scrupulously careful analysis of how the classical mathematician's unstated assumptions make the constructivist view point unintelligible to him. This article is intended for the philosophically minded reader, and requires no mathematical preparation.

### Further Readings. See Bibliography

E. Bishop [1967]; N. Kopell and G. Stolzenberg; G. Stolzenberg

# Mathematical Models, Computers, and Platonism

A S OUR NEXT example, we consider a situation which is very typical; almost a standard situation in applied mathematics.

A mathematician is interested in the solution of a certain differential equation. He knows that this solution $u(t)$ "exists," because standard "existence theorems" on differential equations include his problem.

Knowing that the solution exists, he proceeds to try to find out as much as he can about it. Suppose, for example, that his general theorem tells him that his function $u(t)$ exists uniquely for all $t \geq 0$. His goal is to tabulate the function $u(t)$ as accurately as he can, especially for $t$ close to zero and for $t$ very large (or, as he would say, near infinity).

For $t$ close to zero he uses something called the "Taylor series." He knows a rigorous proof that (for $t$ small) this series converges to the solution of the equation. However, he has no way of proving how many terms of the series he must take in order to get his desired accuracy—say, to within $\frac{1}{1,000,000}$ of the exact value. He adds terms until he finds that the sum is unchanged by adding more terms. At that point he stops. He is guided by common sense, not by rigorous logic. He cannot prove that the neglected high-order terms are, in fact, negligible. On the other hand, he has to stop eventually. So, lacking a completely rigorous argument, he uses a plausible one to make the decision.

For $t$ of moderate size—neither very small nor very large—he calculates $u(t)$ by a recursion scheme, which replaces the differential equation by a succession of algebraic equations. He has great confidence in the accuracy of the result because he is using a differential-equation-solving program that is the most advanced of all available. It has been refined and tested for many years, and is in use in scientific laboratories all over the world. However, there is no rigorous logical proof that the numbers he gets from the machine are correct. First of all, the computing algorithm at the heart of the program cannot be guaranteed to work in all cases—only in all "reasonable" cases. That is to say, the proof that justifies the use of this algorithm assumes that the solution has certain desirable properties which are present "normally" in "problems that usually come up." But there is nothing to *guarantee* this. What if he has an abnormal problem? This abnormality is usually manifested by the calculations breaking down. The numbers "blow up"—become too big for the program to handle—and the program stops running and warns the operator. Undoubtedly, some sufficiently clever person could cook up a differential equation to which this particular program would give reasonable-looking wrong answers.

Moreover, even if the algorithm were rigorously proved to be reliable in our case, the actual machine computation involves both software and hardware. By the "software" we mean the computer program and the whole complex of programmed control systems that permit us to write our programs in ten pages instead of a thousand. By the "hardware" is meant the machine itself, the transistors, memory, wires, and so on.

Software is itself a kind of mathematics. One could demand a rigorous proof that the software does what it is supposed to do. An area is even developing in computer science to provide "proofs of programs." As one might expect, it takes much longer to produce a proof of correctness of the program than to produce the program itself. In the case of the huge compilers that are used in large-scale scientific programming, there is no promised date for the appearance of proofs of correctness; if they ever appear, it is hard to imagine who would read the proofs and check *their* correctness. In the meantime, compilers are used without hesitation. Why? Because they were created by people who were doing their best to make them work correctly; because they have been in use for years, and one presumes there has been time for most of the errors to have been detected and corrected. One hopes that those that remain are harmless. If one wants to be particularly careful, one can do the computation twice, using two different systems programs, on two different machines.

As to the hardware, it usually works properly; one assumes that it is highly reliable, and the probability of failure of any one part is negligible (*not* zero!). Of course, there are very many parts, and it is possible that several could fail. If this happens and the computation is affected, one expects this gross misbehavior to be detected and the computer shut down for repair. But all this is only a matter of likelihood, not certainty.

Finally, what about the function $u(t)$ for large $t$, "near infinity"? Computing with a machine recursively, we can go up to some large value of $t$, but, no matter how large, it is still finite. To finish the study of $u(t)$, letting $t$ approach infinity, it is often possible to use special methods of calculation, so-called "asymptotic methods" which increase in accuracy as $t$ gets larger. Sometimes these methods can be justified rigorously; but they are used often in the absence of such rigorous proof, on the basis of general experience and with an eye on the results to see if they "look reasonable."

If two different methods of asymptotic calculations can be carried out and the results agree, this result is considered to be almost conclusive, even though neither one has been proved correct in a rigorous mathematical sense.

Now, from the viewpoint of the formalist (our imaginary strict, extreme formalist) this whole procedure is sheer nonsense. At least, it isn't mathematics, although maybe it can pass if we call it carpentry or plumbing. Since there are no axioms, no theorems, only "blind calculations" based on fragmentary pieces of arguments, our formalist, if he is true to his philosophy, can only smile pityingly at the foolish and nonsensical work of so-called applied mathematics.

(We must beware here of a verbal trap, caused by the double meaning of the word formalism. Within mathematics itself, formalism often means calculations carried out without error estimates or convergence proofs. In this sense, the numerical and asymptotic methods used in applied mathematics are formal. But in a philosophical context, formalism means the reduction of mathematics to formal deductions from axioms, without regard to meaning.)

Philosophically, the applied mathematician is an uncritical Platonist. He takes for granted that *there is* a function $u(t)$, and that he has a right to use any method he can think of to learn as much as he can about it. He would be puzzled if asked to explain *where* it exists or *how* it exists; but he knows that what he is doing makes sense. It has an inner coherence, and an interconnection with many aspects of mathematics and engineering. If the function $u(t)$ which he attempts to compute, by one means or another—does *not* exist prior to his computations, and independently of them—then his whole enterprise, to compute it, is futile nonsense, like trying to photograph the ectoplasm at a séance.

In many instances, the differential equation whose solution he calculates is proposed as a model for some physical situation. Then, of course, the ultimate test of its utility or validity comes in its predictive or explanatory value to that physical problem. Hence, one must compare these two entities, *each of which has its own objective properties*—the mathematical model, given in our example by a differential equation, and the physical model.

The physical model does *not* correspond exactly to an actual physical object, an observable thing in a particular time and place. It is an idealization or simplification. In any particular time and place, there are infinitely many different kinds of observations or measurements that could be asked for. What is going on at a particular time and place can always be distinguished from what is going on at some other time and place. In order to develop a *theory*, an understanding with some general applicability, the physicist singles out a few particular features as "state-variables" and uses them to represent the actual infinitely complex physical object. In this way he creates a physical model—something which is already a simplification of the physical reality. This *physical model,* being part of a physical theory, is believed or conjectured to obey some mathematical laws. These laws or equations then specify some mathematical objects, the solutions of the mathematics equation—and these solutions are the *mathematical model*. Often the mathematical model one first writes down is too complicated to yield useful information, and so one introduces certain simplifications, "neglecting small terms in the equation," obtaining ultimately a simplified mathematical model which it is hoped (sometimes one can even prove it!) is close in some sense to the original mathematical model.

In any case, one must decide finally whether the *mathematical model* gives an acceptable description of the *physical model*. In order to do this, *each* must be studied, as a distinct reality with its own properties. The study of the mathematical model is done as we have described, with rigorous mathematics as far as possible, with nonrigorous or formal mathematics as far as possible, and with machine computations of many kinds—simulations, truncations, discretizations.

The physical model may be studied in the laboratory, if it is possible to develop it under laboratory conditions. Or if there exists in nature some approximation to it—in the interplanetary plasma, or in deep trenches in the depth of the Atlantic—it may be studied wherever it is best approximated. Or it may be simulated by a computing machine, if we imagine we can tell the machine enough about how our physical model would behave. In this case, we are actually comparing two different mathematical models.

The point is that the Platonic assumption that our mathematical model is a well-defined object seems essential if the whole applied mathematical project is to make any sense.

**Further Readings. See Bibliography**

R. DeMillo, R. Lipton and A. J. Perlis; F. Brooks, Jr.

# Why Should I Believe a Computer?

I N 1976 A RARE event took place. An announcement of the proof of a theorem in pure mathematics actually broke into the news columns of the *New York Times*. The occasion was the proof, by Kenneth Appel and Wolfgang Haken, of the "Four-Color Conjecture." The occasion was newsworthy for two reasons. To begin with, the problem in question was a famous one. The four-color conjecture had been under study for over 100 years. There had been many unsuccessful attacks—now at last it had been proved. But the method of proof in itself was newsworthy. For an essential part of the proof consisted of computer calculations. That is to say, the published proof contained computer programs and the output resulting from calculations according to the programs. The intermediate steps by which the programs were executed were of course not published; in this sense, the published proofs were *permanently and in principle* incomplete.

The four-color problem is, of course, to prove that every map on a flat surface or a sphere can be colored without using more than four different colors. The only requirement is that no two countries sharing a common border should be the same color. If two countries meet only at a single point (as do Utah and New Mexico, for instance, in the United States), then they may be colored the same. The countries can have any shape at all, but each country must consist of one single connected piece.

That four colors are sufficient must have been noticed a long time ago; it was first stated as a mathematical conjecture in 1852 by Francis Guthrie. In 1878 the eminent British mathematician Arthur Cayley proposed it as a problem to the London Mathematical Society, and within a year Arthur Bray Kempe, a London barrister and member of the London Mathematical Society, published a paper that claimed to prove the conjecture.

Kempe attempted to use the method of reductio ad absurdum. To explain his argument, it is sufficient to consider only "normal" maps. A normal map is one where no more than three regions meet at any point, and no region entirely encircles another one. Every map can be associated with a normal map which requires at least as many colors, and it is sufficient to prove the four-color conjecture for normal maps.

Kempe proved, correctly, that in any normal map there is at least one region with five or fewer neighbors. That means that one of these four configurations must appear in any normal map:

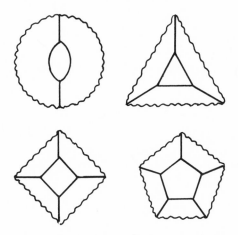

These four diagrams represent the four possible cases of a region with 2, 3, 4 or 5 neighbors. The fact that at least one of these four cases must occur is described by saying that this set of configurations is unavoidable.

Kempe attempted to show how, in each case, one may construct a new map, with fewer countries, which would again be five-chromatic. If this construction can be carried out, one says that the given configuration is reducible. Thus the idea of Kempe's proof is to exhibit an *unavoidable* set of *reducible* configurations. If this can be done, the reductio ad absurdum is immediate. For then we could conclude that, given any five-chromatic map, one could construct from it another five-chromatic map with fewer regions. In a finite number of steps, we would obtain a five-chromatic map with fewer than five regions, which is certainly an absurdity.

Unfortunately, Kempe's argument for reducibility was incorrect in the case of a region with five neighbors. The error was pointed out in 1890 by P. J. Heawood. From 1890 to 1976 the four-color conjecture was one of the outstanding unsolved mathematical problems.

In the end, the proof of Appel and Haken again used the same idea of exhibiting an unavoidable set of reducible configurations. But instead of the four simple configurations in Kempe's proof, the unavoidable set contained thousands of configurations, most of them so complicated that proving reducibility was possible only by use of a high-speed computer.

The use of a computer in this case is in principle quite different from the uses we have described in applied mathematics and in number theory.

In applied mathematics, the computer serves to calculate an approximate answer, when theory is unable to give us an exact answer. We may try to use our theory to prove that the computed answer is in some sense close to the exact answer. But in no way does the theory depend on the computer for its conclusions; rather, the two methods, theoretical and mechanical, are like two independent views of the same object; the problem is to coordinate them.

In the study of distribution of primes or similar number-theoretic problems, the computer serves to generate data. By studying these data, the mathematician may be able to form a conjecture such as the prime number theorem. Of course, he would like to prove the conjecture; but failing that, he can at least *check* it by a second use of the computer, to look at another sample of the natural number system to see if the result predicted by his conjecture is sustained.

In both of these cases, the rigorous mathematics of proof remains uncontaminated by the machine. In the first case, that of the applied mathematician, the machine remains second-best, a substitute to use in that area where theory is unable to go. In the second case, that of the number theorist, the machine is a heuristic helper, which may help us decide what to believe, and even how strongly to believe it, but still does not affect what is *proved*.

In the Haken-Appel four-color theorem, the situation is totally different. They present their work as a definitive, complete, rigorous proof. For that reason, the complex, controversial response which they received gives an unusual insight into what philosophers and mathematicians imagine is meant by this notion, rigorous proof.

In the *Journal of Philosophy* for February, 1979, Stephen Tymoczko considered the philosophical ramifications of Haken-Appel's work. He writes,

> If we accept the four-color theorem as a theorem, then we are committed to changing the sense of "theorem," or more to the point, to changing the sense of the underlying concept of "proof."

From the philosopher's point of view, the use of a computer as an essential part of the proof involves a weakening of the standards of mathematical proof. It introduces grounds for skepticism, and so changes in an essential way the situation, which was previously supposed to involve indubitable conclusions with *no* grounds for skepticism at any stage.

Appel and Haken write,

> A person could carefully check the part of the discharging procedure that did not involve reducibility computations in a month or two, but it does not seem possible to check the reducibility computations themselves by hand. Indeed, the referees of the paper resulting from our work used our complete notes to check the discharging procedure, but they resorted to an independent computer program to check the correctness of the reducibility computations.

So, there is no denying that acceptance of the Haken-Appel theorem involves a certain act of faith. Even if I read and check every line they write, I still have to believe that computer calculations do in fact perform what they are supposed to perform. So my belief in the proof of the four-color theorem depends, not only on my confidence in my own ability to understand and verify mathematical reasoning, it also depends on my belief that computers work and do what they are supposed to do. This is a belief of a totally different order. I have no more grounds for such a belief than for any other belief in the factuality and reliability of "common knowledge"—things everybody knows and which I believe because I accept "what everybody knows."

In this way, mathematical knowledge is reduced to the level of common knowledge. But common knowledge does not claim to be based on rigorous proof or to have the certainty of deductive reasoning to justify it. So the reliance on a computer in the Haken-Appel proof involves sacrificing an essential aspect of mathematical certainty, degrading it to the common level of ordinary knowledge, which is subject to a certain possible skepticism from which mathematical knowledge was always free. So goes the philosopher's critique.

To the mathematician, however, the matter appears in a totally different light. If he is one of that small proportion of mathematicians who feel friendly to computers, who are interested in and can appreciate the art required to take the four-color problem and actually succeed in getting it onto a machine—to such as these, the Haken-Appel theorem will be an inspiration and a vindication. For the majority of mathematicians, however, the response is quite different. When I heard that the four-color theorem had been proved, my first reaction was, "Wonderful! How did they do it?" I expected some brilliant new insight, a proof which had in its kernel an idea whose beauty would transform my day. But when I received the answer, "They did it by breaking it down into thousands of cases, and then running them all on the computer, one after the other," I felt disheartened. My reaction then was, "So it just goes to show, it wasn't a good problem after all."

This reaction is certainly a matter of taste. In this matter my taste is probably that of a passing era. A future generation of mathematicians may well find aesthetic pleasure in computer proofs such as that first achieved by D. H.

Lehmer, and now carried to new heights by Haken and Appel. But this kind of question has to do with whether we like such proofs, with whether we get out of such a proof the insight, pleasure, satisfaction, or whatever it is that we feel a good proof should give.

The philosopher's objection is quite different. It seems to him that there is a degradation in the degree of certainty which violates the nature of mathematics, as he understands it.

To the mathematician, the philosopher's objection seems strangely naive and idealistic, in the bad sense of immature and gullible.

In fact, Prof. Haken was quoted in a newspaper interview as specifically denying that the use of the computer by him and Appel involved any change in the concept of a mathematical proof. He said,

> Anyone, anywhere along the line, can fill in the details and check them. The fact that the computer can run through more details in a few hours than a human could ever hope to do in a lifetime does not change the basic concept of the mathematical proof. What has changed is not the theory but the practice of mathematics.

To the philosopher, there is all the diffence in the world between a proof that depends on the reliability of a machine and a proof that depends only on human reason. To the mathematician, the fallibility of reason is such a familiar fact of life that he welcomes the computer as a more reliable calculator than he himself can hope to be.

In an expository article on their work, Appel and Haken wrote,

> Most mathematicians who were educated prior to the development of fast computers tend not to think of the computer as a routine tool to be used in conjunction with other older and more theoretical tools in advancing mathematical knowledge. Thus they intuitively feel that if an argument contains parts that are not verifiable by hand calculations it is on rather insecure ground. There is a tendency to feel that verification of computer results by independent computer programs is not as certain to be correct as inde-

385

pendent hand checking of the proof of theorems proved in the standard way.

This point of view is reasonable for those theorems whose proofs are of moderate length and highly theoretical. When proofs are long and highly computational, it may be argued that even when hand checking is possible, the probability of human error is considerably higher than that of machine error.

The probability of human error is present even before we bring in the computer! All we can do is to try to minimize it. If a proof is long enough and complicated enough, there is always some room for doubt about its correctness. Using a computer does not eliminate human error, for the computer itself is a man-made object.

In a paper published in *Acta Mathematica* in 1971 and quoted in Yu. I. Manin's text *Introduction to Mathematical Logic*, H. P. F. Swinnerton-Dyer used a computer to calculate the values of a certain determinant arising in the study of homogeneous linear forms. Swinnerton-Dyer made the following comments:

> When a theorem has been proved with the help of a computer, it is impossible to give an exposition of the proof which meets the traditional test—that a sufficiently patient reader should be able to work through the proof and verify that it is correct. Even if one were to print all the programs and all the sets of data used (which in this case would occupy some forty very dull pages) there can be no assurance that a data tape has not been mispunched or misread. Moreover, every modern computer has obscure faults in its software and hardware—which so seldom cause errors that they go undetected for years—and every computer is liable to transient faults. Such errors are rare, but a few of them have probably occurred in the course of the calculations reported here.

Does this mean that the results of the computation should be rejected? Not at all. He continues,

> However, the calculation consists in effect of looking for a rather small number of needles in a six-dimensional haystack: almost all the calculation is concerned with parts of the haystack which in fact contain no needles, and an error

in those parts of the calculation will have no effect on the final results. Despite the possibilities of error, I therefore think it almost certain that the list of permissible $\Delta \leq 17$ is complete; and it is inconceivable that an infinity of permissible $\Delta \leq 17$ have been overlooked.

His conclusion:

Nevertheless, the only way to verify these results (if this were thought worthwhile) is for the problem to be attacked quite independently, by a different machine. This corresponds exactly to the situation in most experimental sciences.

### Further Readings. See Bibliography

K. Appel and W. Haken; Y. I. Manin; T. Tymoczko

# Classification of Finite Simple Groups

WE NOW TURN to another flourishing branch of modern mathematics, where the meaning of *proved* seems to be a bit different from that in the logic texts. This is the theory of finite simple groups. (See Chapter 5, pp. 203.) The classification of finite simple groups is a major problem in algebra in which dramatic progress has been made in recent years. Two aspects of this subject have made it particularly interesting to mathematicians who are not specialists. One of these intriguing aspects is a series of discoveries of "monsters." These are simple groups whose existence was totally unsuspected before 1966, when Z. Janko discovered the first of them, thereby launching the modern theory of sporadic groups. "The existence of these strange objects, discovered at a rate of about one per year, revealed the richness of the subject and lent an air of mystery to the nature of simple groups" (Daniel Gorenstein, *Bull. Amer. Math. Soc.,* January 1979).

Janko's group has 175,560 elements. His discovery was followed by two dozen more; the names of the discoverers and the sizes of the groups can be found in Gorenstein's article. The biggest on the list is "Fischer's monster," which has order $2^{46} \cdot 3^{20} \cdot 5^9 \cdot 7^6 \cdot 11^2 \cdot 13^3 \cdot 17 \cdot 19 \cdot 23 \cdot 29 \cdot 31 \cdot 41 \cdot 47 \cdot 59 \cdot 71$ (approximately $8 \times 10^{53}$).

Actually, at the time Gorenstein's article appeared, this group had not yet been shown to exist, but there was said to be "overwhelming evidence" for its existence.

Workers on finite simple groups now believe they have virtually completed the job of classifying all the finite simple groups. The methodological problems that have arisen in this work are remarkably similar to those discussed by Swinnerton-Dyer in relation to his computer calculation and alluded to earlier.

At the present time the determination of all finite simple groups is very nearly complete. Such an assertion is obviously presumptuous, if not meaningless, since one does not speak of theorems as "almost proved." But the ultimate theorem which will assert the classification of simple groups is unlike any other in the history of mathematics; for the complete proof, when it is attained, will run to well over 5,000 journal pages! Moreover, it is very likely that at the present time more than 80% of these pages exist in either print or preprint form.

—D. Gorenstein

To Swinnerton-Dyer, the traditional test of a proof was that a sufficiently patient reader could work through it and verify that it was correct. When the length of the proof mounts into thousands of pages, the sufficiently patient reader becomes hard to find. We may have to settle for a team of patient readers, and hope that no errors slip by because of problems of communication among members of the team.

But it is not just the length of the proof that causes trouble. There is also its accuracy.

388

This is an appropriate moment to add a cautionary word about the meaning of "proof" in the present context; for it seems beyond human capacity to present a closely reasoned several hundred page argument with absolute accuracy. I am not speaking of the inevitable typographical errors, or the overall conceptual basis for the proof, but of "local" arguments that are not quite right—a misstatement, a gap, what have you. They can almost always be patched up on the spot, but the existence of such "temporary" errors is disconcerting to say the least. Indeed, they raise the following basic question: If the arguments are often ad hoc to begin with, how can one guarantee that the "sieve" has not let slip a configuration which leads to yet another simple group? Unfortunately, there are no guarantees—one must live with this reality. However, there is a prevalent feeling that, with so many individuals working on simple groups over the past fifteen years, and often from such different perspectives, every significant configuration will loom into view sufficiently often and so cannot remain unnoticed for long. On the other hand, it clearly indicates the strong need for continual reexamination of the existing "proofs". This will be especially true on that day when the final classification of simple groups is announced and the exodus, already begun, to more fertile lands takes place. Some of the faithful must remain behind to improve the "text". This will be one of the first major tasks of the "post classification" era.

—D. Gorenstein

For the "hand-made" proof as for the computer proof, one must live with the reality of errors in reasoning. In both cases, one nevertheless has a feeling, based on some overall view of the problem, that the incomplete or erroneous proofs still do give the right answer.

If the program described by Gorenstein is carried out, and if the results are accepted as valid or conclusive by the mathematical community, wherein does this differ from acceptance of computer proofs like that in the four-color theorem?

When a proof is 5000 pages long, and has been pieced together out of contributions by several different mathematicians, then it is clear that the team's claim to have proved their theorem is based in major part on their mutual confidence in each other's competence and integrity; and the acceptance of the work by the mathematical community as a whole is based in large part on its confidence in the members of the team.

This mutual confidence is based on confidence in social institutions and arrangements of the mathematical profession. Acceptance of the computer proof of the four-color theorem is based on confidence that computers do what they are supposed to do. In both cases, the confidence is reasonable and well-warranted. In both cases, there remains room for some doubt, some possibility of error.

The four-color theorem looks exceptional because of the use of computers. The classification of finite simple groups looks exceptional because of the length of its proofs. But no sharp line can be drawn between these examples and the typical proofs and theorems published every month in the mathematics journals.

Mathematicians in every field rely on each other's work, quote each other; the mutual confidence which permits them to do this is based on confidence in the social system of which they are a part. They do not limit themselves to using results which they themselves are able to prove from first principles. If a theorem has been published in a respected journal, if the name of the author is familiar, if the theorem has been quoted and used by other mathematicians, then it is considered established. Anyone who has use for it will feel free to do so.

This mutual confidence is perfectly reasonable and appropriate. But it certainly violates the notion of mathematical truth as indubitable.

### Further Readings. See Bibliography

J. Alper; F. Budden; D. Gorenstein

# Intuition

THE WORD INTUITION, as used by mathematicians, carries a heavy load of mystery and ambiguity. Sometimes it seems to be a dangerous and illegitimate substitute for rigorous proof. In other contexts, it seems to denote an inexplicable flash of insight by which the happy few gain mathematical knowledge which others can attain only by long efforts.

As a first step to exploring the portent of this slippery concept, it will be useful to attempt a list of the various meanings and uses we give to this word.

(1) Intuitive is the opposite of rigorous. This usage is itself not completely clear, for the meaning of "rigorous" itself is never given precisely. We might say that in this usage intuitive means lacking in rigor, and yet the concept of rigor is itself defined intuitively rather than rigorously.

(2) Intuitive means visual. Thus intuitive topology or geometry differ from rigorous topology or geometry in two respects. On the one hand, the intuitive version has a meaning, a referent in the domain of visualized curves and surfaces, which is excluded from the rigorous (i.e., formal or abstract) version. In this respect, the intuitive is superior; it has a quality that the rigorous version lacks. On the other hand, the visualization may lead us to regard as obvious or self-evident statements which are dubious or even false. (The article by Hahn, "The Crises in Intuition" gives a beautiful collection of examples of such statements.)

(3) Intuitive means plausible or convincing in the absence of proof. A related meaning is, "what one might expect to be true in this kind of situation, on the basis of general experience with similar situations or related subjects." "Intuitively plausible" means reasonable as a conjecture, i.e., as a candidate for proof.

(4) Intuitive means incomplete. If one takes limits under

391

the integral sign without using Lebesgue's theorem, if one represents a function by a power series without checking that the function is analytic, then the logical gap is acknowledged by calling the argument intuitive.

(5) Intuitive means relying on a physical model, or on some leading examples. In this sense it is almost the same as heuristic.

(6) Intuitive means holistic or integrative as opposed to detailed or analytic. When we think of a mathematical theory in the large, when we see that a certain statement must be true because of the way it would fit in with everything else we know about it, we are reasoning "intuitively." To be rigorous, we must justify our conclusion deductively, by a chain of reasoning where each step can be defended from criticism, and where the first step is considered known, and the last step is the desired result.

If the chain of reasoning is extremely long and complicated, the rigorous proof may leave the reader still subject to serious doubt and misgiving; in a genuine sense, it may be less convincing than an intuitive argument, which can be grasped as a whole, and which uses implicitly the assumption that mathematics as a whole is coherent and reasonable.

In all these usages the notion of intuition remains rather vague; its aspect changes somewhat from one usage to another. In a preface to a textbook, one author may take pride in avoiding the "merely" intuitive—i.e., the use of figures and diagrams as aids to proof. Another may take pride in emphasizing the intuitive—i.e., in communicating the visual and physical significance of a mathematical theory, or providing the heuristic derivation of a theorem, not only the formal post hoc verification.

With any of these interpretations, the intuitive is to some extent extraneous and unessential. It may be desirable or undesirable—perhaps it would be best to say that it has its desirable and undesirable aspects—and different authors may differ about what place to assign to the intuitive in their writing—but in any case it is optional, like seasoning

392

on a salad. Perhaps it would be foolish and self-defeating, but a teacher *can* teach mathematics and a researcher can write papers without paying attention to the problem of intuition. However, if one is not doing mathematics, but rather is trying to look at people who are doing mathematics and to understand what they are doing, then the problem of intuition becomes central and unavoidable.

We maintain that

(1) All the standard philosophical viewpoints rely in an essential way on some notion of intuition.

(2) None of them even attempt to explain the nature and meaning of the intuition which they postulate.

(3) A consideration of intuition as it is actually experienced leads to a notion which is difficult and complex, but it is not inexplicable or unanalyzable. A realistic analysis of mathematical intuition is a reasonable goal, and should become one of the central features of an adequate philosophy of mathematics.

Let us elaborate each of these three points. By the three main philosophies, we mean constructivism, Platonism or realism, and formalism of one brand or another. For the purpose of the present discussion, we need not make refined distinctions among the various possible versions of Platonism, formalism, and constructivism. It will be sufficient to characterize each viewpoint crudely with a single sentence. Thus, the constructivist regards the natural numbers as the fundamental datum of mathematics, which neither requires nor is capable of reduction to any more basic notion, and from which all meaningful mathematical notions must be constructed.

The Platonist regards mathematical objects, not as things which we construct, but as things already existing, once and for all, in some ideal and timeless (or "tenseless") sense. We do not create, we only discover what is already there, including infinites of any degree of complexity yet conceived or to be conceived by the mind of mathematicians.

Finally, the formalist accepts neither the restrictions of

the constructivist nor the theology of the Platonist. For him, all that matters are the rules of the game—the rules of inference by which we transform one formula to another. Any "meaning" such formulas have is "nonmathematical" and beside the point.

What does each of these three philosophies need from the intuition?

The most obvious difficulty is that besetting the Platonist. If mathematical objects constitute an ideal nonmaterial world, how does the human mind establish contact with this world? Consider the continuum hypothesis. In view of the discoveries of Gödel and Cohen, we know that it can neither be proved nor disproved from any of the systems of axioms by which infinite sets are described in contemporary mathematics.

The Platonist believes that this situation is merely an indication of our ignorance. The continuum is a definite thing, existing independently of the human mind, and it either does or does not contain an infinite subset which is equivalent neither to the set of integers nor to the set of real numbers. Our *intuition* must be developed until it will tell us which of these two is actually the case.

Thus the Platonist requires the intuition to establish the connection between human awareness and mathematical reality. But this intuition is a very elusive thing. The Platonist does not attempt to describe it, let alone analyze its nature. How does one acquire mathematical intuition? Evidently it varies from one person to another, even from one mathematical genius to another; and it has to be developed and refined, since it seems to be inadequate at present. But then by whom, according to what criteria, does one train or develop it?

Are we talking about a mental faculty which can directly perceive an ideal reality just as our physical senses perceive physical reality? Then the intuition becomes a second ideal entity, the counterpart on the subjective level of the ideal mathematical reality on the objective level. We now have two mysteries instead of one, not only the mystery of the relation between the ideal reality of timeless ideas and the mundane reality of change and flux, but also now the relation between the physical mathematician who is born and dies into the world of change and flux, and his intuitive faculty which reaches directly to the hidden reality of the timeless and eternal.

These problems make Platonism a difficult doctrine for any scientifically oriented person to defend.

Mathematical Platonists do not attempt to discuss them. They cannot analyze the notion of mathematical intuition, because their philosophical position turns the intuition into an indispensable but unanalyzable faculty. The intuition is for the Platonist what the "soul" is for the believer in the hereafter. We know it is there but no questions can be asked about it.

The position of the constructivist is different. He is a conscious descendant of Kant, and knows exactly how he relies on intuition. He takes as given (intuitively) the notions of the constructive and of the natural numbers—that is, the notion of an operation which can be iterated, which can always be repeated one more time. This does not seem very problematical, and few people would be inclined to argue about it. Yet it seems that among the followers of Brouwer there have been disagreements, differences of opinion or the right way to proceed, or the right way to be a constructivist. Of course this is only to be expected; every philosophical school has the same experience. But it does create a difficulty for a school which purports to base itself on nothing but a universal and unmistakable intuition. The dogma that the intuition of the natural number system is universal is not tenable in the light of historical, pedagogical, or anthropological experience. The natural number system seems an innate intuition only to mathematicians so sophisticated they cannot remember or conceive of the time before they acquired it; and so isolated that they never have to communicate seriously with people (still no doubt the majority of the human race) who have not internalized this set of ideas and made it intuitive.

What about the formalist? Does not the problem of intuition vanish along with the problem of meaning and truth? Indeed, one can avoid considering intuition as long as one defines mathematics to be nothing other than formal deductions from formal theorems. As A. Lichnerowicz has written,

> Our demands on ourselves have become infinitely larger; the demonstrations of our predecessors no longer satisfy us but the mathematical facts that they discovered remain and we prove them by methods that are infinitely more rigorous and precise, methods from which geometric intuition with its character of badly analyzed evidence has been totally banned.

Geometry is dead as an autonomous branch; it is no more than the study of particularly interesting algebraic-topological structures. But the formalist is able to eliminate intuition only by concentrating all his attention on the refinement of proof and on attaining a dogmatic and irrefutable final presentation.

To the obvious question, why should anyone be interested in these superprecise and superreliable theorems, formalism has no answer, for the interest in them derives from their meaning, and meaning is precisely what the all-out formalist discards as nonmathematical.

If one asks the formalist how our benighted ancestors managed to find correct theorems by incorrect reasoning, he can only answer, "intuition."

Cauchy knew Cauchy's integral theorem, even though he did not "know" (in the formalist's sense of knowing the formal set-theoretic definition)—he did not know the meaning of any of the terms in the statement of the theorem. He did not know what a complex number is, or what an integral is, or what a curve is; still, he knew the right way to evaluate the complex number represented by the integral over this curve! How can this be? Easy. Cauchy was (as we all know) a great mathematician, and therefore he could rely on his intuition.

But what is this intuition? At least the Platonist believes in real objects (ideal in nature) which one can somehow perceive or "intuit." If the formalist does not believe such things exist, what is there to intuit? The only answer he can supply is, unconscious formalizing. Cauchy was such a genius, he subconsciously knew a "correct" proof of the theorem. Which means, no doubt, knowing the correct definitions of all the terms involved in stating the theorem.

This answer is especially interesting to those of us who have the experience of making correct conjectures which we are unable to prove. If our intuitive conjecture is the result of an unconscious calculation, then (a) either the unconscious has a secret method of calculating which is better than any known method, or (b) the proof is there in my head, I just can't get it out to where I can see it.

Formalists who are willing to consider the problem of discovery and to look at mathematics in its historical development have to introduce a mysterious intuition to account for the enormous gap between the account of mathematics (a game played by the rules) and one's real experience of mathematics, where more is often accomplished by breaking rules than obeying them.

Now that we have set up and knocked down our straw men, the time has come to say what the problem is as we see it, and how we propose to look for an answer. The problem is to account for the phenomenon of intuitive knowledge in mathematics, to make it intelligible. This is the basic problem of mathematical epistemology. That is, what do we know, and how do we know it?

We propose to answer this question by another one: what do we teach, and how do we teach it? Or better, what do we try to teach, and how do we find it necessary to teach it?

We try to teach mathematical concepts, not formally (memorizing definitions) but intuitively—by seeing examples, doing problems, developing an ability to think which is the expression of having successfully internalized something. What? An intuitive mathematical idea.

Thus the fundamental intuition of the natural numbers is a shared concept, an idea held in common by everybody who has gone through certain experiences of manipulating coins or bricks, buttons or pebbles, until we can tell (by getting the "right" answers to our questions) that he gets the idea—that even though we will always run out of buttons or coins sooner or later, there is an idea of something like a huge bin of buttons or coins where we will never run out.

That is to say, intuition is not a direct perception of something existing externally and eternally. It is the effect in the mind of certain experiences of activity and manipulation of concrete objects (at a later stage, of marks on paper or even mental images). As a result of this experience, there is something (a trace, an effect) in the pupil's mind which is *his* representation of the integers. But his representation is equivalent to mine, in the sense that we both get the same answer to any question you ask—or if we get different answers, we can compare notes and figure out what's right. We do this, not because we have been taught a set of algebraic rules, but because our mental pictures match each other. If they do not, since I am the teacher and I know my mental picture matches the socially sanctioned one (the one that all the other teachers have), the pupil gets bad marks and is not included in any future discussion of the problem.

We have intuition because we have mental representations of mathematical objects. We acquire these representations, not by memorizing verbal formulas, but by repeated experiences (on the elementary level, experience of manipulating physical objects; on the advanced level, experiences of doing problems and discovering things for ourselves.)

These mental representations are checked for veracity by our teachers and fellow students. If we don't get the right answer, we flunk the course. In this way, different people's representations are always being rubbed against each other to make sure they are congruent with each other.

Of course, we don't know in what way these representations are held in the mind. We know as little about how any other thought or knowledge is held in the mind.

The point is that as shared concepts, as mutually congruent mental representations, they are real objects whose existence is just as "objective" as mother love and race prejudice, as the price of tea or the fear of God. How then do we distinguish mathematics from other humanistic studies? Obviously, there is a fundamental difference between mathematics and, say, literary criticism. If mathematics is a

398

humanistic study with respect to its subject matter, it is like the sciences in its objectivity. Those results about the physical world which are reproducible—which come out the same way every time you ask—are called scientific, and those subjects which have reproducible results are called natural sciences. In the realm of ideas, of mental objects, those ideas whose properties are reproducible are called mathematical objects, and *the study of mental objects with reproducible properties is called mathematics.* Intuition is the faculty by which we can consider or examine these (internal, mental) objects.

To be sure, there is always some possibility for disagreement between intuitions. The process of mutual adjustment to ensure agreement is never finally completed. As new questions are asked, new parts of the structure may come into focus that have never been looked at before.

Sometimes the question has no answer. (There is no reason why the continuum hypothesis must be true or false.)

We know that with physical objects we may ask questions which are inappropriate, which have no answer. For example, what are the exact position and velocity of an electron? Or, how many trees are there growing at this moment in the State of Vermont? For mental objects as for physical ones, it may happen that what seems at first to be an appropriate question is discovered, with great difficulty, to be inappropriate. This doesn't call into question the existence of the particular mental or physical object. There are many other questions which *are* appropriate, to which definite, reliable answers can be given.

The difficulty in seeing what intuition is arises because of the demand that mathematics be infallible. This demand is met by both of the traditional philosophies, formalism and Platonism. Each attempts to create a mathematics which is as superhuman as Plato wanted it to be. But since each of them does this by falsifying the nature of mathematics as it is (in human life, in history) each creates a confusion and a mystery where none need exist.

## Further Readings. See Bibliography

M. Bunge; H. Hahn; A. Lichnerowicz; R. Wilder [1967]

# Four-Dimensional Intuition

A LINE IS one-dimensional. A flat surface is two-dimensional. Solid objects are three-dimensional. But what is the fourth dimension?

Sometimes people say that time is a fourth dimension. In the physics of Einstein's relativity, a four-dimensional geometry is used in which a three-dimensional space and a one-dimensional time coordinate are merged into a single four-dimensional continuum. But we don't want to talk about relativity and space-time. We only want to know if it makes sense to take one more step in the list of geometrical dimensions.

For instance, in two dimensions, we have the familiar figures of the circle and the square. Their three-dimensional analogs are the sphere and the cube. Can we talk about a four-dimensional hypersphere or hypercube, and make sense?

(a)

We can go from a single point up to a cube in three steps. In the first step, we take two points, 1 inch apart, and join them. We get a line interval, a one-dimensional figure. Next, we take two 1-inch line intervals, parallel to each other, 1 inch apart. Connect each pair of end-points, and we get a 1-inch square, a two-dimensional figure. Next, take two 1-inch squares, parallel to each other. Say the first square is directly above the second, 1 inch away. Connect corresponding corners, and we get a 1-inch cube.

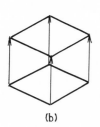

(b)

So, to get a 1-inch hypercube, we must take two 1-inch cubes, parallel to each other, 1 inch apart, and connect vertices. In this way, we should get a 1-inch hypercube, a four-dimensional figure.

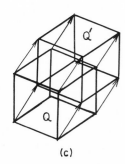

(c)

The trouble is that we have to move in a new direction at each stage. The new direction has to be perpendicular to all the old directions. After we have moved back and forth, then right and left, and finally up and down, we have used up all the directions we have accessible to us. We are three-

400

dimensional creatures, unable to escape from three-dimensional space into the fourth dimension. In fact, the idea of a fourth physical dimension may be a mere fantasy, a device for science fiction. The only argument for it is that we can conceive it; there is nothing illogical or inconsistent about our conception.

We can figure out many of the properties that a four-dimensional hypercube would have, if one existed. We can count the number of edges, vertices, and faces it would have. Since it would be constructed by joining two cubes, each of which has 8 vertices, the hypercube must have 16 vertices. It will have all the edges the two cubes have; it will also have new edges, one for each pair of vertices that have to be connected. This gives $12 + 12 + 8 = 32$ edges. With a little more work, one can see that it will have 24 square faces, and 8 cubical hyperfaces.

The table below shows the number of "parts" of the interval, square, cube, and hypercube. It is a startling discovery that the sum of the parts is always a power of three!

| Dimension | OBJECT | 0-Faces (Vertices) | 1-Faces (Edges) | 2-Faces (Faces) | 3-Faces | 4-Faces |
|---|---|---|---|---|---|---|
| 0 | Point | 1 | | | | |
| 1 | Interval | 2 | 1 | | | |
| 2 | Square | 4 | 4 | 1 | | |
| 3 | Cube | 8 | 12 | 6 | 1 | |
| 4 | Hypercube | 16 | 32 | 24 | 8 | 1 |

In a course on problem-solving for high-school teachers and education students, the gradual discovery of these facts about hypercubes takes a week or two. The fact that we can find out this much definite information about the hypercube seems to mean that it must exist in some sense.

Of course, the hypercube is just a fiction in the sense of physical existence. When we ask how many vertices a hypercube has, we are asking, how many *could* it have, if there were such a thing. It's like the punch line of the old joke—"If you *had* a brother, would he like herring?" The difference is that the question about a nonexistent brother is a foolish question; the question about the vertices of a non-existent cube is not so foolish, since it does have a definite answer.

In fact, by using algebraic methods, defining a hypercube by means of coordinates, we can answer (at least in principle) any question about the hypercube. At least, we can reduce it to algebra, just as ordinary analytic geometry reduces questions about two- or three-dimensional figures to algebra. Then, since algebra in four variables is not essentially more difficult than in two or three, we can answer questions about hypercubes as easily as questions about squares or cubes. In this way, the hypercube serves as a good example of what we mean by mathematical existence. It is a fictitious or imaginary object, but there is no doubt about how many vertices, edges, faces, and hyperfaces it has! (or would have, if one prefers the conditional mode of speaking about it.)

The objects of ordinary three- or two-dimensional geometry are also mathematical objects, which is to say, imaginary or fictitious; yet they are closer to physical reality, unlike the hypercube which we cannot construct.

The mathematical three-cube is an ideal object, but we can look at a wooden cube and use it to determine properties of the three-cube. The number of edges of the three-cube is 12; so is the number of edges of a sugar cube 12. We can get a lot of information about two- and three-dimensional geometry by drawing pictures or building models and then inspecting our pictures or models. While it is possible to go wrong by misusing a picture or model, it is rather difficult to do so. It takes ingenuity to invent a situation where one could go wrong in this way. As a general rule, the use of pictures and models is helpful, even essential in understanding two- or three-dimensional geometry.

Reasoning based on models and figures, either actual ones or mental images of them, would be called intuitive reasoning, as opposed to formal or rigorous reasoning.

When it comes to four-dimensional geometry, it might seem that since we ourselves are mere three-dimensional creatures, we are excluded by nature from the possibility of reasoning intuitively about four-dimensional objects. And yet, it is not so. Intuitive grasp of four-dimensional figures is not impossible.

At Brown University Thomas Banchoff, a mathematician, and Charles Strauss, a computer scientist, have made computer-generated motion pictures of a hypercube moving in and out of our three-dimensional space. To understand what they have done, imagine a flat, two-dimensional creature who lived at the surface of a pond and could see only other objects on the surface (not above or below). This flat fellow would be limited to two physical dimensions, just as we are limited to three. He could become aware of three dimensional objects only by way of their two-dimensional intersections with his flat world. If a solid cube passes from the air into the water, he sees the cross-sections that the cube makes with the surface as it enters the surface, passes through it, and finally leaves it.

If the cube passed through repeatedly, at many different angles and directions, he would eventually have enough information about the cube to "understand" it even if he couldn't escape from his two-dimensional world.

The Strauss-Banchoff movies show what we would see if a hypercube passed through our three-space, at one angle or another. We would see various more or less complex configurations of vertices and edges. It is one thing to describe what we would see by a mathematical formula. It is quite another to see a picture of it; and still better to see it in motion. When I saw the film presented by Banchoff and Strauss, I was impressed by their achievement,* and by the sheer visual pleasure of watching it. But I felt a bit disappointed; I didn't gain any intuitive feeling for the hypercube.

A few days later, at the Brown University Computing Center, Strauss gave me a demonstration of the interactive graphic system which made it possible to produce such a film. The user sits at a control panel in front of a TV screen. Three knobs permit him to rotate a four-dimen-

---

* This film, incidentally, won Le Prix de la Récherche Fondamentale au Festival de Bruxelles, 1979.

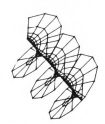

*The complex exponential function (a four-dimensional object) looked at from several points of view.*

*Courtesy: Banchoff & Strauss Productions.*

sional figure on any pair of axes in four-space. As he does so, he sees on the screen the different three-dimensional figures which would meet our three-dimensional space as the four-dimensional figure rotates through it.

Another manual control permits one to take this three-dimensional slice and to turn it around at will in three-space. Still another button permits one to enlarge or shrink the image; the effect is that the viewer seems to be flying away from the image, or else flying toward and actually into the image on the screen. (Some of the effects in Star Wars of flying through the battle-star were created in just this way, by computer graphics.)

At the computing center, Strauss showed me how all these controls could be used to get various views of three-dimensional projections of a hypercube. I watched, and tried my best to grasp what I was looking at. Then he stood up, and offered me the chair at the control.

I tried turning the hypercube around, moving it away, bringing it up close, turning it around another way. Suddenly I could *feel* it! The hypercube had leaped into palpable reality, as I learned how to manipulate it, feeling in my fingertips the power to change what I saw and change it back again. The active control at the computer console created a union of kinesthetics and visual thinking which brought the hypercube up to the level of intuitive understanding.

In this example, we can start with abstract or algebraic understanding alone. This can be used to design a computer system which can simulate for the hypercube the kinds of experiences of handling, moving and seeing real cubes that give us our three-dimensional intuition. So four-dimensional intuition is available, for those who want it or need it.

The existence of this possibility opens up new prospects for research on mathematical intuition. Instead of working with children or with ethnographic or historical material, as we must do to study the genesis of elementary geometric intuition (the school of Piaget), one could work with adults, either trained mathematically or naive, and attempt to document by objective psychological tests the develop-

404

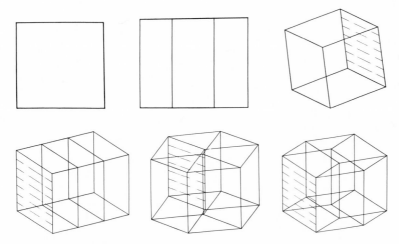

*Six views of a hypercube extracted from a general computer graphics system for the real time display of four dimensional "wire frame" objects.*

*Courtesy: Banchoff and Strauss Productions.*

ment of four-dimensional intuition, possibly sorting out the roles played by the visual (passive observation) and the kinesthetic (active manipulation.) With such study, our understanding of mathematical intuition should increase. There would be less of an excuse to use intuition as a catch-all term to explain anything mysterious or problematical.

Looking back at the epistemological question, one wonders whether there really ever was a difference in principle between four-dimensional and three-dimensional. We can develop the intuition to go with the four-dimensional imaginary object. Once that is done, it does not seem that much more imaginary than "real" things like plane curves and surfaces in space. These are all ideal objects which we are able to grasp both visually (intuitively) and logically.

### Further Readings. See Bibliography

H. Freudenthal [1978]; J. Piaget, [1970, 71]; T. Banchoff and C. M. Strauss

# True Facts About Imaginary Objects

THE UNSPOKEN assumption in the three traditional foundationist viewpoints is that mathematics must be a source of indubitable truth.

The actual experience of all schools—and the actual daily experience of mathematicians—shows that mathematical truth, like other kinds of truth, is fallible and corrigible.

Do we really have to choose between a formalism that is falsified by our everyday experience, and a Platonism that postulates a mythical fairyland where the uncountable and the inaccessible lie waiting to be observed by the mathematician whom God blesses with a good enough intuition? It is reasonable to propose a different task for mathematical philosophy, not to seek indubitable truth, but to give an account of mathematical knowledge as it really is—fallible, corrigible, tentative, and evolving, as is every other kind of human knowledge. Instead of continuing to look in vain for foundations, or feeling disoriented and illegitimate for lack of foundations, we have tried to look at what mathematics really is, and account for it as a part of human knowledge in general. We have tried to reflect honestly on what we do when we use, teach, invent or discover mathematics.

The essence of mathematics is its freedom, said Cantor. Freedom to construct, freedom to make assumptions. These aspects of mathematics are recognized in constructivism and formalism. Yet Cantor was a Platonist, a believer in a mathematical reality that transcends the human mind. These constructions, these imagined worlds, then impose *their* order on us. We have to recognize their objectivity; they are partly known, partly mysterious and hard to know; partly, perhaps, unknowable. This is the truth that the Platonist sees.

What happens when we observe two contradictory facts

in the world, both of which are undeniably true? We are forced to change our way of looking at the world, we are forced to find a viewpoint from which the facts are not contradictory but compatible.

Another way of saying the same thing is this. If a fact goes against common sense, and we are nevertheless compelled to accept and deal with this fact, we learn to alter our notion of common sense.

There are two pieces of information which we have about the nature of mathematics.

Fact 1 is that mathematics is a human invention. Mathematicians know this, because they do the inventing.

The arithmetic and elementary geometry that everybody knows seem to be God-given, for they are present everywhere, and seemingly always were. The latest algebraic gadget used by topologists, the newest variation of pseudodifferential operators, have been invented recently enough so that we know the names and addresses of the inventors. They are still shiny from the mint. But we can see the line of descent. The family resemblance is unmistakable, from the latest new thing to the most ancient. Arithmetic and geometry came from the same place as homotopy theory—from the human brain. Every day, millions of us labor to instil these into other human brains.

Fact 2 is that these things we bring into the world, these geometric figures and arithmetical functions and algebraic operators, are mysterious to us, their creators. They have properties which we discover only by dint of great effort and ingenuity; they have other properties which we try in vain to discover; they have properties which we do not even suspect. The whole activity of mathematical problem-solving is evidence for Fact 2.

Formalism is built on Fact 1. It recognizes that mathematics is the creation of the human mind. Mathematical objects are imaginary.

Platonism is built on Fact 2. The Platonist recognizes that mathematics has its own laws which we have to obey. Once we construct a right triangle, with sides $a$, $b$, and hypotenuse $h$, then $a^2 + b^2 = h^2$ whether we like it or not. I don't know whether 1375803627 is a prime number or not, but I

407

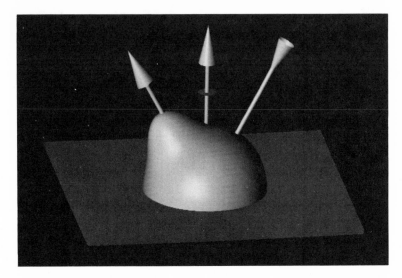

do know that it is not up to me to chose which it is; that is already decided as soon as I write the number down.

These imaginary objects have definite properties. There *are* true facts about imaginary objects.

From the Platonist point of view, Fact 1 is unacceptable. Since mathematical objects are what they are, in defiance of our ignorance or preferences, they must be real in a sense independent of human minds. In some way we cannot explain or understand, they exist, outside of the material world, and outside the human mind.

From the constructivist viewpoint, Fact 2 is unacceptable. Since mathematics is our creation, nothing in mathematics is true until it is known—in fact, until it is proved by constructive methods.

As for the formalist, he escapes the dilemma by simply denying everything. There are no mathematical objects, so there are no problems about the nature of mathematical objects.

If we are willing to forget about Platonism, constructivism, and formalism, we can take as our starting point the two facts that we learn from mathematical experience:

Fact 1: Mathematics is our creation; it is about ideas in our minds.

408

Fact 2: Mathematics is an objective reality, in the sense that mathematical objects have definite properties, which we may or may not be able to discover.

If we believe our own experience and accept these two facts, then we have to ask how they can be reconciled; how we can look at them as compatible, not contradictory. Or better said, we have to see what presumptions we are making that force us to find something incompatible or contradictory about these two facts. Then we can try to dispose of these presuppositions, in order to develop a point of view broad enough to accept the reality of mathematical experience.

We are accustomed, in a philosophical context, to think of the world as containing only two kinds of material—matter, meaning physical substance, what you study in a physics lab—and mind, meaning my mind or your mind, the private psyche each of us has someplace inside our skulls. But these two categories are inadequate. Just as inadequate as the four categories, earth, air, fire, and water of ancient Greece were inadequate for physics.

Mathematics is an objective reality that is neither subjective nor physical. It is an ideal (i.e., nonphysical) reality that is objective (external to the consciousness of any one person). In fact, the example of mathematics is the strongest, most convincing proof of the existence of such an ideal reality.

This is our conclusion, not to truncate mathematics to fit into a philosophy too small to accommodate it—rather, to demand that the philosophical categories be enlarged to accept the reality of our mathematical experience.

The recent work of Karl Popper provides a context in which mathematical experience fits without distortion. He has introduced the terms World 1, 2, and 3, to distinguish three major levels of distinct reality.

World 1 is the physical world, the world of mass and energy, of stars and rocks, blood and bone.

The world of consciousness emerges from the material world in the course of biological evolution. Thoughts, emotions, awareness are nonphysical realities. Their existence is inseparable from that of the living organism, but they are different in kind from the phenomena of physiology and anatomy; they have to be understood on a different level. They belong to World 2.

In the further course of evolution, there appear social consciousness, traditions, language, theories, social institutions, all the nonmaterial culture of mankind. Their existence is inseparable from the individual consciousness of the members of the society. But they are different in kind from the phenomena of individual consciousness. They have to be understood on a different level. They belong to World 3. Of course, this is the world where mathematics is located.

Mathematics is not the study of an ideal, preexisting nontemporal reality. Neither is it a chess-like game with made-up symbols and formulas. Rather, it is the part of human studies which is capable of achieving a science-like consensus, capable of establishing *reproducible* results. The existence of the subject called mathematics is a fact, not a question. This fact means no more and no less than the existence of modes of reasoning and argument about ideas which are compelling and conclusive, "noncontroversial when once understood."

Mathematics does have a subject matter, and its statements are meaningful. The meaning, however, is to be found in the shared understanding of human beings, not in an external nonhuman reality. In this respect, mathematics is similar to an ideology, a religion, or an art form; it deals with human meanings, and is intelligible only within the context of culture. In other words, mathematics is a humanistic study. It is one of the humanities.

The special feature of mathematics that distinguishes it from other humanities is its science-like quality. Its conclusions are compelling, like the conclusions of natural science. They are not simply products of opinion, and not subject to permanent disagreement like the ideas of literary criticism.

As mathematicians, we know that we invent ideal objects, and then try to discover the facts about them. Any philosophy which cannot accommodate this knowledge is too small. We need not retreat to formalism when attacked by philosophers. Neither do we have to admit that our belief in the objectivity of mathematical truth is Platonic in the sense of requiring an ideal reality apart from human thought. Lakatos' and Popper's work shows that modern philosophy is capable of accepting the truth of mathematical experience. This means accepting the legitimacy of mathematics as it is: fallible, correctible, and meaningful.

## Further Readings. See Bibliography

K. Popper and J. Eccles

# Glossary

**algorithm**  A fixed process which if carried out systematically produces a desired result. Thus, the "Euclidean algorithm" is a set of rules which when applied to two integers produces their greatest common divisor.

**analysis**  The modern outgrowth of differential and integral calculus.

**argument**  A particular value at which a function is evaluated. Thus if the function is $y = x^2$, the argument $x = 7$ yields the functional value $y = 7^2 = 49$. The independent variable of a function.

**axiom**  A statement which is accepted as a basis for further logical argument. Historically, an axiom was thought to embody a "self-evident" truth or principle.

**axiom of choice**  A principle which validates a certain type of mathematical construction. The axiom of choice asserts that given a collection of sets, one may form a set consisting of precisely one element from each set of the given collection.

**bit**  One digit in the binary representation of a number. Thus 1101 is a four-bit number. The fundamental unit of formalized information.

**binary notation**  The representation of integers in terms of powers of 2, much used in computers.

| DECIMAL | BINARY | | EXPLANATION |
|---------|--------|---|-------------|
| 1 | 1 | $1 =$ | $(1 \times 1)$ |
| 2 | 10 | $2 =$ | $(2 \times 1) + (1 \times 0)$ |
| 3 | 11 | $3 =$ | $(2 \times 1) + (1 \times 1)$ |
| 4 | 100 | $4 =$ | $(4 \times 1) + (2 \times 0) + (1 \times 0)$ |
| 5 | 101 | $5 =$ | $(4 \times 1) + (2 \times 0) + (1 \times 1)$ |
| 6 | 110 | $6 =$ | $(4 \times 1) + (2 \times 1) + (1 \times 0)$ |
| 7 | 111 | $7 =$ | $(4 \times 1) + (2 \times 1) + (1 \times 1)$ |

**byte**  eight binary digits (bits)

**combinatorics**  The mathematical discipline which "counts without counting" in an attempt to answer questions of the form: In how many ways can . . .? Example: In how many ways can five husbands and five wives be arranged around a circular table so that no wife sits next to her husband?

412

**complex number**  A number of the form $a + bi$ where $a$ and $b$ are real numbers and $i = \sqrt{-1}$ (or $i^2 = -1$). The systematic study of numbers of this form is called the "theory of functions of a complex variable" and is much employed in both pure and applied mathematics.

**constructivism** (Earlier term: intuitionism)  The doctrine which asserts that only those mathematical objects have real existence and are meaningful which can be "constructed' from certain primitive objects in a finitistic way. Associated with the Dutch mathematician L. E. J. Brouwer and his followers.

**continuum hypothesis**  In Cantorian set theory, the cardinal number of a set designates its "manyness." The cardinality of the set of integers 1, 2, 3, . . . is designated by $\aleph_0$. The cardinality of the set of real numbers is $2^{\aleph_0}$. The continuum hypothesis asserts that there is no set whose cardinal falls between $\aleph_0$ and $2^{\aleph_0}$.

**curve fitting**  Finding a "simple" formula or a "simple" geometric curve which approximates physical or statistical data. Kepler's discovery that planetary orbits are elliptical is a classic example of curve fitting.

**Diophantine equation**  An equation which is required to be solved by integers. Thus, the Diophantine equation $2x^2 - 3y^2 = 5$ has the integer solution $x = 2$, $y = 1$.

**Dirichlet problem**  Given a region $R$ in the $x - y$ plane and a function $f$ defined on the boundary $B$ of $R$. The Dirichlet problem asks for a function $u(x, y)$ which satisfies the partial differential equation $\dfrac{\partial^2 u}{\partial x^2} + \dfrac{\partial^2 u}{\partial y^2} = 0$ in $R$ and which takes on the values $f$ on $B$.
The Dirichlet problem is of fundamental importance in theoretical hydro and aerodynamics, elasticity, electrostatics, etc.

**Fermat's last problem/theorem**  The statement that the equation $x^n + y^n = z^n$ has no positive integer solution $x$, $y$, $z$ if $n$ is an integer greater than 2. Fermat's last problem is currently one of the most famous of the unsolved mathematical problems.

**fixed point**  A point which is left unchanged by a transformation. Thus, if a disc is rotated about its center, the center is a fixed point.

**formalism**  The position that mathematics consists merely of formal symbols or expressions which are manipulated or combined according to preassigned rules or agreements. Formalism makes no inquiry as to the meaning of the expressions.

**four-color theorem**  It is desired to color a political map on a plane so that countries sharing a common boundary are colored differently. Experience of map-makers has shown that four colors suffice. This assertion in its precise formulation was an unsolved problem in mathematics for over a hundred years. It has now been "proved" by an assist from the computer.

**Fourier analysis**  The mathematical strategy wherein a periodic curve (function) is decomposed into elementary periodic curves (sines and cosines).

**function**  Generally: an association between the elements of two sets.

413

More narrowly: a formula by which such an association can be computed; a curve, a rule or a "black box" which produces a fixed output for a given input. Example: $y = x^2$. Input: $x$; output: $x^2$.

**function space** A set of functions, membership in which is restricted by entry rules. A famous example is the set of all functions defined on the interval $-\pi \leq x \leq \pi$ which are measurable in the sense of Lebesgue and such that $\int_{-\pi}^{\pi}|f|^2 dx < \infty$. A less fancy example of a function space might be the set of all parabolic functions $y = ax^2 + bx + c$.

**fundamental theorem of algebra** The assertion that an equation of the form $a_0 x^n + a_1 x^{n-1} + \cdots + a_n = 0$, where the numbers $a_0, \ldots, a_n$ may be complex, has at least one complex value $x$ which satisfies it.

**Goldbach's conjecture** The conjecture that every even number is the sum of two prime numbers.

**group** A group is an algebraic structure for which the following requirements (axioms) are fulfilled. There is a set of elements. The elements are combinable two by two by an operation "·" yielding another element of the set. The process of combination satisfies the "associative law": $a \cdot (b \cdot c) = (a \cdot b) \cdot c$. The set contains an "identity" element $e$ which under combination satisfies $a \cdot e = e \cdot a = a$ for all $a$ in the set. For each element $a$ in the set, there is an inverse element $a^{-1}$ which under combination satisfies $a \cdot a^{-1} = a^{-1} \cdot a = e$. A very simple example of a group is the two numbers 1 and $-1$ combined by multiplication. The element 1 is the identity. The inverses of 1 and $-1$ are 1 and $-1$. The *order* of a group is the number of elements in it.

**integers** The positive integers are 1, 2, 3, . . . The negative integers are $-1, -2, -3, \ldots$ Then there is 0, the zero integer.

**irrational number** A real number that is not a rational number (fraction). One of the first irrational numbers to be discovered was $\sqrt{2} = 1.414 \ldots$ "Most" real numbers are irrational, in a sense made precise by Georg Cantor.

**lemma** A preparatory theorem. Possibly of slight interest in its own right, but used as a stepping stone in the proof of a more substantial theorem.

**logicism** The position that mathematics is identical to symbolic logic. One of the early advocates of this position was Bertrand Russell.

**matrix** A rectangular array of items (generally numbers).

$$\begin{pmatrix} 2 & 1 \\ 2 & 3 \end{pmatrix} \quad \begin{pmatrix} 1 & 2 & 3 \\ 4 & 5 & 9 \end{pmatrix} \text{ are matrices.}$$

**modulo** When integers are taken "modulo 3," one neglects multiples of 3 and considers only the remainder. Thus, 7 (modulo 3) = 1 because $7 = (3 \cdot 2) + 1$. Similarly for moduli other than 3. A clock registers time modulo 12, not accumulated time. The odometer on a car registers mileage modulo 100,000 miles.

**natural number** Any of the integers 1, 2, . . . . The positive integers.

**non-Euclidean geometry**  A geometry which is based upon axioms that contradict those set forth by Euclid. Especially, a geometry in which Euclid's Fifth Axiom is replaced by alternatives. The fifth axiom asserts that through a point not on a line there exists one and only one parallel line. One particular non-Euclidean geometry asserts that there is no such parallel.

**non-standard analysis**  A mathematical number system (and related calculus), made logically rigorous in the 1960's by Abraham Robinson, which admits infinitely small quantities.

**Platonism**  As used in this book, Platonism is the position that the whole of mathematics exists eternally, independently of man, and the job of the mathematician is to discover these mathematical truths.

**power series**  An infinite series of a special type wherein successive powers of a variable are multiplied by certain numbers and then added.

Examples:   $1 + x + x^2 + x^3 + \ldots$

$$1 + x + \frac{x^2}{1 \cdot 2} + \frac{x^3}{1 \cdot 2 \cdot 3} + \frac{x^4}{1 \cdot 2 \cdot 3 \cdot 4} + \ldots$$

The general form of a power series in the variable $x$ is

$$a_0 + a_1x + a_2x^2 + a_3x^3 + \cdots$$

**prime number theorem**  A statement about the frequency with which the prime numbers occur in the sequence of positive integers. This theorem was conjectured in the early 1800s but was not firmly established until the 1890s. More precisely, if $\pi(n)$ designates the number of primes not greater than $n$, $\pi(n)$ is approximately equal to $\dfrac{n}{\log n}$, the approximation becoming better as $n$ becomes larger.

**rational number**  Any number that is the ratio of two integers: $\frac{1}{1}$, $-\frac{6}{7}$, $\frac{21}{108}$, $\frac{4627}{1039}$. A fraction.

**real number**  Any finite or infinite decimal. Examples:

1,
2817.
−30.00792
81.1111 . . .
.12345678911011121314
3.14159 . . .

Any rational or irrational number.

**residue**  The remainder upon division of one integer by another. The residue of 10 modulo (i.e., upon division by) 6 is 4.

**Russell's paradox**  Popularly: the Cretan who said that all Cretans are liars—was he lying or telling the truth?

In a mathematical context: the set of all sets that are not members of themselves. This concept embodies a contradiction.

**sequence**  One thing after another in a set ordered as are the positive integers. Generally, a sequence of numbers. Thus: the sequence of even numbers is 2, 4, 6, 8, . . . The sequence of square numbers is 1, 4, 9, . . .

**series** Generally, but not always, an *infinite sum*. A mathematical process which calls for an infinite number of additions. Thus, $1 + \frac{1}{2} + \frac{1}{4} + \frac{1}{8} + \cdots$ and $1 - \frac{1}{3} + \frac{1}{5} - \frac{1}{7} + \cdots$ and $1 - 1 + 1 - 1 + 1 \ldots$ are famous infinite series.

**word length** The number of bits normally processed by a computer as a single unit.

# Bibliography

ADLER, ALFRED: Mathematics and Creativity. The New Yorker Magazine, February 19, 1972

AHRENS, W.: Mathematiker Anektoden, ed. L. J. Cappon. Chapel Hill: University of North Carolina Press 1959

ALEXANDROFF, A. D., KOLMOGOROFF, A. N., LAWRENTIEFF, M. A. (eds.): Mathematics: Its Content, Methods and Meaning. Cambridge: M.I.T. Press 1963

ALPER, J. L.: Groups and Symmetry. In: Mathematics Today, ed. L. A. Steen, pp. 65–82. New York: Springer-Verlag 1978

ANDERSON, D. B., BINFORD, T. O., THOMAS, A. J., WEYBRAUCH, R. W., WILKS, V. A.: After Leibniz: Discussions on Philosophy and Artificial Intelligence. Stanford Artificial Intelligence Laboratory Memo AIM-229 March 1974

ANONYMOUS: Federal Funds for Research and Development, Fiscal Years 1977, 1978, 1979. Vol. 27. Detailed Statistical Tables, Appendix C. NSF-78-312. Washington, D.C.: National Science Foundation 1978

APPEL, K., HAKEN, W.: The Four-Color Problem. In: Mathematics Today, ed. L. A. Steen, pp. 153–190. New York: Springer-Verlag 1978

ARAGO, F. J.: Éloge historique de Joseph Fourier. Mém. Acad. Roy. Sci. 14, 69–138. English translation in: Biographies of Distinguished Scientific Men, London, 1857

ARCHIBALD, R. C.: Outline of the History of Mathematics. Slaught Memorial Paper. Buffalo: Mathematical Association of America 1949

ARIS, R.: Mathematical Modelling Techniques. San Francisco: Pitman 1978

AUBREY, JOHN: Brief Lives. Edited by Andrew Clark. Oxford: Oxford University Press 1898

AUBREY, JOHN: Aubrey's Brief Lives. Edited by Oliver Lawson Dick; foreword by Edmund Wilson. Ann Arbor: Ann Arbor Paperbacks 1962

BANCHOFF, T., STRAUSS, C. M.: On Folding Algebraic Singularities in

Complex 2-Space. Talk and movie presented at meeting of the American Mathematical Society. Dallas, Texas January 1973

BARBEAU, E.J., LEAH, P. J.: Euler's 1760 Paper on Divergent Series. Historia Mathematica 3, 141–160 (1976)

BARKER, STEPHEN F.: Philosophy of Mathematics. Englewood Cliffs: Prentice-Hall 1964

BARWISE, JON (ed.): Handbook of Mathematical Logic. Amsterdam: North-Holland 1977

BASALLA, GEORGE: The Rise of Modern Science: Internal or External Factors. Boston: D. C. Heath and Co., 1968

BAUM, ROBERT J. (ed.): Philosophy and Mathematics from Plato to the Present. Freeman, San Francisco 1973

BELL, E. T.: Men of Mathematics. New York: Simon and Schuster 1937

BELL, E. T.: The Development of Mathematics. New York: McGraw-Hill 1949

BELL, M. S. (ed.): Studies in Mathematics, Vol. XVI: Some Uses of Mathematics: a Sourcebook for Teachers and Students of School Mathematics. Stanford: School Mathematics Study Group 1967

BELLMAN, RICHARD: A Collection of Modern Mathematical Classics: Analysis. New York: Dover 1961

BENACERRAF, PAUL, PUTNAM, HILARY (eds.): Philosophy of Mathematics: Selected Readings. Englewood Cliffs: Prentice-Hall 1964

BERNAL, J. D.: The Social Function of Science. New York: Macmillan 1939

BERNSTEIN, DOROTHY L.: The Role of Applications in Pure Mathematics. American Mathematical Monthly. 86, 245–253 (1979)

BETH, EVERT W., PIAGET, JEAN: Mathematics, Epistemology and Psychology. Translated by W. Mays. New York: Gordon and Breach 1966

BIRKHOFF, GARRETT: Mathematics and Psychology. SIAM Review. 11, 429–469 (1969)

BIRKHOFF, GARRETT (ed.): A Source Book in Classical Analysis. Cambridge: Harvard University Press 1973

BIRKHOFF, GARRETT: Applied Mathematics and Its Future. In: Science and Technology in America, pp. 83–103. Washington, D.C.: National Bureau of Standards, Special Publication No. 465 1977

BISHOP, E.: Foundations of Constructive Analysis. New York: McGraw-Hill 1967

BISHOP, E.: Aspects of Constructivism. Las Cruces: New Mexico State University 1972

BISHOP, E.: The Crisis in Contemporary Mathematics. Historia Mathematica. 2, 507–517 (1975)

BLANCHÉ, ROBERT: Axiomatization. In: Dictionary of the History of Ideas. Vol. I Scribner's 1973

BLOCH, MARC: The Historian's Craft. New York: Alfred A. Knopf 1953

BLOOR, DAVID: Wittgenstein and Mannheim on the Sociology of Mathematics. Studies in the History and Philosophy of Science. 2, 173–191. New York: Macmillan 1973

418

BOCHNER, SALOMON: The Role of Mathematics in the Rise of Science. Princeton: Princeton University Press 1966

BOCHNER, SALOMON: Mathematics in Cultural History. In: Dictionary of the History of Ideas. New York: Charles Scribner's Sons 1973

BOLZANO, B.: Paradoxes of the Infinite (1850), D. A. Steele, ed. London: Routledge and Kegan Paul 1950

BOREL, EMILE: L'imaginaire et le réal en mathématiques et en physique. Paris: Editions Alvin Michel 1952

BOOSS, BERNHELM, NISS, MOGENS (eds.): Mathematics and the Real World. Basel: Birkhauser Verlag 1979

BOWNE, GWENDOLYN D.: The Philosophy of Logic, 1880–1908. The Hague: Mouton 1966

BOYER, C. B.: A History of Mathematics. New York: John Wiley & Sons 1968

BRIDGMAN, P. W.: The Way Things Are. Cambridge: Harvard University Press 1959

BROOKS, FREDERICK P., JR.: The Mythical Man-Month: Essays on Software Engineering. Reading, Mass.: Addison-Wesley 1975

BRUNER, JEROME S.: On Knowing: Essays for the Left Hand. New York: Atheneum 1970

BRUNER, JEROME S.: The Process of Education. Cambridge: Harvard University Press 1960

BRUNO, GIORDANO: Articuli centum et sexaginta adversos huius tempestatis mathematicos atque philosophos. Prague 1588

BRUNSCHVICG, LÉON: Les étapes de la philosophie mathématique. Paris: Alcan 1912

BUDDEN, F. J.: The Fascination of Groups. Cambridge: Cambridge University Press 1972

BUNGE, MARIO: Intuition and Science. Englewood Cliffs: Prentice-Hall 1962

BUNT, L. N. H., JONES, P. S., BEDIENT, J. D.: The Historical Roots of Elementary Mathematics. Englewood Cliffs: Prentice Hall 1976

BURGESS, J. P.: Forcing. In: Jon Barwise (ed.), Handbook of Mathematical Logic, pp. 403–452. Amsterdam: North-Holland 1977

CAJORI, F.: The Early Mathematical Sciences in North and South America. Boston: R. G. Badger 1928

CAJORI, F.: History of Mathematical Notations. Chicago: The Open Court Publishing Co. 1928–29

CHIHARA, CHARLES S.: Ontology and the Vicious Circle Principle. Ithaca, N.Y.: Cornell University Press 1973

CHINN, W. G., STEENROD, N. E.: First Concepts of Topology. Washington, D.C., The New Mathematical Library, Mathematical Association of America 1966

CLARK, G. N.: Science and Social Welfare in the Age of Newton, Oxford: Oxford University Press 1949

COHEN, M. R., DRABKIN, I. E.: Source Book in Greek Science. Cambridge: Harvard University Press 1958

# Bibliography

COHEN, P. J.: Set Theory and the Continuum Hypothesis. New York: W. A. Benjamin 1966

COHEN, P. J.: Comments on the Foundation of Set Theory. In: Dana Scott (ed.), Axiomatic Set Theory, pp. 9–15. Providence, R.I.: American Mathematical Society 1971

COHEN, P. J., HERSH, R.: Non-Cantorian Set Theory. In: Mathematics in the Modern World. San Francisco: W. H. Freeman 1968

COPI, IRVING M., GOULD, JAMES A. (eds.): Readings on Logic. New York: Macmillan 1967

COURANT, R., ROBBINS, H.: What is Mathematics? New York: Oxford University Press 1948

CROSSLEY, JOHN N., et al.: What is Mathematical Logic? Oxford: Oxford University Press 1972

CROWE, M. J.: Ten "Laws" Concerning Patterns of Change in the History of Mathematics. Historia Mathematica. 2, 161–166 (1975)

CUDHEA, DAVID: Artificial Intelligence. The Stanford Magazine, spring/summer (1978)

CURRY, HASKELL B.: Some Aspects of the Problem of Mathematical Rigor. Paper presented at the Meeting of the American Mathematical Society, New York, October 26, 1940

CURRY, HASKELL B.: Outlines of a Formalist Philosophy of Mathematics. Amsterdam: North-Holland, 1951

DANTZIG, TOBIAS: Henri Poincaré: Critic of Crises. Reflections on his Universe of Discourse. New York: Charles Scribner's Sons 1954

DANTZIG, TOBIAS: Number, the Language of Science. New York: Macmillan 1959

DAUBEN, J. W.: The Trigonometric Background to Georg Cantor's Theory of Sets. Arch. History of the Exact Sciences 7, 181–216 (1971)

DAVIS, CHANDLER: Materialist Mathematics. Boston Studies in the Philosophy of Science. 15, 37–66. Dordrecht: D. Reidel 1974

DAVIS, H. T.: Essays in the History of Mathematics. Evanston, Illinois, mimeographed 1949

DAVIS, MARTIN: Applied Nonstandard Analysis. New York: John Wiley & Sons 1977

DAVIS, MARTIN: The Undecidable. Hewlett, N.Y.: Raven Press 1965

DAVIS, MARTIN: Unsolvable Problems. In: Jon Barwise (ed.): Handbook of Mathematical Logic. Amsterdam: North-Holland 1977

DAVIS, MARTIN, HERSH, REUBEN: Nonstandard Analysis. Scientific American. June, 1972, pp. 78–84.

DAVIS, MARTIN, HERSH, REUBEN: Hilbert's Tenth Problem. Scientific American. November, 1973, pp. 84–91.

DAVIS, N. P.: Lawrence and Oppenheimer. New York: Simon and Schuster 1968

DAVIS, P. J.: Leonhard Euler's Integral: An Historical Profile of the Gamma Function. American Mathematical Monthly. 66, 849–869 (1959)

DAVIS, P. J.: The Criterion Makers: Mathematics and Social Policy. American Scientist. 50, 258A–274A (1962)

DAVIS, P. J.: Number. Scientific American. September, 1964. Reprinted in: Mathematics: An Introduction to its Spirit and Use. San Francisco: W. H. Freeman 1978

DAVIS, P. J.: Numerical Analysis. In: The Mathematical Sciences, pp. 128–137. Cambridge: M.I.T. Press 1969

DAVIS, P. J.: Fidelity in Mathematical Discourse: Is 1 + 1 Really 2? American Mathematical Monthly. 78, 252–263 (1972)

DAVIS, P. J.: Simple Quadratures in the Complex Plane. Pacific Journal of Mathematics. 15, 813–824 (1965)

DAVIS, P. J.: Visual Geometry, Computer Graphics, and Theorems of Perceived Type. In: Proceedings of Symposia in Applied Mathematics, vol. 20. Providence, R.I.: American Mathematical Society 1974

DAVIS, P. J.: Towards a Jamesian History of Mathematics. Invited address at the Winter Meeting of the American Mathematical Society, January 22, 1967, San Antonio, Texas

DAVIS, P. J.: Mathematics by Fiat? The Two Year College Mathematics Journal. June, 1980

DAVIS, P. J.: Circulant Matrices. New York: John Wiley & Sons 1979

DAVIS, P. J., ANDERSON, J. A.: Non-Analytic Aspects of Mathematics and Their Implication for Research and Education. SIAM Review. 21, 112–127 (1979)

DAVIS, P. J., CERUTTI, ELSIE: FORMAC Meets Pappus: Some Observations on Elementary Analytic Geometry by Computer. American Mathematical Monthly. 75, 895–905 (1969)

DEE, JOHN: Monas Hieroglyphica 1564

DEE, JOHN: The Mathematical Praeface to the Elements of Geometrie of Euclid of Megara (1570). With an introduction by Allen G. Debus. New York: Neale Watson Academic Publications 1975

DEMILLO, R. A., LIPTON, R. J., PERLIS, A. J.: Social Processes and Proofs of Theorems and Programs. Communications of the ACM. 22, 271–280 (1979)

DERTOUZOS, M. L., MOSES, J. (eds.): The Computer Age: A Twenty-Year View. Cambridge: M.I.T. Press 1979

DESARMANIEN, J., KUNG, J. P. S., ROTA, G.-C.: Invariant Theory, Young Bitableaux and Combinatories. Advances in Mathematics. 27, 63–92 (1978)

DICKSON, L. E.: History of the Theory of Numbers, vol. 2. New York: G. E. Stechert 1934

DICTIONARY OF THE HISTORY OF IDEAS: New York: Charles Scribner's Sons 1973

DIEUDONNÉ, J.: Modern Axiomatic Methods and the Foundations of Mathematics. In: Great Currents of Mathematical Thought, Vol. 2, pp. 251–266. New York: Dover 1971

DIEUDONNÉ, J.: The Work of Nicholas Bourbaki. American Mathematical Monthly. 77, 134–145 (1970)

421

*Bibliography*

DIEUDONNÉ, J.: Should We Teach Modern Mathematics? American Scientist. 61, 16–19 (1973)

DIEUDONNÉ, J.: Panorame des mathématiques pures: le choix bon'bachique. Paris: Bordas, Dunod, Gauthier-Villars 1977

DI SESSA, A.: Turtle Escapes the Plane: Some Advanced Turtle Geometry. Artificial Intelligence Memo. 348, Artificial Intelligence Laboratory, M.I.T., Boston, Mass. December 1975

DRESDEN, ARNOLD: Mathematical Certainty. Scientia. 45, 369–374 (1929)

DREYFUS, HERBERT: What Computers Can't Do: a Critique of Artificial Reason. New York: Harper & Row 1972

DUHEM, P.: The Aim and Structure of Physical Theory. First edition 1906. Princeton: Princeton University Press 1954

DUMMETT, MICHAEL: Elements of Intuitionism. Oxford: The Clarendon Press. 1977

DUMMETT, MICHAEL: Reckonings: Wittgenstein on Mathematics. Edited by C. Diamond; notes of R. Bosanquet, N. Malcolm, R. Rhees, Y. Smithies. Cambridge 1939

DUNMORE, PAUL V.: The Uses of Fallacy. New Zealand Mathematics Magazine. 1970

DUNNINGTON, G. W.: C. F. Gauss. New York: Exposition Press 1955

DUPREE, A. HUNTER: Science in the Federal Government. Cambridge: Harvard University Press 1957

DYCK, MARTIN: Novalis and Mathematics. Chapel Hill: University of North Carolina Press 1960

EDWARDS, H. M.: Riemann's Zeta Function. New York: Academic Press 1974

EDMUNDSON, H. P.: Definitions of Random Sequences. TR-360, Computer Science Department, University of Maryland, College Park, Maryland. March, 1975

EUCLID: The Thirteen Books of Euclid's Elements. Introduction and Commentary by T. L. Heath. New York: Dover 1956

EVES, H., NEWSOM, C. V.: An Introduction to the Foundations and Fundamental Concepts of Mathematics. New York: Holt, Rinehart and Winston 1965

FANG, J. AND TAKAYAMA, K. P.: Sociology of Mathematics and Mathematicians. Hauppauge, N.Y.: Paideia Press 1975.

FEFERMAN, SOLOMON: The Logic of Mathematical Discovery vs. the Logical Structure of Mathematics. Departure of Mathematics, Stanford University. 1976

FEFERMAN, SOLOMON: What Does Logic Have to Tell Us About Mathematical Proofs? Mathematical Intelligence 2, No. 4.

FERGUSON, E. S.: The Mind's Eye: Nonverbal Thought in Technology. Science. 197, 827–836 (1977)

FISHER, CHARLES S.: The Death of a Mathematical Theory: A Study in the Sociology of Knowledge. Arch. History of the Exact Sciences. 3, 137–159 (1966)

FITZGERALD, ANNE, MACLANE, SAUNDERS (eds.): Pure and Applied Mathematics in the People's Republic of China. Washington, D.C.: National Academy of Sciences 1977

FRAENKEL, A. A.: The Recent Controversies About the Foundations of Mathematics. Scripta Mathematica. 13, 17–36 (1947)

FRAME, J. S.: The Working Environment of Today's Mathematician. In: T. L. Saaty and F. J. Weyl (eds.), The Spirit and Uses of the Mathematical Sciences. New York: McGraw-Hill 1969

FRANK, PHILIPP: The Place of Logic and Metaphysics in the Advancement of Modern Science. Philosophy of Science. 5, 275–286 (1948)

FRENCH, PETER J.: John Dee. London: Routledge and Kegan Paul 1972

FREUDENTHAL, HANS: The Concept and Role of the Model in Mathematics and Social Sciences. Dordrecht: Reidel 1961

FREUDENTHAL, HANS: Symbole. In: Encyclopaedia Universalis. Paris. 1968

FREUDENTHAL, HANS: Mathematics as an Educational Task. Dordrecht: Reidel 1973

FREUDENTHAL, HANS: Weeding and Sowing. Dordrecht: Reidel 1978

FRIEDMAN, JOEL I.: On Some Relations Between Leibniz' Monodology and Transfinite Set Theory (A Complement to the Russell Thesis). In: Akten des II Internationalen Leibniz-Kongresses. Wiesbaden: Franz Steiner 1975

FRIEDMAN, JOEL I.: Some Set-Theoretical Partition Theorems Suggested by the Structure of Spinoza's God. Synthese. 27, 199–209 (1974)

GAFFNEY, M. P., STEEN, L. A.: Annotated Bibliography of Expository Writings in the History of the Mathematical Sciences. Washington, D.C.: Mathematical Association of America 1976

GARDNER, H.: The Shattered Mind. New York: Knopf 1975

GARDNER, MARTIN: Aha! Insight. San Francisco: W. H. Freeman 1978

GILLINGS, R. A.: Mathematics in the Times of the Pharaohs. Cambridge: M.I.T. Press 1972

GODEL, K.: What is Cantor's Continuum Problem? In: P. Benacerraf and H. Putnam (eds.): Philosophy of Mathematics, Selected Readings, pp. 258–273. Englewood Cliffs: Prentice-Hall 1964

GOLDSTEIN, IRA, PAPERT, SEYMOUR: Artificial Intelligence Language and the Study of Knowledge. M.I.T. Artificial Intelligence Laboratory, Memo. 337, Boston March 1976

GOLDSTINE, HERMAN H.: The Computer from Pascal to von Neumann. Princeton: Princeton University Press 1972

GOLDSTINE, HERMAN H.: A History of Numerical Analysis. New York: Springer 1977

GOLOS, E. B.: Foundations of Euclidean and Non-Euclidean Geometry. New York: Holt, Rinehart and Winston 1968

GONSETH, FERDINAND: Philosophie mathématique. Hermann, Paris 1939

GOOD, I. J., CHURCHHOUSE, R. F.: The Riemann Hypothesis and Pseu-

423

dorandom Features of the Möbius Sequence. Mathematics of Computation. 22, 857–864 (1968)

GOODFIELD, JUNE: Humanity in Science: A Perspective and a Plea. The Key Reporter. 42, Summer 1977

GOODMAN, NICHOLAS D.: Mathematics as an Objective Science. American Mathematical Monthly. 86, 540–551 (1979)

GORENSTEIN, D.: The Classification of Finite Simple Groups. Bulletin of the American Mathematical Society. N.S. 1, 43–199 (1979)

GRABINER, JUDITH V.: Is Mathematical Truth Time-Dependent? American Mathematical Monthly. 81, 354–365 (1974)

GRATTAN-GUINNESS, I.: The Development of the Foundations of Mathematical Analysis from Euler to Riemann. Cambridge: M.I.T. Press 1970

GREENBERG, MARVIN J.: Euclidean and Non-Euclidean Geometries: Development and History. San Francisco: W. H. Freeman 1974

GREENWOOD, T.: Invention and Description in Mathematics. Meeting of the Aristotelian Society 1/20 1930

GRENANDER, ULF: Mathematical Experiments on the Computer. Division of Applied Mathematics, Brown University, Providence, R.I. 1979

GRIFFITHS, J. GWYN: Plutarch's De Iside et Osiride. University of Wales Press 1970

GROSSWALD, E.: Topics from the Theory of Numbers. New York: Macmillan 1966

GUGGENHEIMER, H.: The Axioms of Betweenness in Euclid. Dialectica. 31, 187–192 (1977).

HACKING, IAN: Review of I. Lakatos' Philosophical Papers. British Journal of the Philosophy of Science (to appear)

HADAMARD, JACQUES. The Psychology of Invention in the Mathematical Field. Princeton: Princeton University Press 1945

HAHN, HANS: The Crisis in Intuition. In: J. R. Newman (ed.), The World of Mathematics, pp. 1956–1976. New York: Simon and Schuster 1956

HALMOS, P.: Mathematics as a Creative Art. American Scientist. 56, 375–389 (1968)

VON HARDENBERG, FRIEDRICH (Novalis): Tagebücher. München: Hanser 1978

HARDY, G. H.: Mathematical Proof. Mind. 38, 1–25 (1929)

HARDY, G. H.: A Mathematician's Apology. Cambridge: Cambridge University Press 1967

HARTREE, D. R.: Calculating Instruments and Machines. Urbana: University of Illinois Press 1949

HEATH, T. L.: A History of Greek Mathematics. Oxford: The Clarendon Press 1921

HEATH, T. L.: Euclid's Elements. Vol. I. New York: Dover 1956

HEISENBERG, WERNER: The Representation of Nature in Contemporary

Physics. In: Rollo May (ed.), Symbolism in Religion and Literature. New York: Braziller 1960

HENKIN, LEON A.: Are Logic and Mathematics Identical? The Chauvenet Papers, Vol. II. Washington, D.C.: The Mathematical Association of America 1978

HENRICI, PETER: Reflections of a Teacher of Applied Mathematics. Quarterly of Applied Mathematics. 30, 31–39 (1972)

HENRICI, PETER: The Influence of Computing on Mathematical Research and Education. In: Proceedings of Symposia in Applied Mathematics, Vol. 20. Providence: American Mathematical Society 1974

HERSH, REUBEN: Some Proposals for Reviving the Philosophy of Mathematics. Advances in Mathematics. 31, 31–50 (1979)

HERSH, REUBEN: Introducing Imre Lakatos. Mathematical Intelligencer. 1, 148–151. (1978)

HESSEN, B.: The Social and Economic Roots of Newton's Principia. New York: Howard Fertig 1971

HILBERT, D.: On the infinite. In: P. Benacerraf and H. Putnam (eds.), Philosophy of Mathematics: Selected Readings, pp. 134–151. Englewood Cliffs: Prentice-Hall 1964

HONSBERGER, R.: Mathematical Gems, II. Washington, D.C.: Mathematical Association of America 1973

HORN, W.: On the Selective Use of Sacred Numbers and the Creation in Carolingian Architecture of a new Aesthetic Based on Modular Concepts. Viator. 6, 351–390 (1975)

HOROVITZ, J.: Law and Logic. New York: Springer 1972

HOUSTON, W. ROBERT (ed.): Improving Mathematical Education for Elementary School Teachers. East Lansing, Michigan: Michigan State University 1967

HOWSON, A. G. (ed.): Developments in Mathematical Education. Cambridge: Cambridge University Press 1973

HRBACEK, K., JECH, T.: Introduction to Set Theory. New York: Marcel Dekker 1978

HUNT, E. B.: Artificial Intelligence. New York: Academic Press 1975

HUNTLEY, H. E.: The Divine Proportion. New York: Dover 1970

HUSSERL, EDMUND: The Origins of Geometry. Appendix VI in Edmund Husserl, The Crisis of European Science. Translated by David Carr. Evanston: Northwestern University Press 1970

ILIEV, L.: Mathematics as the Science of Models. Russian Mathematical Surveys. 27, 181–189 (1972)

JACOB, FRANÇOIS: Evolution and Tinkering. Science. 196, 1161–1166 (1977)

JAMES, WILLIAM: Great Men and Their Environment. In: Selected Papers on Philosophy, pp. 165–197. London: J. M. Dent and Sons 1917

JAMES, WILLIAM: Psychology (Briefer Course). New York: Collier 1962

# Bibliography

JAMES, WILLIAM: The Varieties of Religious Experiences. Reprint: New York: Mentor Books 1961

JOSTEN, C. H.: A Translation of Dee's 'Monas Hieroglyphica' with an Introduction and Annotations. Ambix. 12, 84–221 (1964)

JOUVENEL, BERTRAND DE: The Republic of Science. In: The Logic of Personal Knowledge: Essays to M. Polányi. London: Routledge and Kegan Paul 1961

JUNG, C. J.: Man and His Symbols. Garden City, New York: Doubleday 1964

JUSTER, NORTON: The Phantom Tollbooth. New York: Random House 1961

KAC, MARK: Statistical Independence in Probability, Analysis and Number Theory. Carus Mathematical Monographs No. 12. Washington, D.C.: Mathematical Association of America 1959

KANTOROWICZ, ERNST: Frederick the Second. London: Constable 1931

KASNER, E., NEWMAN, J.: Mathematics and the Imagination. New York: Simon and Schuster 1940

KATZ, AARON: Toward High Information-Level Culture. Cybernetica. 7, 203–245 (1964)

KERSHNER, R. B. AND WILCOX, L. R.: The Anatomy of Mathematics. New York: Ronald Press 1950

KESTIN, J.: Creativity in Teaching and Learning. American Scientist. 58, 250–257 (1970)

KLEENE, S. C.: Foundations of Mathematics. In: Encyclopaedia Britannica, 14th edition, volume 14, pp. 1097–1103. Chicago 1971

KLENK, V. H.: Wittgenstein's Philosophy of Mathematics. The Hague: Nijhoff 1976

KLIBANSKY, R. (ed.): La philosophie contemporain. Vol. 1. Florence: UNESCO 1968

KLINE, M. (ed.): Mathematics in the Modern World. Readings from Scientific American. San Francisco: W. H. Freeman 1968

KLINE, M.: Logic Versus Pedagogy. American Mathematical Monthly. 77, 264–282 (1970)

KLINE, M.: Mathematical Thought from Ancient to Modern Times. Oxford: Oxford University Press 1972

KLINE, M.: Why the Professor Can't Teach. New York: St. Martin's Press 1977

KNEEBONE, I. G. T., CAVENDISH, A. P.: The Use of Formal Logic. The Aristotelian Society Supplementary Vol. 45 1971

KNOWLTON, K.: The Use of FORTRAN-Coded EXPLOR for Teaching Computer Graphics and Computer Art. In: Proceedings of the ACM SIGPLAN Symposium on Two-Dimensional Man-Machine Communication, Los Alamos, New Mexico, October 5–6, 1972

KNUTH, D. E.: Mathematics and Computer Science: Coping with Finiteness. Science. 192, 1235–1242 (1976)

KOESTLER, ARTHUR: The Sleepwalkers. New York: Macmillan 1959

KOESTLER, ARTHUR: The Act of Creation. London: Hutchinson 1964

KOLATA, G. BARI: Mathematical Proof: the Genesis of Reasonable Doubt. Science. 192, 989–990 (1976)

KOLMOGOROV, A. D.: Mathematics. In: Great Soviet Encyclopaedia, third edition. New York: Macmillan 1970

KOPELL, N., STOLTZENBERG, G.: Commentary on Bishop's Talk. Historia Mathematica. 2, 519–521 (1975)

KORNER, S.: On the Relevance of Post-Gödelian Mathematics to Philosophy. In: I. Lakatos (ed.), Problems in the Philosophy of Mathematics, pp. 118–133. Amsterdam: North-Holland 1967

KOVALEVSKAYA, SOFYA: A Russian Childhood. Translated by Beatrice Stillman. New York: Springe-Verlag 1978

KUHN, T. S.: The Structure of Scientific Revolutions. Chicago: University of Chicago Press 1962

KUHNEN, K.: Combinatorics. In: Jon Barwise (ed.), Handbook of Mathematical Logic, pp. 371–401. Amsterdam: North-Holland 1977

KUNTZMANN, JEAN. Où vont les mathématiques? Paris: Hermann 1967

KUYK, WILLEM: Complementarity in Mathematics. Dordrecht: Reidel 1977

LAKATOS, I.: Infinite Regress and the Foundations of Mathematics. Aristotelian Society Supplementary Volume 36, pp. 155–184 (1962)

LAKATOS, I.: A Renaissance of Empiricism in the Recent Philosophy of Mathematics? In: I. Lakatos (ed.): Problems in the Philosophy of Mathematics, pp. 199–203. Amsterdam: North-Holland 1967

LAKATOS, I.: Proofs and Refutations. J. Worral and E. Zahar, eds. Cambridge: Cambridge University Press 1976

LAKATOS, I.: Mathematics, Science and Epistemology. Cambridge: Cambridge University Press 1978

LAKATOS, I., MUSGRAVE, A. (eds.): Problems in the Philosophy of Science. Amsterdam: North-Holland 1968

LANGER, R. E.: Fourier Series. Slaught Memorial Paper. American Mathematical Monthly. Supplement to Volume 54, pp. 1–86 (1947)

LASSERRE, FRANÇOIS: The Birth of Mathematics in the Age of Plato, Larchmont, N.Y.: American Research Council 1964

LEBESGUE, HENRI: Notices d'histoire des mathématiques. L'enseignement mathématique. Geneva 1958

LEHMAN, H.: Introduction to the Philosophy of Mathematics. Totowa, N.J.: Rowman and Littlefield 1979

LEHMER, D. N.: List of Prime Numbers from 1 to 10,006,721. Washington, D.C.: Carnegie Institution of Washington Publication No. 163 1914

LEITZMANN, W.: Visual Topology. Translated by M. Bruckheimer. London: Chatto and Windus 1965

LEVINSON, NORMAN: Wiener's Life. Bulletin of the American Mathematical Society. 72, 1–32 (1966) (a special issue on Norbert Wiener)

LIBBRECHT, ULRICH: Chinese Mathematics in the Thirteenth Century. Cambridge: M.I.T. Press 1973

LICHNEROWICZ, ANDRÉ: Rémarques sur les mathématiques et la realité.

427

In: Logique et connaissance scientifique. Dijon: Encyclopédie de la Pléiade, 1967

LITTLETON, A. C., YAMEY, B. S. (eds.): Studies in the History of Accounting. Homewood, Illinois: R. D. Irwin 1956

LITTLEWOOD, J. E.: A Mathematician's Miscellany. London: Methuen and Co. 1953

MACFARLANE, ALEXANDER: Ten British Mathematicians. New York: John Wiley and Sons 1916

MAIMONIDES, MOSES: Mishneh Torah. Edited and translated by M. H. Hyamson. New York: 1937

MANIN, Y. I.: A Course in Mathematical Logic. New York: Springer-Verlag 1977

MARSAK, LEONARD M.: The Rise of Science in Relation to Society. New York: Macmillan 1966

MAZIARZ, EDWARD A., GREENWOOD, THOMAS: Greek Mathematical Philosophy. New York: Ungar 1968

MEDAWAR, PETER B.: The Art of the Solvable. London: Methuen 1967 (In particular: "Hypothesis and Imagination.")

MEHRTENS, HERBERT: T. S. Kuhn's Theories and Mathematics. Historia Mathematica. 3, 297–320 (1976)

MERLAN, PHILIP: From Platonism to Neoplatonism. The Hague: Martinus Nijhoff 1960

MESCHKOWSKI, H.: Ways of Thought of Great Mathematicians. Translated by John Dyer-Bennet. San Francisco: Holden-Day 1964

MEYER ZUR CAPELLEN, W.: Mathematische Maschinen and Instrumente. Berlin: Akademie Verlag 1951

MICHENER, EDWINA R.: The epistemology and associative representation of mathematical theories with application to an interactive tutor system. Doctoral thesis, Department of Mathematics, M.I.T., Cambridge, Mass. 1977

MIKAMI, YOSHIO: The Development of Mathematics in China and Japan. Abhandlung zur Geschichte der Mathematischen Wissenschaften. 30, 1–347 (1913). Reprinted Chelsea, New York

MINSKY, M.: Computation: Finite and Infinite Machines. Englewood Cliffs: Prentice-Hall 1967

VON MISES, R.: Mathematical Theory of Probability and Statistics. New York: Academic Press 1964

MOLLAND, A. G.: Shifting the Foundations: Descartes' Transformation of Ancient Geometry. Historia Mathematica. 3, 21–79 (1976)

MONK, J. D.: On the Foundations of Set Theory. American Mathematical Monthly. 77, 703–711 (1970)

MORITZ, ROBERT E.: Memorabilia Mathematica. New York: Macmillan 1914

MOYER, R. S., LANDAUER, T. K.: Time Required for Judgments of a Numerical Inequality. Nature. 215, 1519–1529 (1967)

MOYER, R. S., LANDAUER, T. K.: Determinants of Reaction Time for

428

Digit Inequality Judgments. Bulletion of the Psychonomic Society. 1, 167–168 (1973)

MURRAY, F. J.: Mathematical Machines. New York: Columbia University Press 1961

MURRAY, F. J.: Applied Mathematics: An Intellectual Orientation. New York: Plenum Press 1978

MUSGRAVE, ALAN: Logicism Revisited. British Journal of the Philosophy of Science. 28, 99–127 (1977)

NALIMOV, V. V.: Logical Foundations of Applied Mathematics. Dordrecht: Reidel 1974

NATIONAL RESEARCH COUNCIL (ed.): The Mathematical Sciences. Cambridge: M.I.T. Press 1969

NEEDHAM, JOSEPH: Science and Civilization in China. Vol. III. Cambridge: Cambridge University Press 1959

NEUGEBAUER, O.: Babylonian Astronomy: Arithmetical Methods for the Dating of Babylonian Astronomical Texts. In: Studies and Essays to Richard Courant on his Sixtieth Birthday, pp. 265–275. New York: 1948

NEUGEBAUER, O.: The Exact Sciences in Antiquity. New York: Dover 1957

NEUGEBAUER, O., VON HOESEN, H. B.: Greek Horoscopes. Philadelphia: American Philosophical Society 1959

NEUGEBAUER, O., SACHS, A. J.: Mathematical Cuneiform Texts. New Haven: American Oriental Society and American Schools of Oriental Research 1945

VON NEUMANN, J.: The Mathematician. In: Works of the Mind, Robert B. Heywood (ed.). Chicago: University of Chicago Press 1947

NEWMAN, J. R. (ed.): The World of Mathematics. Four volumes. New York: Simon and Schuster 1956

NOVY, LUBOS: Origins of Modern Algebra. Translated by Jaroslav Taver. Leiden: Noordhoff International Publishing 1973

PACIOLI, LUCA: De Divina Proportione. 1509; reprinted in 1956

PAPERT, SEYMOUR: Teaching Children to be Mathematicians vs. Teaching About Mathematics. Memo. No. 249, Artificial Intelligence Laboratory, M.I.T., Boston, Mass. July, 1971

PAPERT, SEYMOUR: The Mathematical Unconscious. In Judith Wechsler (ed.), On Aesthetics in Science, pp. 105–121. Cambridge: M.I.T. Press 1978

PIERPONT, JAMES: Mathematical Rigor, Past and Present. Bulletin of the American Mathematical Society. 34, pp. 23–53 (1928)

PHILLIPS, D. L.: Wittgenstein and Scientific Knowledge. New York: Macmillan 1977

PIAGET, JEAN: Psychology and Epistemology. Translated by Arnold Rosin. New York: Grossman 1971

PIAGET, JEAN: Genetic Epistemology. Translated by Eleanor Duckworth. New York: Columbia University Press 1970

429

PINGREE, DAVID: Astrology. In: Dictionary of the History of Ideas. New York: Charles Scribner's Sons 1973

POINCARÈ, HENRI: Mathematical Creation. Scientific American. 179, 54–57 (1948); also in M. Kline (ed.), Mathematics and the Modern World, pp. 14–17, San Francisco: W. H. Freeman (1968); and in J. R. Newman (ed.), The World of Mathematics, Vol. 4, pp. 2041–2050, New York: Simon and Schuster, 1956

POINCARÉ, HENRI: The Future of Mathematics. Revue genérale des sciences pures et appliquées. Vol. 19. Paris 1908

POLÁNYI, M.: Personal Knowledge: Towards a Post-Critical Philosophy. Chicago: University of Chicago Press 1960

POLÁNYI, M.: The Tacit Dimension. New York: Doubleday 1966

PÓLYA, G.: How to Solve It. Princeton: Princeton University Press 1945

PÓLYA, G.: Patterns of Plausible Inference. Two volumes. Princeton: Princeton University Press 1954

PÓLYA, G.: Mathematical Discovery. Two volumes. New York: John Wiley & Sons 1962

PÓLYA, G.: Some Mathematicians I Have Known. American Mathematical Monthly. 76, 746–752 (1969)

PÓLYA, G.: Mathematical Methods in Science. Washington, D.C.: Mathematical Association of America 1978

PÓLYA, G., KILPATRICK, J.: The Stanford Mathematics Problem Book. New York: Teacher's College Press 1974

POPPER, KARL R.: Objective Knowledge. Oxford: The Clarendon Press 1972

POPPER, KARL R., ECCLES, J. C.: The Self and Its Brain. New York: Springer International 1977

PRATHER, R. E.: Discrete Mathematical Structures for Computer Science. Boston: Houghton Mifflin 1976

PRENOWITZ, W., JORDAN, M.: Basic Concepts of Geometry. New York: Blaisdell-Ginn 1965

PRIEST, GRAHAM: A Bedside Reader's Guide to the Conventionalist Philosophy of Mathematics. Bertrand Russell Memorial Logic Conference, Uldum, Denmark, 1971. University of Leeds, 1973

PUTNAM, H.: Mathematics, Matter and Method. London and New York: Cambridge University Press 1975

RABIN, MICHAEL O.: Decidable Theories. In Jon Barwise (ed.), Handbook of Mathematical Logic. Amsterdam: North-Holland 1977

RABIN, MICHAEL O.: Probabilistic Algorithms. In J. F. Traub (ed.), Algorithms and Complexity: New Directions and Recent Results. New York: Academic Press 1976

REBIÈRE, A.: Mathématiques et mathématicien. Second edition. Paris: Libraire Nony et Cie. 1893

RIED, CONSTANCE: Hilbert. New York: Springer-Verlag 1970

RÉNYI, ALFRED: Dialogues on Mathematics. San Francisco: Holden-Day 1967

RESCH, R. D.: The topological design of sculptural and architectural systems. In: AFIPS Conference Proceedings. Vol. 42, pp. 643–650 (1973)

RESTLE, F.: Speed of Adding and Comparing Numbers. Journal of Experimental Psychology. 83, 274–278 (1970)

ROBINSON, A.: Nonstandard Analysis. Amsterdam: North Holland 1966

ROBINSON, A.: From a Formalist's Point of View. Dialectica. 23, 45–49 (1969)

ROBINSON, A.: Formalism 64. In Proceedings, International Congress for Logic, Methodology and Philosophy of Science, 1964, pp. 228–246.

ROSS, S. L.: Differential Equations. New York: Blaisdell 1964

ROTA, G.-C.: A Husserl Prospectus. The Occasional Review, No. 2, 98–106 (Autumn 1974)

ROUSE BALL, W. W.: Mathematical Recreations and Essays. 11th edition; revised by H. S. M. Coxeter. London: Macmillan 1939

RUBENSTEIN, MOSHE F.: Patterns of Problem Solving. Englewood Cliffs: Prentice-Hall 1975

RUSSELL, BERTRAND: The Principles of Mathematics. Cambridge: Cambridge University Press 1903

RUSSELL, BERTRAND: A History of Western Philosophy. New York: Simon and Schuster 1945

RUSSELL, BERTRAND: Human Knowledge, Its Scope and Its Limits. New York: Simon and Schuster 1948

RUSSELL, BERTRAND: The Autobiography of Bertrand Russell. Boston: Little, Brown 1967

RUSSELL, BERTRAND, WHITEHEAD, A. N.: Principia Mathematica, Cambridge: Cambridge University Press, 1910.

SAADIA GAON (Saadia ibn Yusuf): The Book of Beliefs and Opinions. Translated by S. Rosenblatt. New Haven: Yale University Press 1948

SAATY, T. L., WEYL, F. J.: The Spirit and Uses of the Mathematical Sciences. New York: McGraw-Hill 1969

SAMPSON, R. V.: Progress in the Age of Reason: the Seventeenth Century to the Present Day. Cambridge: Harvard University Press 1956

SHAFAREVITCH, I. R.: Über einige Tendenzen in der Entwicklung der Mathematik. Jahrbuch der Akademie der Wissenschaften in Göttingen, 1973. German, 31–36. Russian original, 37–42

SCHATZ, J. A.: The Nature of Truth. Unpublished manuscript.

SCHMITT, F. O. WORDEN, F. G. (eds.): The Neurosciences: Third Study Program. Cambridge: M.I.T. Press 1975

SCHOENFELD, ALAN H.: Teaching Mathematical Problem Solving Skills. Department of Mathematics, Hamilton College, Clinton, N.Y., 1979

SCHOENFELD, ALAN H.: Problem Solving Strategies in College-Level Mathematics. Physics Department, University of California (Berkeley), 1978

SCOTT, D. (ed.): Axiomatic Set Theory. Proceedings of Symposia in Pure Mathematics. Providence: American Mathematical Society 1967

SEIDENBERG, A.: The ritual origin of geometry. Archive for the History of the Exact Sciences. 1, 488–527 (1960–1962)

SEIDENBERG, A.: The ritual origin of counting. Archive for the History of the Exact Sciences. 2, 1–40 (1962–1966)

SHOCKLEY, J. E.: Introduction to Number Theory. New York: Holt, Rinehart and Winston 1967

SINGER, CHARLES: A Short History of Scientific Ideas. Oxford: Oxford University Press 1959

SJÖSTEDT, C. E.: Le Axiome de Paralleles. Lund: Berlingska 1968

SLAGLE, J. R.: Artificial Intelligence: The Heuristic Programming Approach. New York: McGraw-Hill 1971

SMITH, D. E., MIKAMI, YOSHIO: Japanese Mathematics. Chicago: Open Court Publishing Co. 1914

SMITH, D. E.: A Source Book in Mathematics. New York: McGraw-Hill 1929

SNAPPER, ERNST: What is Mathematics? American Mathematical Monthly. 86, 551–557 (1979)

SPERRY, R. W.: Lateral Specialization in the Surgically Separated Hemispheres. In: The Neurosciences: Third Study Program, F. O. Schmitt and F. G. Worden (eds.). Cambridge: M.I.T. Press 1975

STABLER, E. R.: Introduction to Mathematical Thought. Reading, Mass.: Addison-Wesley 1948

STEEN, L. A.: Order from Chaos. Science News. 107, 292–293 (1975)

STEEN, L. A. (ed.): Mathematics Today. New York: Springer-Verlag 1978

STEINER, GEORGE: After Babel. New York: Oxford University Press 1975

STEINER, GEORGE: Language and Silence. New York: Atheneum 1967

STEINER, MARK: Mathematical Knowledge. Ithaca, N.Y.: Cornell University Press 1975

STIBITZ, G. R.: Mathematical Instruments. In: Encyclopaedia Britannica, 14th edition, vol. 14, pp. 1083–1087. Chicago 1971

STOCKMEYER, L. J., CHANDRA, A. K.: Intrinsically Difficult Problems. Scientific American. 140–149 May, (1979)

STOLZENBERG, GABRIEL: Can an Inquiry into the Foundations of Mathematics Tell Us Anything Interesting About Mind? In: George Miller (ed.), Psychology and Biology of Language and Thought. New York: Academic Press

STRAUSS, C. M.: Computer-encouraged serendipity in pure mathematics. Proceedings of the IEEE. 62, (1974)

STROYAN, K. D., LUXEMBURG, W. A. U.: Introduction to the Theory of Infinitesimals. New York: Academic Press 1976

STRUIK, D. J.: A Concise History of Mathematics. New York: Dover 1967

STRUIK, D. J.: A Source Book in Mathematics, 1200–1800. Cambridge: Harvard University Press 1969

SZABÓ, ÁRPÁD: The Transformation of Mathematics into a Deductive

Science and the Beginnings of its Foundations on Definitions and Axioms. Scripta Mathematica. 27, 28–48A, 113–139 (1964)

TAKEUTI, G., ZARING, W. M.: Introduction to Axiomatic Set Theory. New York: Springer 1971

TAVISS, IRENE (ed.): The Computer Impact. Englewood Cliffs: Prentice-Hall 1970

TAYLOR, JAMES G.: The Behavioral Basis of Perception. New Haven: Yale University Press 1948

THOM, R.: Modern Mathematics: An Educational and Philosophical Error? American Scientist. 59, 695–699 (1971)

THOM, R.: Modern Mathematics: Does it Exist? In: A. G. Howson (ed.), Developments in Mathematical Education, pp. 194–209. London and New York: Cambridge University Press 1973

TRAUB, J. F.: The Influence of Algorithms and Heuristics. Department of Computer Science, Carnegie-Mellon University, Pittsburgh, Pa. 1979

TUCKER, JOHN: Rules, Automata and Mathematics. The Aristotelian Society, February 1970.

TYMOCZKO, THOMAS: Computers, Proofs and Mathematicians: A Philosophical Investigation of the Four-Color Proof. Mathematics Magazine. 53, 131–138 (1980)

TYMOCZKO, THOMAS: The Four-Color Problem and its Philosophical Significance. Journal of Philosophy. 76, 57–83 (1979)

ULAM, S.: Adventures of a Mathematician. New York: Scribners 1976

VAN DER WAERDEN, B. L.: Science Awakening. Groningen: P. Noordhoff 1954

WANG, HAO: From Mathematics to Philosophy. London: Routledge and Kegan Paul 1974

WECHSLER, JUDITH (ed.): On Aesthetics in Science. Cambridge: M.I.T. Press 1978

WEDBERG, ANDERS: Plato's Philosophy of Mathematics. Westport, Conn.: Greenwood Press 1977

WEINBERG, JULIUS: Abstraction in the Formation of Concepts. In: Dictionary of the History of Ideas, Vol. 1. Charles Scribner's Sons 1973

WEISS, E.: Algebraic Number Theory. New York: McGraw-Hill 1963

WEISS, GUIDO L.: Harmonic Analysis. The Chauvenet Papers, Vol. II, p. 392. Washington, D.C.: The Mathematical Association of America 1978

WEISSGLASS, JULIAN: Higher Mathematical Education in the People's Republic of China. American Mathematical Monthly. 86, 440–447 (1979)

WEISSINGER, JOHANNES: The Characteristic Features of Mathematical Thought. In: T. L. Saaty and F. J. Weyl (eds.), The Spirit and Uses of the Mathematical Sciences, pp. 9–27. New York: McGraw-Hill 1969

WEYL, HERMANN: God and the Universe: The Open World. New Haven: Yale University Press 1932

WEYL, HERMANN: Philosophy of Mathematics and Natural Science.

Translated by Olaf Helmer. Princeton: Princeton University Press 1949

WHITE, LYNN, JR.: Medieval Astrologers and Late Medieval Technology. Viator. 6, 295–308 (1975)

WHITE, L. A.: The Locus of Mathematical Reality. Philosophy of Science. 14, 289–303 (1947). Reprinted in The World of Mathematics, J. R. Newman (ed.), Volume 4, pp. 2348–2364. New York: Simon and Schuster 1956

WHITEHEAD, A. N.: Science and the Modern World. New York: Macmillan 1925

WHITEHEAD, A. N.: Mathematics as an Element in the History of Thought. In: J. R. Newman (ed.), The World of Mathematics, Volume 1, pp. 402–416. New York: Simon and Schuster 1956

WIGNER, EUGENE P.: The Unreasonable Effectiveness of Mathematics in the Natural Sciences. Communications in Pure and Applied Mathematics. 13, 1–14 (1960)

WILDER, RAYMOND L.: The Nature of Mathematical Proof. American Mathematical Monthly. 51, 309–323 (1944)

WILDER, RAYMOND L.: The Foundations of Mathematics. New York: John Wiley & Sons 1965

WILDER, RAYMOND L.: The Role of Intuition. Science. 156, 605–610 (1967)

WILDER, RAYMOND L.: The Evolution of Mathematical Concepts. New York: John Wiley & Sons 1968

WILDER, RAYMOND L.: Hereditary Stress as a Cultural Force in Mathematics. Historia Mathematica. 1, 29–46 (1974)

WITTGENSTEIN, L.: On Certainty. New York: Harper Torchbooks 1969

WRONSKI, J. M.: Oeuvres mathématiques. Reprinted Paris: J. Hermann 1925

YATES, FRANCES A.: Giordano Bruno and the Hermetic Tradition. Chicago: University of Chicago Press 1964

YUKAWA, HIDEKI: Creativity and Intuition. Tokyo, New York, San Francisco: Kodansha International 1973

ZAGIER, DON: The First 50 Million Prime Numbers. The Mathematical Intelligencer. 0, 7–19 (1977)

ZIMAN, JOHN: Public Knowledge: The Social Dimension of Science. Cambridge: Cambridge University Press 1968

ZIPPIN, LEO. Uses of Infinity. Washington, D.C.: Mathematical Association of America 1962

# Index

437